Advances in

MICROBIAL PHYSIOLOGY

Advances in
MICROBIAL PHYSIOLOGY

Edited by

A. H. ROSE

Department of Microbiology
University of Newcastle upon Tyne
England

and

J. F. WILKINSON

Department of General Microbiology
University of Edinburgh
Scotland

VOLUME 1

1967

ACADEMIC PRESS · LONDON and NEW YORK

ACADEMIC PRESS INC. (LONDON) LTD.
BERKELEY SQUARE HOUSE
BERKELEY SQUARE
LONDON, W.1

U.S. Edition published by
ACADEMIC PRESS INC.
111 FIFTH AVENUE
NEW YORK, NEW YORK 10003

Library of Congress Catalog Card Number: 67-19850

PRINTED IN GREAT BRITAIN BY
SPOTTISWOODE, BALLANTYNE AND CO. LTD.
LONDON AND COLCHESTER

Contributors to Volume I

N. O. KJELDGAARD, *University Institute of Microbiology, Copenhagen, Denmark.*

H. LARSEN, *Department of Biochemistry, The Technical University of Norway, Trondheim, Norway.*

R. I. MATELES, *Department of Nutrition and Food Science, Massachusetts Institute of Technology, Cambridge, Massachusetts, U.S.A.*

W. G. MURRELL, *Commonwealth Scientific and Industrial Research Organization, Division of Food Preservation, Ryde, New South Wales, Australia.*

J. R. POSTGATE, *University of Sussex, Falmer, Sussex, England.*

G. N. WOGAN, *Department of Nutrition and Food Science, Massachusetts Institute of Technology, Cambridge, Massachusetts, U.S.A.*

Preface

The explosion in biological research which has taken place over the past two decades has inevitably induced a considerable degree of specialization among those engaged in this research. At the same time, there has been a move towards integration among the biological sciences, largely as a result of the realization that there exist many basic similarities among living organisms, particularly at the physiological level. These two trends have placed an even greater premium on the availability of review articles with which biologists, who are researching in one area, can become acquainted with progress that is being made in areas outside their immediate sphere of interest. The physiology of micro-organisms is now being studied by individuals representing widely diverse specializations in the chemical as well as the biological sciences. It was with the belief that work being carried out under these many specializations could profitably be brought together in the form of review articles that we agreed to edit the present series.

In *Advances in Microbial Physiology*, we aim to include articles covering as wide a range as possible of the specialized interests that constitute microbial physiology. Clearly there must be limits to our field of interest, and we have confined it to unicellular micro-organisms. Except in so far as they pertain to the physiology of the host micro-organism, the articles will not deal primarily with viruses and virus multiplication. Moreover, while we aim to include articles that deal with all of the major groups of unicellular micro-organisms, the concentration of research on bacteria—and indeed on just a very few species of bacteria —makes it difficult and probably undesirable to strike a true balance. Nevertheless, we hope that these reviews will be of value to all biologists who are interested in the physiological activities of micro-organisms.

A. H. ROSE
J. F. WILKINSON

January, 1967

Contents

Contributors to Volume I v

Editors' Preface vii

Viability Measurements and the Survival of Microbes under Minimum Stress. JOHN R. POSTGATE

I. Introduction 1
II. Methodology 2
A. Definitions and Principles 2
B. Indirect Assessment of Viability 4
C. Direct Assessment of Viability 5
D. Cryptic Growth 8
III. Results 9
A. Starvation and Microbial Populations 9
B. Survival of Old Cultures: the "Death Phase" 16
C. Mortality during Growth 16
D. Senescence of Microbes 17
IV. Prospects 18
V. Acknowledgement 20
References 21

Aflatoxins. RICHARD I. MATELES and GERALD N. WOGAN

I. Introduction 25
A. Biological Effects of Aflatoxins. 25
B. Chemical Characteristics of Aflatoxins 27
II. Microbiological Aspects of Aflatoxin Production 28
A. Organisms 28
B. Media and Culture Conditions 31
III. Biosynthesis of Aflatoxins 32
IV. Acknowledgements 36
References 36

Regulation of Nucleic Acid and Protein Formation in Bacteria.
NIELS OLE KJELDGAARD

I. Introduction . . . 39

II. Regulation and Bacterial Growth . . . 40
A. DNA . . . 42
B. RNA . . . 43
C. Proteins . . . 46
D. States of Transition . . . 47

III. Regulation of DNA Formation . . . 50
A. Molecular Weight of Bacterial DNA . . . 50
B. The Replication Cycle . . . 52
C. Induction of DNA Replication . . . 55
D. Genetic Studies of DNA Replication . . . 60

IV. Regulation of RNA Formation . . . 62
A. Synthesis of RNA . . . 62
B. Modes of Regulation . . . 63
C. Role of Amino Acids . . . 64

V. Regulation of Enzyme Synthesis . . . 76
A. The Operator Model . . . 76
B. Regulatory Genes . . . 83
C. Models of Regulation . . . 88

References . . . 90

Biochemical Aspects of Extreme Halophilism. HELGE LARSEN

I. Introduction . . . 97

II. Extreme Halophiles and Their Growth Relations . . . 98

III. Metabolic Apparatus of the Extreme Halophiles . . . 101
A. Intracellular Salt Content . . . 101
B. Salt Relations of Individual Enzymes . . . 103
C. Salt Relations of the Ribosomes . . . 107
D. Salt Relations of the Uptake Mechanism . . . 108
E. Nucleic Acids. . . . 108
F. Metabolic Pathways . . . 110

IV. Cell Envelope of the Halobacteria . . . 111
A. Requirement for Salt to Maintain the Cell Envelope . . . 112
B. Structure and Chemical Composition. . . . 114
C. Function of Salts in Maintaining the Envelope . . . 123

V. On the Rise of Extreme Halophilism . . . 126

References . . . 130

The Biochemistry of the Bacterial Endospore. W. G. MURRELL

I. Introduction 133
A. Definitions 135
II. Cytological Development of the Spore 135
A. Nuclear Movements Associated with Sporogenesis . . 136
B. Spore Septum Formation 138
C. Envelopment of the Spore Protoplast by the Mother Cell . 139
D. Cortex Formation 140
E. Coat Formation 142
F. Exosporium 144
G. Maturation or Ripening 144
III. Chemical Composition of the Mature Spore 144
A. Structures 145
B. Chemical Fractions 164
C. Inorganic Composition 167
D. Dipicolinic Acid and Calcium 174
IV. Biochemistry of Spore Formation 184
A. Metabolic Changes during Spore Formation in *Bacillus* Species 185
B. Metabolic Changes during Sporulation in *Clostridium* Species . 205
C. Nucleic Acid Changes during Sporulation 209
D. Synthesis of Spore Components 216
V. Genetic Control of Spore Morphogenesis and Induction of Spore Formation 230
VI. Maturation and the Physico-Chemical State of the Mature Resting Spore 235
A. Maturation 235
B. State of the Mature Spore 236
VII. Heat Resistance 240
VIII. Conclusions 243
IX. Acknowledgement 244
References 244

Author Index 253

Subject Index 263

ABBREVIATIONS AND SYMBOLS

Certain abbreviations in this book are used without definition. These abbreviations were recommended by the IUPAC-IUB Combined Commission on Biochemical Nomenclature, and have been reproduced in the *Biochemical Journal* **102**, 15 (1967), *Biochemistry* **5**, 1445 (1966), *Biochimica et Biophysica Acta* **108**, 1 (1965) and the *Journal of Biological Chemistry* **241**, 527 (1966). Enzymes are referred to by the trivial names recommended in the Report of the Commission of Enzymes of the IUB (Pergamon Press, Oxford; 1961). All temperatures recorded in this volume are in degrees Centigrade.

Viability Measurements and the Survival of Microbes Under Minimum Stress

JOHN R. POSTGATE

University of Sussex, Falmer, Sussex, England

I. Introduction 1
II. Methodology 2
  A. Definitions and Principles 2
  B. Indirect Assessment of Viability 4
  C. Direct Assessment of Viability 5
  D. Cryptic Growth 8
III. Results 9
  A. Starvation and Microbial Populations 9
  B. Survival of Old Cultures: the "Death Phase" 16
  C. Mortality during Growth 16
  D. Senescence of Microbes 17
IV. Prospects 18
V. Acknowledgement 20
  References 21

I. Introduction

The title of this review is obliquely phrased, because it is concerned with the "spontaneous" death of vegetative microbes. A moment's reflection will show that microbes, being uni- or non-cellular, have no simple analogue of the "natural" senescence and death undergone by multicellular organisms (though in molecular terms it is possible that processes analogous to the senescence of multicellular organisms could be demonstrated in individual microbes). Most studies on microbial survival have been concerned with the viability of populations after they have been subjected to some kind of stress: a population is examined after it has been heated, frozen, dried; exposed to adverse pH, pressure, salinity; to toxic chemicals or lethal radiation; or to biological antagonists (anti-sera, phage, phagocytes). Experiments on the viability of microbes after such treatments in fact constitute a great part of the literature of experimental microbiology, particularly that concerned with sterilization and disinfection or with the preservation of stock cultures, but such experiments require a control in which the viability of a comparable population not subjected to the overt stress is assessed. Such "stress-free" controls have a "spontaneous" death-rate which may or may not act additively with respect to the stress imposed, and it is largely with the death of such "un-stressed" populations that this review is concerned.

In practice, stress-free conditions are difficult to realize, and the investigator must usually adopt a form of treatment in which as mild and simple stress as possible is imposed on the population. Three conditions are generally available for such studies: (1) Exposure of microbial populations to non-nutrient environments in conditions in which growth is in principle possible, but is prevented by withholding one or more essential nutrients. (2) Allowing a culture to age until multiplication has ceased. (3) Growing a continuous culture at so slow a rate that the spontaneous death-rate of the population makes a contribution to the dynamics of the steady state.

In contrast to the voluminous literature available on spore survival, remarkably little work has been published relevant to the survival of populations of vegetative organisms in such conditions, though experiments of this kind have obvious relevance to hygiene, water bacteriology, and microbial ecology. So far, published work has largely illustrated the methodological difficulties of the study of microbial death. Therefore, before reviewing such published data as exist, a critical consideration of the techniques used to assess microbial survival is desirable.

II. Methodology

A. Definitions and Principles

Viability, as used in this review, has the sense of the ratio of the number of viable units in a microbial population to the total number of microbes present. Thus a yeast with a bud is a single unit, viable or non-viable as the case may be, and the question whether it represents one or two individuals will not be discussed. Some authorities have also used the term viability in a clonal sense; a certain clone is less viable than another if the generations include a higher proportion of non-viable progeny. This sense of the term, though legitimate, can lead to ambiguity in the present context and will not be used.

Viable microbes, for microbiologists, are microbes capable of dividing to form one or more live daughter cells when provided with a favourable environment. What constitutes a favourable environment can rarely be stated precisely, and must therefore be derived from a consensus of informed opinions. The usage of the term thus differs from that current among scientists concerned with higher organisms, where a creature may be viable without being capable of multiplication. It is possible, in special conditions, to obtain microbes having many of the characters of normal, living organisms but unable to divide and form a colony. Such microbes are committed, as it were, to death and will be termed "moribund". Because of the need to use colony formation as a test, the viability of an individual microbe can only be rigidly determined retrospectively: as soon as its viability is established, it ceases to be the same individual. Since bacteria, for example, do not necessarily multiply the moment they are given the opportunity to do so, an element of ambiguity arises in deciding whether a given cell is dead or in a state of division lag. Undoubtedly, in fact, organisms from a dying population may continue to die during a lag phase, so that an assessment of the viability of such a population may actually measure the resultant of two theoretical viabilities: the viability at the time of sampling supplemented by the mortality of those organisms that died during the lag phase. These elements of uncertainty in the assessment of viability are often trivial, but they can assume importance on occasions. Death during the lag phase takes place among the survivors of frozen populations (Postgate and Hunter, 1963b), and has been proposed as a partial explanation of "metabolic injury", a phenomenon in which the recovery of viable bacteria from frozen populations may be greater the "richer" the recovery medium used. The possibility that comparable metabolic injury may precede death by starvation must be borne in mind, though Postgate and Hunter (1962) did not observe it in their experiments. Bacteria subject to stress become hypersensitive to secondary stresses, such as toxicity of certain medium components (see Jacobs and Harris, 1960, 1961), and the possibility of comparable hypersensitivity among starved populations must also be considered. Starved *Escherichia coli* K–12, for example, show increased sensitivity to infection by λ bacteriophage (Arber, 1963). Sykes and Tempest (1965) studied a slow-growing continuous culture of magnesium-limited *Pseudomonas* which showed nearly 100% viability when the sample for assay was diluted in a cold saline solution, but gave considerable evidence of lysis when diluted in distilled water; normal populations were indifferent to the diluent used. Postgate and Hunter (1962) routinely centrifuged their populations for starvation studies and observed that the survival patterns of the populations were influenced by whether distilled water or a saline solution was used to resuspend the bacteria.

Viability is normally assessed by growth tests. Because such tests are slow and can give ambiguous results, many indirect procedures have been proposed for determining viability. Though these can have value for the study of populations subject to major stresses, they are uniformly unsuitable for the examination of starved populations. A critical summary of the major indirect procedures follows.

B. Indirect Assessment of Viability

1. *Staining and Dye-Uptake Tests*

Vital staining, either with fluorescent dyes (Strugger, 1948) or ordinary dyes such as methylene blue (see Gilliland, 1959), depends on the assumption that microbes become permeable to dye as they die. Thus they fail to score moribund cells as dead. This point is almost certainly unimportant when the stress applied is heating or disinfection with, for example, a quaternary bactericide; it is far from trivial when a mild stress such as chilling or starvation is applied. Postgate and Hunter (1962) showed that the permeability barriers of starved *Aerobacter aerogenes* persisted for many hours after loss of viability; Postgate *et al.* (1961) observed that methylene blue, as a vital stain, under-estimated the viability of old baker's yeast by a considerable margin, doubtless because it administered the *coup de grâce* to the moribund members of the population (see, for example, Ketterer, 1956). Gilliland (1959), on the other hand, found that brewer's yeast was tougher, and that vital staining with methylene blue over-estimated the viability of old yeast populations. Razumovskaya and Osipova (1958) reported a lack of correlation between the ability of *Acetobacter* cells to multiply and their permeability to a fluorescent vital stain. Differential staining (e.g. White, 1947) likewise depends on assumed differences in the inherent stainability of dead and live cells, which are expected to persist after the surviving organisms have been killed during the staining procedure.

Permeable microbes take up more dye than impermeable ones, and the assumption that permeability and death are coincident has led to proposals for the determination of the gross viability of populations by dye uptake tests (Borzani and Vairo, 1958, 1960). Like vital staining, these procedures are excluded for use with unstressed or mildly stressed populations because stainability and viability do not necessarily correlate. Fluorescence when stained by dyes of the anilino-naphthalene-sulphonic acid class has, in fact, been used to demonstrate persistence of osmotic integrity in certain classes of moribund cells (Postgate and Hunter, 1962; Mathews and Sistrom, 1960; Strange and Postgate, 1964).

2. *Optical Tests*

Mager *et al.* (1956) described the "optical effect", whereby the optical density of a suspension of live bacteria is greater in a dilute salt solution than in distilled water by a factor of 30–100% (the percent change is now known to depend on the geometry of the apparatus used to measure extinction; see Powell, 1963). Heat- or phenol-killed bacteria showed no optical effect. Though proposed by its discoverers as a rapid method of assessing viability, the method is in fact useless for this purpose unless the populations have been violently damaged. On the other hand, the optical effect provides an invaluable test of the integrity of the osmotic barrier in cell suspensions (Postgate and Hunter, 1962; Strange, 1964) and is very useful in permeability studies.

Immersion refractometry is based on the principle that breakdown of the osmotic barrier leads to a decline in refractive index and consequent loss of contrast under the phase-contrast microscope. Barer *et al.* (1953), Fikhman (1959a, b) and Fikhman and Pryadkina (1961) adopted this phenomenon using gelatin solutions of various refractive indices to abolish light-scattering by one class of cells ("dead") while leaving others visible ("live"). The method suffers from the same formal objections as others based on permeability changes.

3. *Leakage of Pool Materials*

Koch (1959) estimated death during growth of a culture by measuring leakage of purines from labelled bacteria. Since bacteria appear to undergo quite dramatic changes in permeability during active growth, without loss of viability (see, for example, Strange and Postgate, 1964), this test is of doubtful validity.

4. *Enzyme Activity*

Loss of viability in practice is often paralleled by a decline in dehydrogenase enzyme activity, and the ability to reduce a dye such as tetrazolium or methylene blue has been proposed as the measure of the viability of microbial populations (e.g. Delpy *et al.*, 1956). In specialized circumstances, such procedures may work satisfactorily, even with unstressed populations, but their lack of sound theoretical basis makes them unsuitable as a tool for studying the physiology of microbial survival.

C. Direct Assessment of Viability

1. *Plate and Total Counting*

It is probably desirable here to repeat certain textbook truisms concerning the accuracy of plate and total counts. Plate counts record the numbers of organisms in a known volume able to form a macrocolony on the medium provided and, since their distribution is Poissonian, the standard deviation is the square root of the number counted. Thus it is necessary to count about 300 colonies, but little gain in accuracy arises from counting more. Many errors arise in the serial dilution process, and it is usually preferable to perform a single 1 in 10^4 dilution (e.g.

rinsing a 10 μl. capillary of culture into 100 ml. of diluent) than to do four 1 in 10 dilutions. The necessity to use a non-toxic and osmotically suitable diluent is widely realized, but the fact that cold diluents can cause errors due to cold shock in sensitive populations (e.g. Gorrill and McNiel, 1960; Strange and Dark, 1962) seems less well known. Replicate plate counts on a population of discrete bacteria should always fall within the fiducial limits of a Poissonian distribution and, if they do not, some aspect of the technique should be suspect. For completeness, a dutiful reviewer should cite de Silva's (1953) claim to be able to influence plate counts of typhoid bacilli by mental concentration.

On the other hand, microscopic total counts are normally subject to a systematic error that exaggerates the count. The depth of the chamber varies considerably according to the manner in which it is set up, and for complete accuracy the depth of the chamber should be checked interferometrically each time it is used (Topley and Wilson, 1955; Norris and Powell, 1961). If this precaution is not taken, cultures that are 100% viable often give ratios of total to viable bacteria indicating viabilities of 50–80%, and historically this technical error led to a widespread belief that normal bacterial cultures contained many dead organisms (e.g. Wilson, 1922, 1926). This view is no longer tenable, but the existence of the technical error should alert one to view with suspicion statements of the form: "plate and microscopic counts were equal".

Weibull (1960) reported an ingenious technique for total counts: the suspension under test is set in agar between an ordinary slide and coverslip and whole microscopic fields are counted throughout their depth. The areas of such fields are known from the optical characteristics of the objective, and their depths may be found using the micrometer scale of the fine focusing adjustment, after due allowance for the refractive index of the agar environment. In practice, the accuracy of the method is limited by the extent to which sloping coverslips can be avoided and the exactness with which the micrometer scale can be used. Collins and Kipling (1957) evaporated a known volume of stained aqueous suspension to near-dryness in the presence of about 5 μl. of glycerol; the bacteria concentrated in the tiny glycerol droplet in which, provided clumping did not occur, every cell could be counted. Gabe (1957) counted small volumes of suspensions in flat-walled capillaries; special techniques are needed to manufacture these capillaries. The problems of total counts have been discussed further by Meynell and Meynell (1965). Provided certain conditions regarding orifice size and interference by dust are observed, electronic particle counters such as the Coulter counter (Kubitschek, 1960) can be used to obtain very exact and reproducible total counts of large (>1 μ^3) bacteria. This instrument, however, measures a conductivity signal that is determined by the cell volume as it flows through a tunnel, and the user should remember that live and dead microbes of the same nominal volume may not have the same effective volume: the writer has observed an apparent net shrinkage of *Aerobacter aerogenes* on adding formaldehyde to a live population. With yeasts or erythrocytes, such alterations are usually readily allowed for in adjusting the instrument's setting; with bacteria, however, the particles are so small that the counter is, perforce, operating near to the electronic noise level and such counts need far more care than those with larger microbes.

2. *Viability by Micro-culture*

Powell (1956, 1958) and Quesnel (1963) used micro-culture to study the multiplication of individual bacteria, and incidentally, to record viability. Valentine and Bradfield (1954) adapted slide culture for the rapid assessment of viability by adding urea to the population and assuming that elongated forms appearing after brief incubation were derived from originally viable cells; Gilliland (1959) used slide culture in a haemocytometer for assessing the viability of yeast populations; Postgate *et al.* (1961) described a simple method of slide culture on agar for the assessment of viability and reported on its statistical accuracy in practice; Bretz (1962) described a similar procedure using blotting paper in place of agar; Jebb and Tomlinson (1960) cultured mycobacteria in agar films suspended in wire loops in order to determine clonal multiplication rates, a procedure which could obviously be adapted for viability determinations. Kogut *et al.* (1965) used micro-culture to follow growth of bacterial clones.

Slide culture is accurate within the range 5–100% viability, and since that is the operational range of most workers concerned with the physiology of death, it is often the method of choice. It is relatively rapid and, though it does not give either a total or a viable count, it gives the ratio of these with Poissonian precision. Results obtained with it differ slightly from conventional plate counts because pairs are normally scored as representing one live parent, though both may be dead; the error so introduced is normally trivial. It is unsuitable for use with organisms that lyse or that form filamentous networks; these and other limitations were discussed by Postgate *et al.* (1961).

3. *An Oblique "Direct" Method*

Wade and Morgan (1954), using the principle that bacteria accumulate "fluctuating RNA" when about to divide, devised a rational vital stain whereby populations were inoculated for a brief period in defined conditions and those that had amassed RNA were distinguished by a staining procedure from those that had not. Though successful in model mixtures of dead and live organisms, this procedure was unsatisfactory with starved populations of *Aerobacter aerogenes* when tested jointly by H. E. Wade and myself: too many organisms of ambiguous staining reaction were observed.

D. Cryptic Growth

When members of populations of starved microbes die, the survivors are no longer starved. Leakage and lysis products from the dead organisms may support growth of the survivors. This phenomenon, known as cryptic growth (see Ryan, 1959), can interfere with studies on the physiology of starvation. Postgate and Hunter (1962) recorded that death of fifty members of a starved population of *Aerobacter aerogenes*, derived from a chemostat, allowed the doubling of one survivor, a figure closer to four or five deaths per multiplication probably applied to organisms from batch cultures. Cryptic growth can be prevented by adding a non-toxic inhibitor of multiplication, such as penicillin or chloramphenicol; it is particularly important to allow for cryptic growth in studies in which a carbon substrate is added to a starved population (e.g. those concerned with maintenance energy or substrate-accelerated death), because then situations can arise in which death of one cell supports the growth of one and sometimes, in the writer's experience, more than one survivor. Slide culture can give qualitative evidence of cryptic growth because the distribution of division lags in the populations become highly irregular: cryptically grown newcomers among starved populations have typically a short lag phase and show large micro-colonies. Systematic differences in colony sizes after plate-counting also suggest cryptic growth.

Cryptic growth interfered with attempts by Postgate and Hunter (1964) to demonstrate substrate-accelerated death with *Bacillis subtilis*, *Pseudomonas ovalis* and *Candida utilis*. Jannasch (1965) found it interfered with determinations of the lowest substrate concentration accepted by *Spirillum serpens*, and he had recourse to continuous culture for that measurement. Its importance in the design of experiments is not appreciated sufficiently widely. For example, Theil and Zamenhof (1963), in a discussion of DNA turnover in a mutant of *Escherichia coli*, dismiss cryptic growth on the grounds that the species shows no tendency to lysis. In fact, all studies on cryptic growth have used non-lysing bacteria for the simple reason that slide culture would be inapplicable if lysis of the dying population were extensive. Many experiments on "maintenance energy" could be interpreted as yielding the resultant of cryptic growth and substrate-accelerated death; this point is discussed further below.

III. Results

A. Starvation of Microbial Populations

The virtues of saline solutions versus distilled water as an environment for prolonging survival of starving bacteria, and the influence of pH value and of certain cations on their survival, occupied certain microbiologists during the earlier decades of this century (e.g. Shearer, 1916, 1917; Cohen, 1922; Winslow and Falk, 1923a, b). Cohen (1922) cited earlier work, and Postgate and Hunter (1962) briefly discussed work after 1922. In view of the pronounced effect nutritional status is now known to have on survival, much of the earlier work is of limited general significance, though two facts still stand: cations such as Mg^{2+} or Ca^{2+} prolong the survival of coliform bacteria, and a stable pH value of about 6·0 favours survival even though growth is optimum at pH 7–7·5. Between 1923 and 1960, only occasional papers on storage of bacterial suspensions were published; the speculations of Hinshelwood (1957), followed by the work of Harrison (1960) and of Strange *et al.* (1961), represented a revival of interest in the systematic study of microbial death by starvation. The state of the subject up to 1962 was reviewed by Postgate and Hunter (1963a); this article will take that review for granted and deal mainly with work published between 1962 and the spring of 1966. A symposium on "Microbial Physiology and Continuous Culture" was held in March, 1966 at the Microbiological Research Establishment (M.R.E.), Porton, England and will be published by H.M. Stationery Office, London, (Editor: E. O. Powell). It included material relevant to certain aspects of this review which will be mentioned, though not discussed in detail.

1. *Protection from Death by Starvation*

Lovett (1964) showed that suspensions of *Aerobacter aerogenes*, starved in buffered 75% deuterium oxide, survived longer than control suspensions in aqueous buffers at their growth temperature. The death curves were sigmoid, and the mode of action of D_2O was to prolong the flat part of the sigmoid curve; once death started, the linear death-rates were similar in both environments. Lovett attributed the protective effect of D_2O to a delaying action on the endocellular ribonuclease (RNase), thus delaying the net degradation of RNA that accompanies death of carbon-limited *A. aerogenes* (Strange *et al.*, 1961); release of substances absorbing at 260 mμ was repressed in D_2O. *Pseudomonas fluorescens* 8248, a strain that contains no endocellular RNase, did not show prolonged survival in D_2O-based buffer. In an unpublished experiment J. R. Hunter, using the techniques described by Postgate and Hunter (1962), could observe no preservative effect of N-(6)-benzyladenine (100 μg./ml.) on the survival of *A. aerogenes*; this compound has been held to delay senescence in plants (Salunkhe *et al.*, 1962).

Magnesium-deficient bacteria tend to be particularly mortal and, conversely, Mg^{2+} and, to a lesser extent, Ca^{2+} and Fe^{2+} have protective effects against starvation as well as against other mild stresses. These points were mentioned in earlier reviews. Strange and Shon (1964) showed that *Aerobacter aerogenes*, grown in a carbon-limited mannitol medium, carried reserves of Mg^{2+} through washing by centrifugation, and that this material influenced their survival properties. Washing the organisms at 20° in a saline-phosphate buffer depleted them of adsorbed Mg^{2+}, and increased their sensitivity to a mild heat stress as compared with bacteria washed in distilled water. Na^{+} and K^{+} antagonized the resorption of Mg^{2+} by depleted bacteria. A depletion of this kind could account for the influence of washing conditions on death by starvation reported by Postgate and Hunter (1962), but their phenomenon occurred in an opposite sense: bacteria washed in distilled water died in comparable test conditions more rapidly than those washed in saline. The importance of the ionic environment in survival experiments is indicated by a number of incidental reports in recent papers, some of which are not easily reconciled. Phosphate-limited *A. aerogenes* show K^{+}-accelerated death by starvation (Strange and Dark, 1965) but carbon-limited bacteria were earlier reported to be immune to K^{+} (Postgate and Hunter, 1962). Mg^{2+} protects against some kinds of substrate-accelerated death (see below) but not others. Mn^{2+} or Co^{2+} can replace Mg^{2+} to some extent as protective agents against mild heat stress, though they are inactive with starving populations (Strange and Shon, 1964; Postgate and Hunter, 1962). Ethylenediamine tetraacetic acid (EDTA), which presumably acts by influencing the ionic environment, can have markedly different effects on the survival properties of various sub-strains of *A. aerogenes* (see "Substrate-Accelerated Death", below). Further evidence in the rôle of Mg^{2+}, and indications that K^{+} is important, were discussed at the M.R.E. symposium referred to earlier. High concentrations of both Na^{+} and Cl^{-} are essential for survival of a strong halophile and these ions are probably involved in the structure of the cell wall (Mohr and Larsen, 1963).

2. *Substrate-Accelerated Death*

Postgate and Hunter (1964) gave a detailed account of "substrate-accelerated death", a phenomenon whereby the substrate that limited the growth of a population accelerated its death in starvation conditions: for example, a population of coliform organisms that ceased growth because of exhaustion of glucose in the medium (e.g. a well-aerated broth culture; Freter and Ozawa, 1963) dies much faster when starved in phosphate buffer containing glucose than in phosphate buffer alone (Strange *et al.*, 1961). Substrate-accelerated death had the following characteristics according to Postgate and Hunter (1964).

(1) A high degree of specificity. Glucose or ribose did not accelerate the death of glycerol-limited *Aerobacter aerogenes*, though they accelerated death of glucose- or ribose-limited populations. Ammonium- or phosphate-limited populations showed ammonium- or phosphate-accelerated death; sulphate- or magnesium-limited populations did not show correspondingly accelerated deaths.

(2) Intermediates in the metabolism of glycerol (e.g. pyruvate, succinate) accelerated death of glycerol-limited organisms to various extents.

(3) Magnesium ions, 2,-4dinitrophenol or azide prevented glycerol-accelerated death. 2,4-Dinitrophenol was toxic in the absence of the lethal substrate; the interaction of lethal substrate and drug thus provided an unusual example of negative synergy.

(4) Glycerol-accelerated death was associated with excretion of materials from the cells, but several processes directly or obliquely associated with death from starvation, such as polymer breakdown or osmotic breakdown, were not augmented during substrate-accelerated death.

(5) The survivors of substrate-accelerated death showed long division lags reminiscent of organisms recovering from a repressed state.

Postgate and Hunter (1964) used a strain of *Aerobacter aerogenes* which had been in continuous culture for some years and which had developed several morphological divergencies from the parent stock. In a formal repetition of part of the work, using putatively the same strain and restricting the period of continuous culture to not more than 4 weeks, Strange and Dark (1965) reported several divergencies from the findings of Postgate and Hunter. These were:

(1) Specificity as between carbon substrates did not occur: glucose and ribose, for example, accelerated death of glycerol-limited bacteria.

(2) Ammonium-accelerated death of ammonium-limited cells did not occur.

(3) Apparent phosphate-accelerated death of phosphate-limited cells could be attributed to K^+ ion toxicity; KH_2PO_4 but not NaH_2PO_4 accelerated death in this buffer. Postgate and Hunter (1964) had only tested carbon-limited cells for toxicity by K^+ and (in contrast to Strange and Dark) found none.

(4) Several differences of detail occurred: the effect of bacterial concentration on glycerol-accelerated death was the opposite of that found

by Postgate and Hunter; EDTA, which was protective when added to "saline-tris" buffer, was toxic in the hands of Strange and Dark.

Strange and Dark (1965) attributed glycerol-accelerated death in part to a toxic product of glycerol metabolism which they did not identify; magnesium was not wholly protective against it. The divergencies among the findings of Strange and Dark compared with those of Postgate and Hunter arose to some extent from their use of different sub-strains of *Aerobacter aerogenes*. A further study by Strange and Hunter (1966), using Postgate and Hunter's original variant strain, showed unequivocal ammonium-accelerated death while confirming the insusceptibility of the parent strain, and of *Escherichia coli*, to this stress. Ammonium-accelerated death was augmented by SO_4^{2-} and was not prevented by Mg^{2+}; susceptibility to ammonium-accelerated death was associated with restricted polysaccharide synthesis in conditions of nitrogen limitation. The specificity claimed by Postgate and Hunter for substrate-accelerated death may thus be less clear with other strains than it originally appeared, but the phenomenon is undoubtedly widespread and is not limited to energy-yielding substrates. Strange and Shon (1964) reported substrate-accelerated death among populations of *A. aerogenes* subjected to a mild heat stress; Smith and Wyss (1965) observed it in *Azotobacter vinelandii*.

A report by Pittillo and Narkates (1964) claimed that various bacteria, including *Escherichia coli*, died more rapidly when starved in saline containing folic acid than without this vitamin. Saline is a notoriously haphazard storage environment (see Cohen, 1922; Postgate and Hunter, 1962) and in the absence of data concerning pH value, population densities, aeration and trace metal contamination it is difficult to assess their data.

3. *Suicidal Mutants and Substrate-Accelerated Death*

Certain biochemical mutant strains of microbe show suicidal properties: they die when deprived of their required metabolite in otherwise nutrient conditions. They were discussed briefly in an earlier review (Postgate and Hunter, 1963a). Pronounced suicidal behaviour has been observed in microbes requiring biotin, inositol, succinate, thymine, or diaminopimelic acid; "thymine-less death" has become of importance in the study of cellular control processes (see Hayes, 1965).

Suicidal behaviour and substrate-accelerated death may have features in common. If the organisms used to study thymine-less death, for example, were in fact from a carbon-limited population, then the test conditions used to evoke thymine-less death (incubation with substrate in conditions in which growth can nevertheless not take place) might be precisely those that would cause substrate-accelerated death. Although much information is available concerning the influence of nutrients such as amino acids and purines on thymine-less death, critical data on the nutritional status of bacteria undergoing this stress are not available in most publications in this field. The relationship between these two phenomena needs further examination; several critical experiments suggest themselves, such as a study of the susceptibility of the wild type strains to substrate-accelerated death and the relative protective effects of components of media in which metabolite-less death may be observed. While it is possible that substrate-accelerated death has mechanistic features in common with thymine-less death and with suicidal behaviour in response to deprivation of such metabolites as biotin, succinate or inositol, it is clear that death resulting from deprivation of diaminopimelic acid has an entirely different origin. Imperfect cell-wall synthesis leading to lysis appears to provide a complete interpretation of this phenomenon (Meadow *et al.*, 1957; Rhuland, 1957) and is supported by the fact that osmotically strong environments prevent lysis but cause spheroplast formation (McQuillen, 1958).

4. *The Question of Maintenance Energy*

McGrew and Mallette (1962) and Mallette (1963) determined the amount of carbon required for maintenance of viability in starved *Escherichia coli* on the presumption that, if sufficiently small amounts of substrate were added continuously to a starved suspension, it should be possible to sustain viability without permitting growth. They succeeded in prolonging viability (according to plate counts) of such suspensions without a parallel increase in turbidity but, as Dawes and Ribbons (1964) pointed out, cryptic growth may well have taken place during their experiments. One must add the further criticisms that these experiments took no account of (1) the decline in turbidity that accompanies the starvation and death of coliform bacteria, (2) the possibility that substrate-accelerated death took place, and (3) the considerable sensitivity of growing *E. coli* to cold shock (see Strange and Dark, 1962), to which their organisms appear routinely to have been subjected. Such considerations also apply to the experiments of Mallette *et al.* (1964) and Bohinski and Mallette (1965) intended to demonstrate absences of maintenance requirements for phosphate and sulphate in *E. coli*; the publication of McGrew and Mallette (1965), citing incorporation of labelled glucose as evidence for maintenance, could readily be interpreted in terms of death and cryptic growth. While these criticisms do not necessarily obviate the conclusions reached by Mallette and his coworkers, they reflect on the rigidity of the experimental design and could account for some of the curious responses, recorded by these authors, that sulphate- and phosphate-limited populations showed to small amounts of added substrate. Criticisms of this kind apply to the design of many published studies concerned with endogenous metabolism and maintenance energy.

5. *Rôle of Cell Polymers in Survival*

The question whether "reserve" materials, such as polysaccharide, actually function for the maintenance of a microbe when starved was raised by Wilkinson (1959, 1963), who found the position unclear. Turnover of cell constituents is well known to occur in "resting" bacteria (see, for example, Mandelstam, 1963) and the question arises whether the degradation of polymers known to occur on starvation assists in prolonging viability. Strange *et al.* (1961) provided a partial answer by their demonstration that protein- or polysaccharide-rich *Aerobacter aerogenes* maintained viability while utilizing their excess of these materials. Such populations had survival curves of a sigmoid form, the flat part corresponding to consumption of these "reserves". RNA is also, to some extent, expendable, in that up to 50% of the ribosomal RNA of some organisms can be mobilized without death taking place; RNA degradation always accompanies death by starvation and if RNA is the only expendable polymer, death is most rapid. RNA degradation appears to be a rather critical process in the survival of *A. aerogenes*, but no absolute correlation between rate of RNA breakdown and death-rate exists. Harrison and Lawrence's (1963) starvation-resistant mutants had an augmented rate of RNA metabolism compared with the wild type; this point was discussed further by Postgate and Hunter (1963a). A claim has been made that a starved uracil-less mutant of *Escherichia coli* did not degrade its ribosomes (Nakada and Smith, 1962) but no viability tests were recorded. Campbell *et al.* (1963), in experiments unaccompanied by viability determinations, demonstrated RNA and protein as expendable polymers in *Pseudomonas aeruginosa*.

Dawes and Ribbons (1965) reported that starved *Escherichia coli* degraded intercellular polysaccharide, and that release of ammonia, indicating protein breakdown, did not take place until the polysaccharide was exhausted; nevertheless, protein turnover took place during the starvation period though death was delayed. As Strange and his colleagues had found for *Aerobacter aerogenes*, RNA degradation took place during the starvation period; no lipid metabolism was observed. Once death had started, it was more rapid in anaerobic than aerobic conditions. Strange (1966) presented evidence that adaptively formed β-galactosidase was mobilized more rapidly than the bulk of cell protein when *E. coli* was starved; thiomethylgalactoside antagonized this process. Sierra and Gibbons (1962) presented evidence that mobilization of stored poly-β-hydroxybutyrate preserved the viability of the halophile, *Micrococcus denitrificans*.

A considerable volume of work has been published on the endogenous metabolism of bacteria. Recent work in this field was reviewed by Dawes and Ribbons (1962, 1964) and a valuable symposium (Lamana, 1963) was published. This work will not be reviewed in detail here because correlations with viability have not generally been published or, if present, have been cursory. The general pattern is to the effect that starved organisms break down all their major polymeric constituents, except, perhaps, lipid and DNA, including such materials as poly-β-hydroxybutyric acid or sulphur when present. The rates and orders in which they conduct these breakdowns depend on the type of organism and its nutritional status. It would be interesting to know whether the polymetaphosphate deposits, which may be induced in certain bacteria by sulphur starvation (Harold and Sylvan, 1962), can function as energy reserves for starved populations. It is possible that the absence of substrate-accelerated death in Postgate and Hunter's (1964) S-limited populations was a result of protection by stored polyphosphates. Pine (1963), in experiments unaccompanied by viability measurements, illustrated a methodological difficulty in experiments involving gross analysis for loss of cell polymers on starvation: an alcohol-soluble protein, thought to be a reserve material in *Escherichia coli*, apparently disappeared on sulphur starvation. In fact it merely conjugated with polymetaphosphate formed by the bacteria in response to sulphur starvation.

McCarthy (1962) claimed that magnesium-starved *Escherichia coli*, in an otherwise complete medium, degraded their ribosomes to particles of low sedimentation coefficients, but did not die on incubation for up to 40 hr., according to plate counts. The conclusion that they did not die is surprising in view of the fragility that Mg^{2+}-limited bacteria seem generally to show, because it is probable that McCarthy's populations became Mg^{2+}-limited during his experiments. It is regrettable that a more rigid assessment of viability was not used, because magnesium starvation provides ideal circumstances for cryptic growth and it is possible that the experiments concerned a dynamic steady state of dying and multiplying cells analogous to the magnesium-starved cultures described by Postgate and Hunter (1963a); Dr. D. Kennell mentioned the appearance of non-viable bacteria in comparable experiments reported at the M.R.E. symposium in March, 1966.

Pardee and his colleagues, in a sequence of papers cited by Rachmeler and Pardee (1963), showed that bacteria heavily labelled in their DNA died more rapidly, when cold-stored in a protective environment containing glycerol, than those heavily labelled in their RNA or protein. Loss of viability was attributed to breakage of the polymer chain by radioactive decay. Their work advances, in their own words, "recognition that intact DNA is required for bacterial viability and enzyme synthesis"; one notes respectfully that the converse proposition, that provided DNA is unchanged a cell will be viable, does not follow. Ogg and his colleagues (Ogg and Zelle, 1957; Zelle and Ogg, 1957; Olsen and Ogg, 1963) claim to have induced diploidy in *Escherichia coli* by treatment with camphor, and to have shown that radiation resistance is thereby enhanced. As a means of studying the rôle of DNA in death by starvation, this approach has obvious possibilities but has not, to the writer's knowledge, been used.

6. *The Population Effect*

Postgate and Hunter (1963a) showed rigidly that Harrison's (1960) population effect, whereby sparse populations die more rapidly than denser ones, is not an experimental artifact. Population effects of divergent characters occur during substrate-accelerated death (see above). No further work on the intrinsic factors responsible for the population effect in starvation has been published.

B. Survival of Old Cultures: the "Death Phase"

The textbook curve normally includes, after the lag, logarithmic and stationary phases, a death phase which is often termed "exponential". In practice, the survival patterns of such populations are complex and are determined by their nutritional status. Postgate and Hunter (1963a) published survival curves for nitrogen-, carbon- and magnesium-limited flask cultures of *Aerobacter aerogenes* and observed that, though a majority of the Mg^{2+}-starved organisms were very mortal, the population as a whole reached a steady state of 2–4% viability which is maintained for many days. They attributed this phenomenon to cryptic growth.

That morphological changes occur in old bacterial cultures has been known for many years (e.g. Knaysi, 1951). Chatterjee and Williams (1962) described sequential morphological changes in ageing cultures of *Bacillus anthracis*. Imshenetskii and Zhiltsova (1965) recorded that young cells of various coliform bacteria and *Bacillus mycoides* contained stainable "nucleoids" which disappeared on ageing.

C. Mortality during Growth

The belief (see Wilson, 1922) that normal growing cultures contain a proportion of non-viable cells arose, as indicated earlier, from an experimental error in microscopic counting. Normal cultures of most bacteria are 95–100% viable (Jensen, 1928; Kelly and Rahn, 1932; Ziegler and Halvorson, 1935; Powell, 1958; Valentine and Bradfield, 1954; Postgate *et al.*, 1961; Quesnel, 1963). However, a few bacteria do die, and these can sometimes be recognized by their loss of contrast. With coliform bacteria, it is fair to say that an organism of low contrast is usually dead though one with strong contrast is not necessarily alive. Filamentous forms of coliform bacteria are usually incapable of further division in the writer's experience; Hoffman and Frank (1963) reported that heat-induced filaments of *Escherichia coli* may be considered dead after the second mean division time though, before then, they may sometimes be observed to fragment and to behave normally. Quesnel (1963) studied the growth of clones derived from individual *E. coli* and observed that more non-viable forms appeared in the second generation than at other times. He offered delayed trauma as a result of subculture as an interpretation of this finding. Delayed traumata of a comparable kind occur when organisms are treated with streptomycin (Kogut *et al.*, 1965).

Postgate and Hunter (1962) grew continuous cultures of *Aerobacter aerogenes* at so slow a rate that steady states with a proportion of dead organisms were obtained. Later (Postgate and Hunter, 1963a) they obtained some evidence that partly moribund magnesium-limited continuous cultures showed a population effect: the viability was greater the denser the population. The finding that continuous cultures reach such steady states conflicts with reports that, below a certain flow rate, chemostat cultures "go into lag" or otherwise become unstable (Novick, 1958; Schulze and Lipe, 1964). Dr. D. Herbert reported some further experiments on partly moribund continuous cultures at the M.R.E. symposium in March, 1966, and reported that the transition to this state was accompanied by morphological elongation of *A. aerogenes* and that these changes were reversible. It is a pity that the interesting experiments of Marr *et al.* (1963) on maintenance energy, using slow continuous cultures, were not accompanied by viability measurements.

D. Senescence of Microbes

Bacteria and most other microbes have no tendency to clonal senescence and, normally, clones can continue to develop indefinitely, at least in principle. In this respect, they contrast with tissue cultured from higher organisms: such cultures seem to senesce and die out over approximately fifty doublings unless they are of neoplastic origin (Hayflick, 1965). The nearest approach to an experimentally recognizable senescent state in bacteria is only reached in response to starvation: in that phase between death as expressed by loss of viability and by breakdown of the osmotic regulating mechanism (the moribund state) when the microbe is presumably a functioning biological entity, but one incapable of multiplication. Marcou, Rizet and their colleagues in a series of papers cited by Rizet *et al.* (1958), and summarized by Rizet and Marcou (1959), described the curious behaviour of a fungus, *Podospora anserina*, which apparently showed clonal senescence: after protracted logarithmic growth, the mycelium died, and subcultures from aged mycelia showed decreased longevity. Senescence appeared to be determined by the concentration of an infective particulate factor in the cytoplasm (Marcou and Schecroun, 1959).

Senescence among individuals in a growing population of unicellular microbes is, in principle, possible and has been observed in yeasts (Barton, 1950; Mortimer and Johnston, 1959). Beran and his colleagues (see Streibelova and Beran, 1963) used primulin staining and fluorescence microscopy to confirm that mother cells became scarred as a result of budding-off daughter cells and that overlap of scars on a given individual did not occur. Thus, after budding about twenty-five daughters, the cell surface became wholly scarred and non-viable in the sense that it could generate no further progeny. The proportion of such "25-scarred" cells in a freely growing culture is exceedingly small, but Liebelova *et al.* (1964) have been able to fractionate cells with low levels of scarring by centrifugation in dextran gradients, thus showing that an experimental approach to senescence in yeast cultures is possible. Yeasts characterized by apical budding showed comparable scarring with limitations on overlap, according to Dr. Beran's report at the M.R.E. Symposium (March, 1966) suggesting that senescence of individuals may occur generally in cultures of yeasts.

IV. Prospects

The study of the survival of vegetative microbes under minimum stress has some practical consequences that need little emphasis. Any process involving the intentional or unintentional storage of microbes, from sewage processing through food and water microbiology to fermentation and industrial production, will be influenced by the survival characters of the population in the environment being used; an understanding of the nutritional factors underlying such character should enable one in practice to enhance or decrease the mortality of such populations at will. Certain research practices also need reconsideration. The practice of chilling samples of cultures, or even stock cultures, accelerates their death (Strange and Dark, 1962; Postgate and Hunter, 1962) and promotes leakage of materials that are potential substrates for cryptic growth. The assessment of viability, as discussed already, is only rigidly satisfactory if performed by direct growth experiments. Cryptic growth and substrate-accelerated death are of obvious importance in the study of maintenance energy experiments and since, in most cases, it is viability that is being maintained, the importance of meticulous measurements of viability in such studies cannot be over-emphasized. Postgate and Hunter (1963a) mentioned some ecological consequences of the discovery of partly moribund slow continuous cultures; a recent paper by MacLeod (1965) comments on the excess of visible over viable bacteria normally found in the sea, indicating that the sea itself may be analogous to such a moribund continuous culture.

Studies on microbial survival in stress-free conditions, when a microbial population presumably dies from purely innate causes, could provide information on the fragility of the molecular architecture of the microbial cell. But stress-free conditions are probably impossible to obtain experimentally. Postgate and Hunter's (1962) relatively crude tests showed that washing procedures, the presence of trace elements in the distilled water, and even failure to alter the pH from the growth value, introduced mild stresses that influenced survival considerably. Some of these stresses can be overcome by suitable experimental design, but the work of Quesnel (1963), for example, suggests that, when such unintentional stresses are recognized and eliminated, even so simple an operation as transferring a loop of culture to a new environment can introduce a trauma which will lead to lowered viability among the first generation progeny. Logically, as in most of this review, the scientist is forced back to considering the question of what factors influence microbial survival under minimum stress, and the answer, of course, will be that it depends what that stress, however minimal it may be, is. Therefore such research can provide its most clear-cut data on the stresses to which microbes are or can become sensitive. Substrate-accelerated death, for example, is a very curious response to the situation imposed on the population, and the fact that it can be reproducibly demonstrated with ammonium-limited as well as carbon-limited cells, if only so far in one strain of organism, implies, as Postgate and Hunter (1964) pointed out, that it is likely to be intimately involved in the control mechanisms governing the cell's synthesis of constitutive material. The stress implicit in the population effect is one of the more baffling of the problems raised recently: what is it about a sparse population that causes each individual to die more rapidly than it would if it had more neighbours? Much of the work in this field has been done with strains of *Aerobacter aerogenes*, some of which were admittedly (and perhaps fortunately) peculiar. Extension of work of this kind to other organisms, particularly those that do not form polysaccharide, or which deposit "reserves" of fat, polymetaphosphate, sulphur or poly-β-hydroxybutyrate, would be very rewarding. In non-growing insect tissue (Maynard Smith, 1962), senescence and ultimate death may well be associated with cumulative damage to the cell's DNA, comparable to that induced by low doses of radiation. Starved, non-growing bacteria show no evidence for involvement of degradative damage to DNA on death; indeed, an increase in the DNA content of starved, logarithmic-phase *Escherichia coli* has been claimed (Brdrar *et al.*, 1965). The most lethal process conducted by the organism seems to be auto-degradation of RNA. Nevertheless, gross chemical analysis may well not detect the sort of fine damage to DNA that could cause loss of viability, and a follow-up of the work of Ogg and his colleagues mentioned earlier, showing that camphor-induced polyploidy in *E. coli* caused enhanced radiation resistance, might greatly enhance our understanding of microbial survival under the relatively mild stress of starvation.

Research on microbial death has so far provided little information that is relevant to the senescence and death of higher organisms, but Beran's demonstration of an analogy to senescence among multiplying yeasts opens fascinating possibilities for the study of "natural" death among microbes. It was discovered in yeasts because a method became available for distinguishing mother from daughter cells by scarring, which led to the observation that scarring imposed a restriction on further budding at that site. Though no satisfactory method exists at present for distinguishing mother from daughter organisms among bacteria, and though there is no obvious analogy between the budding of yeasts and the fission of bacteria, it is nevertheless perfectly possible that an analogous situation exists and that it accounts for the small percentage of non-viable cells observed even among the "healthiest" bacterial populations. In such circumstances microbial clones can be said to survive indefinitely because only the mothers, and not the daughters, suffer the restrictions on further multiplication caused by scarring or some analogous trauma.

V. Acknowledgement

I am grateful to Dr. K. Stacey who read and criticized the manuscript of this review.

References

Arber, W. (1963). *In* "Symbiotic Associations", (P. S. Nutman and B. Mosse, eds.) p. 12, *13th. Symp. Soc. gen. Microbiol.* University Press, Cambridge.
Barer, R., Ross, K. F. A. and Tkaczyk, S. (1953). *Nature, Lond.* **171**, 720.
Barton, A. A. (1950). *J. gen. Microbiol.* **4**, 84.
Bohinski, R. C. and Mallette, M. F. (1965). *Canad. J. Microbiol.* **11**, 663.
Borzani, W. and Vairo, M. L. R. (1958). *J. Bact.* **76**, 251.
Borzani, W. and Vairo, M. L. R. (1960). *J. Bact.* **80**, 574.
Brdrar, B., Kos, E. and Drakulis, M. (1965). *Nature, Lond.* **208**, 303.
Bretz, H. W. (1962). *J. Bact.* **84**, 1115.
Campbell, J. J. R., Gronlund, A. F. and Duncan, M. G. (1963). *Ann. N.Y. Acad. Sci.* **102**, 669.
Chatterjee, B. R. and Williams, R. P. (1962). *J. Bact.* **84**, 340.
Cohen, B. (1922). *J. Bact.* **7**, 183.
Collins, V. G. and Kipling, C. (1957). *J. appl. Bac.* **20**, 257.
Dawes, E. A. and Ribbons, D. W. (1962). *Annu. Rev. Microbiol.* **16**, 241.
Dawes, E. A. and Ribbons, D. W. (1964). *Bact. Rev.* **28**, 126.
Dawes, E. A. and Ribbons, D. W. (1965). *Biochem. J.* **95**, 332.
Delpy, L. P., Béranger, G. and Kaweh, M. (1956). *Ann. Inst. Pasteur* **91**, 112.
Fikhman, B. A. (1959a). *C.R. Acad. Sci. USSR* **124**, 1141.
Fikhman, B. A. (1959b). *J. Microbiol. Epidemiol. Immunobiol.* **30**, 121.
Fikhman, B. A. and Pryadkina, M. D. (1961). *J. Microbiol. Epidemiol. Immunobiol.* **32**, 462.
Freter, R. and Ozawa, A. (1963). *J. Bact.* **86**, 904.
Gabe, D. R. (1957). *Mikrobiologiya* **26**, 109.
Gilliland, R. B. (1959). *J. Inst. Brew.* **65**, 424.
Gorrill, R. H. and McNiel, E. M. (1960). *J. gen. Microbiol.* **22**, 437.
Harold, F. M. and Sylvan, S. (1962). *Procs. 8th intern. Congr. Microbiol., Montreal.* p. 36.
Harrison, A. P. (1960). *Proc. Roy. Soc. B.* **152**, 418.
Harrison, A. P. and Lawrence, F. R. (1963). *J. Bact.* **85**, 742.
Hayes, W. (1965). *In* "Function and Structure in Micro-organisms", (M. R. Pollock and M. H. Richmond, eds.) p. 294, *15th. Symp. Soc. gen. Microbiol.* University Press, Cambridge.
Hayflick, L. (1965). *Exp. Cell Res.* **37**, 614.
Hinshelwood, C. N. (1957). *In* "The Biology of Ageing", (W. B. Yapp and G. H. Bourne, eds.) p. 1. *6th. Symp. Inst. Biol.*, Institute of Biology, London.
Hoffman, H. and Frank, M. E. (1963). *J. Bact.* **85**, 1221.
Imshenetskii, A. A. and Zhiltsova, G. K. (1965). *Mikrobiologiya* **34**, 300.
Jacobs, S. E. and Harris, N. D. (1960). *J. appl. Bact.* **23**, 294.
Jacobs, S. E. and Harris, N. D. (1961). *J. appl. Bact.* **24**, 172.
Jannasch, H. W. (1965). *Biotechnol. Bioeng.* **7**, 279.
Jebb, W. H. H. and Tomlinson, A. H. (1960). *J. gen. Microbiol.* **22**, 93.
Jensen, K. A. (1928). *Z. Bakt.* (I Abt.), **107**, 1.
Kelly, C. D. and Rahn, O. (1932). *J. Bact.* **23**, 147.
Ketterer, H. (1956). *Brauwissenschaft* **9**, 59.
Knaysi, G. (1951). "Elements of Bacterial Cytology", 2nd Ed., Comstock Publishing Co., New York.
Koch, A. L. (1959). *J. Bact.* **77**, 623.
Kogut, M., Lightbown, J. W. and Issacson, P. (1965). *J. gen. Microbiol.* **39**, 165.
Kubitschek, H. E. (1960). *Research, Lond.* **13**, 128.

Lamana, C. (1963). *Ann. N.Y. Acad. Sci.* **102**, 515.
Liebelova, J., Beran, K. and Streibelova, E. (1964). *Folia Microbiol.* **9**, 205.
Lovett, S. (1964). *Nature, Lond.* **203**, 429.
MacLeod, R. A. (1965). *Bact. Rev.* **29**, 9.
Mager, J., Kuczynski, M., Schatzberg, G. and Avi-dor, Y. (1956). *J. gen. Microbiol.* **14**, 69.
Mallette, M. F. (1963). *Ann. N.Y. Acad. Sci.* **102**, 521.
Mallette, M. F., Cowan, C. I. and Campbell, J. J. R. (1964). *J. Bact.* **87**, 779.
Mandelstam, J. (1963). *Ann. N.Y. Acad. Sci.* **102**, 621.
Marcou, D. and Schecroun, J. (1959). *C.R. Acad. Sci., Paris*, **248**, 280.
Marr, A. G., Nilson, E. H. and Clark, D. J. (1963). *Ann. N.Y. Acad. Sci.* **102**, 536.
Mathews, M. M. and Sistrom, W. R. (1960). *Arch. Mikrobiol.* **35**, 139.
Maynard Smith, J. (1962). *Proc. Roy. Soc. B.* **157**, 115.
McCarthy, B. J. (1962). *Biochim. Biophys. Acta* **55**, 880.
McGrew, S. B. and Mallette, M. F. (1962). *J. Bact.* **83**, 844.
McGrew, S. B. and Mallette, M. F. (1965). *Nature, Lond.* **209**, 1096.
McQuillen, K. (1958). *Biochim. Biophys. Acta* **27**, 410.
Meadow, P., Hoare, D. S. and Work, E. (1957). *Biochem. J.* **66**, 270.
Meynell, G. G. and Meynell, E. (1965). "Theory and Practice in Experimental Bacteriology", 12–24 pp. University Press, Cambridge.
Mohr, V. and Larsen, H. (1963). *J. gen. Microbiol.* **31**, 267.
Mortimer, R. K. and Johnston, J. R. (1959). *Nature, Lond.* **183**, 1751.
Nakada, D. and Smith, I. (1962). *Biochim. Biophys. Acta* **61**, 414.
Norris, K. P. and Powell, E. O. (1961). *J.R. micr. Soc.* **80**, 107.
Novick, A. (1958). *In* "Continuous Cultivation of Micro-organisms", (I. Malek, ed.), p. 29, Czechoslovak Academy of Sciences, Prague.
Ogg, J. E. and Zelle, M. R. (1957). *J. Bact.* **74**, 477.
Olsen, R. H. and Ogg, J. E. (1963). *J. Bact.* **86**, 494.
Pine, M. J. (1963). *J. Bact.* **85**, 301.
Pittillo, R. F. and Narkates, A. J. (1964). *Canad. J. Microbiol.* **10**, 345.
Postgate, J. R. and Hunter, J. R. (1962). *J. gen. Microbiol.* **29**, 233. Corrigenda: (1964). *J. gen. Microbiol.* **34**, 473.
Postgate, J. R. and Hunter, J. R. (1963a). *J. appl. Bact.* **26**, 295.
Postgate, J. R. and Hunter, J. R. (1963b). *J. appl. Bact.* **26**, 405.
Postgate, J. R. and Hunter, J. R. (1964). *J. gen. Microbiol.* **34**, 459.
Postgate, J. R., Crumpton, J. E. and Hunter, J. R. (1961). *J. gen. Microbiol.* **24**, 15.
Powell, E. O. (1956). *J. gen. Microbiol.* **14**, 153.
Powell, E. O. (1958). *J. gen. Microbiol.* **18**, 382.
Powell, E. O. (1963). *J. Sci. Food. Agric.* (1) 1.
Quesnel, L. B. (1963). *J. appl. Bact.* **26**, 127.
Rachmeler, M. and Pardee, A. B. (1963). *Biochim. Biophys. Acta* **68**, 62.
Razumovskaya, Z. G. and Osipova, I. V. (1958). *Microbiologiya* **27**, 727.
Rhuland, L. E. (1957). *J. Bact.* **73**, 778.
Rizet, G. and Marcou, D. (1959). *Procs. 8th. intern Congr. Bot., Paris*, p. 121.
Rizet, G., Marcou, D. and Schecroun, J. (1958). *Bull. Soc. Francais Physiol. vegetale* **4**, 136.
Ryan, F. J. (1959). *J. gen. Microbiol.* **21**, 530.
Salunkhe, D. K., Dhaliwal, A. S. and Boe, A. A. (1962). *Nature, Lond.* **195**, 724.
Schulze, K. L. and Lipe, R. S. (1964). *Arch. Mikrobiol.* **48**, 1.
Shearer, C. (1916). *Lancet* **2**, 902.
Shearer, C. (1917). *Proc. Roy. Soc. B.* **89**, 440.

Sierra, G. and Gibbons, N. E. (1962). *Canad. J. Microbiol.* **8**, 255.
de Silva, R. (1953). *Abs. 6th. int. Congr. Microbiol., Rome* **1**, 191.
Smith, D. D. and Wyss, O. (1965) *Bact. Proc.*, A.24.
Strange, R. E. (1964). *Nature, Lond.* **203**, 1304.
Strange, R. E. (1966). *Nature. Lond.* **209**, 428.
Strange, R. E. and Dark, F. A. (1962). *J. gen. Microbiol.* **29**, 719.
Strange, R. E. and Dark, F. A. (1965). *J. gen. Microbiol.* **39**, 215.
Strange, R. E. and Hunter, J. R. (1966). *J. gen. Microbiol.* **44**, 255.
Strange, R. E. and Postgate, J. R. (1964). *J. gen. Microbiol.* **36**, 393.
Strange, R. E. and Shon, M. (1964). *J. gen. Microbiol.* **34**, 99.
Strange, R. E., Dark, F. A. and Ness, A. G. (1961). *J. gen. Microbiol.* **25**, 61.
Streibelova, E. and Beran, K. (1963). *Exp. Cell Res.* **30**, 603.
Strugger, S. (1948). *Canad. J. Res.* **26**, 188.
Sykes, J. and Tempest, D. W. (1965). *Biochim. Biophys. Acta* **103**, 93.
Theil, E. C. and Zamenhof, S. (1963). *Nature, Lond.* **199**, 599.
"Topley and Wilson's Principles of Bacteriology and Immunity" (1955). 4th ed. vol. 1, (G. S. Wilson and A. A. Miles, eds.) p. 115, Arnold, London.
Valentine, R. C. and Bradfield, J. R. G. (1954). *J. gen. Microbiol.* **11**, 349.
Wade, H. E. and Morgan, D. M. (1954). *Nature, Lond.* **174**, 920.
Weibull, C. (1960). *J. Bact.* **79**, 155.
White, P. B. (1947). *J. Path. Bact.* **59**, 334.
Wilkinson, J. F. (1959). *Exp. Cell Res.* Supp(7), 111.
Wilkinson, J. F. (1963). *J. gen. Microbiol.* **32**, 171.
Wilson, G. S. (1922). *J. Bact.* **7**, 405.
Wilson, G. S. (1926). *J. Hyg., Camb.* **25**, 150.
Winslow, C.-E. A. and Falk, I. S. (1923a). *J. Bact.* **8**, 215.
Winslow, C.-E. A. and Falk, I. S. (1923b). *J. Bact.* **8**, 237.
Zelle, M. R. and Ogg, J. E. (1957). *J. Bact.* **74**, 485.
Ziegler, N. R. and Halvorson, H. O. (1935). *J. Bact.* **29**, 609.

Aflatoxins*

Richard I. Mateles and Gerald N. Wogan

Department of Nutrition and Food Science,
Massachusetts Institute of Technology,
Cambridge, Massachusetts, U.S.A.

I. Introduction 25
A. Biological Effects of Aflatoxins 25
B. Chemical Characteristics of Aflatoxins 27
II. Microbiological Aspects of Aflatoxin Production 28
A. Organisms 28
B. Media and Culture Conditions 31
III. Biosynthesis of Aflatoxins 32
IV. Acknowledgements 36
References 36

I. Introduction

The aflatoxins are a series of secondary metabolites produced by some strains of *Aspergillus flavus* as well as certain other fungi associated with the spoilage of agricultural commodities. The compounds were discovered in England in 1961 as a result of widespread episodes of unexplained mortality encountered in poultry flocks, swine and cattle (Allcroft and Carnaghan, 1963). Investigations into the cause of the toxicity revealed that the compounds were contained in certain batches of heavily mould-damaged peanut meals.

The fungal origin of these toxic substances was suspected when highly toxic meals were found to contain large quantities of fungal hyphae (Austwick and Ayerst, 1963). Isolation of the fungi and subculture of the isolates on non-toxic substrates resulted in production of the toxins. The causative fungus was identified as *A. flavus* and the toxin produced was named "aflatoxin" to denote its origin (Sargeant *et al.*, 1961).

A. Biological Effects of Aflatoxins

Extensive studies with purified aflatoxins have revealed their high order of biological potency in many animal species. Their discovery in

* Contribution No. 931 from the Department of Nutrition and Food Science, Massachusetts Institute of Technology, Cambridge, Massachusetts.

agricultural commodities used as animal and human foodstuffs, and the subsequent demonstration of various biological effects resulting from ingestion of contaminated diets, emphasized the potential public health hazards which might arise from contamination of the food supply by these so-called "mycotoxins". These factors have recently stimulated considerable research activity dealing with many aspects of the aflatoxins as well as other toxic mould metabolites.

The toxic properties of the aflatoxins manifest themselves differently depending on the test system, dose and duration of exposure. Thus, they have been shown to be lethal to animals and animal cells in culture when administered acutely at certain dose levels and to cause histological changes in animals when smaller doses are administered subacutely. Chronic exposure for extended periods has resulted in the induction of malignant tumours in several animal species.

The pathology and toxicology of aflatoxin poisoning in animals has been extensively reviewed (Allcroft and Carnaghan, 1963; Wogan, 1965, 1966) and require only brief summary for the present purpose. The compounds are acutely toxic to most animal species, aflatoxin B_1 having the greatest lethal potency. Although there is some species variation in susceptibility, the LD_{50} for a single dose of the compound is in the range of 0·5–10 mg./kg. body weight for most experimental animals. The toxic effects of aflatoxin B_1 have also been investigated in animal cells *in vitro*. In cell cultures, lethality has been reported at concentrations of 1–5 μg./ml. of medium (Gabliks *et al.*, 1965) whereas inhibition of growth and mitotic rate has been reported at concentrations of the order of 0·03 μg./ml. (Legator and Withrow, 1964; Daniels, 1965).

The potency of the aflatoxins as acute poisons is thus clearly indicated. Significant effects also result from prolonged administration of sublethal quantities of the compounds to animals. Early investigations of aflatoxin-contaminated peanut meals revealed that rats fed diets containing toxic peanut meal developed malignant liver tumours within a six-month period (Lancaster *et al.*, 1961). This finding was the first indication that the aflatoxins are carcinogenic to rats, and this observation has since been amply confirmed (Barnes and Butler, 1964; Butler and Barnes, 1964; Dickens and Jones, 1964). In experiments with purified aflatoxin B_1, significant tumour incidence has been observed in rats fed diets containing less than 0·3 p.p.m. of the compound. Liver carcinoma has also been induced by feeding contaminated diets to ducks (Carnaghan, 1965), and to the rainbow trout (Ashley *et al.*, 1965). By virtue of the low doses required for tumour induction, aflatoxin B_1 is among the most potent hepatocarcinogens yet discovered.

B. Chemical Characteristics of Aflatoxins

The proposed structures of the four known aflatoxins are shown in Fig. 1 (Asao *et al.*, 1963, 1965; van Dorp *et al.*, 1963; Chang *et al.*, 1963). The results of X-ray analysis of aflatoxin G_1 (Cheung and Sim, 1964) confirmed the structures shown in Fig. 1 but not the alternative configurations proposed by van der Merwe and his coworkers (1963).

Aflatoxins are extractable from mould culture filtrates by moderately polar solvents such as chloroform, methanol or acetone. In most cases, the extracts contain mixtures of the four aflatoxins although their relative concentrations depend upon factors such as the fungal strain,

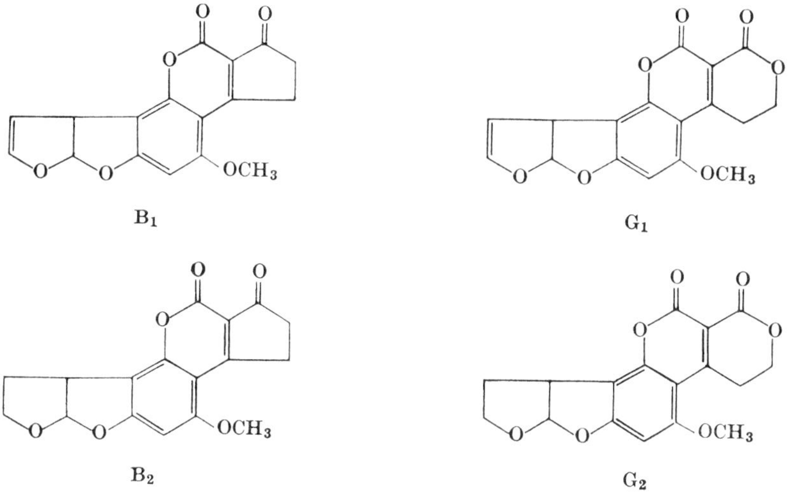

Fig. 1. Chemical formulae of the four known aflatoxins.

and the culture medium and conditions as discussed subsequently. Individual components are isolated chromatographically, and the designations B and G refer to the characteristic blue or green fluorescence exhibited upon exposure to ultraviolet radiation. The subscripts (1 and 2) refer to their relative chromatographic mobilities. Aflatoxins B_2 and G_2 are also known as dihydro-aflatoxins B and G, names which indicate the saturation in the terminal furan ring. Some relevant data concerning the physical properties of these compounds are summarized in Table 1.

Thin-layer chromatography has been found to be the most convenient method for the separation of aflatoxins in mixtures. A widely used system, of which many modifications exist, comprises development of silica gel (Kieselgel G) plates with chloroform containing 2–3% methanol (de Iongh *et al.*, 1962, 1964). Another type of system which has given excellent results in our hands uses a liquid–liquid partition chromatogram prepared with kieselguhr (Kieselguhr G) impregnated with a formamide–water solution and developed with benzene saturated with formamide (Adye and Mateles, 1964).

Determinations of aflatoxin concentration after separation on chromatograms can be accomplished by two techniques. The first comprises visual comparison of the fluorescence intensity of aflatoxin spots on a chromatogram with the intensity of a parallel spot of a known amount of reference aflatoxin on the same plate. This method has good sensitivity since the minimum amount of the compound visible in a compact area is of the order of 0·5 mμg. Alternatively, when individual chromatogram

TABLE I. Selected Physical and Chemical Data on Aflatoxins

Aflatoxin	Empirical formula	Mol. wt.	Molar extinction coefficient at 265 mμ	Molar extinction coefficient at 363 mμ	Fluorescence emission max. (mμ)
B_1	$C_{17}H_{12}O_6$	312	13,400	21,800	425
B_2	$C_{17}H_{14}O_6$	314	9,200	14,700	425
G_1	$C_{17}H_{12}O_7$	328	10,000	16,100	450
G_2	$C_{17}H_{14}O_7$	330	11,200	19,300	450

areas contain more than 1 or 2 μg., the supporting material can be removed from the plate, the compound eluted with solvent, and the extinction of the solution determined at the absorption maximum (363 mμ). The concentration of aflatoxin is then calculated utilizing the molar extinction coefficients listed in Table I.

Although the latter method is less sensitive and hence less suitable for screening samples in which aflatoxins may be present in only relatively small amounts, it is much less subject to either quenching or enhancement of fluorescence by other compounds in the extract being chromatographed, and is the method of choice for studies on biosynthesis where relatively large amounts of aflatoxins will be present in the culture filtrates.

II. Microbiological Aspects of Aflatoxin Production

A. ORGANISMS

The fungi which produce aflatoxins appear to lie almost exclusively within the *Aspergillus flavus* group (Raper and Fennell, 1965). Table II shows the results of various surveys undertaken to provide information on the frequency of occurrence of aflatoxin-producing strains. From these and other unpublished surveys, it appears that 10–30% of strains of *A. flavus* or *A. parasiticus* isolated from mouldy peanuts or other agricultural products are potentially able to produce aflatoxins, although in many cases the substrates from which the isolates were made contained no measurable amounts of aflatoxins.

TABLE II. Aflatoxin-Producing Ability of Fungi

Species	Aflatoxin producers	Source of cultures	Reference
Aspergillus flavus	0/13	C.B.S.[a]	de Vogel *et al.*, 1965
Aspergillus flavus	2/2	Uganda nuts	de Vogel *et al.*, 1965
Aspergillus flavus	3/4	Brazil peanuts	de Vogel *et al.*, 1965
Aspergillus flavus	1/1	Nigeria peanuts	de Vogel *et al.*, 1965
Aspergillus flavus	26/93	Q.M.C.C.[b]	Parrish *et al.*, 1966
Aspergillus flavus	9/59	Peanuts	Austwick and Ayerst, 1963
Aspergillus flavus-oryzae	0/5	C.B.S.	de Vogel *et al.*, 1965
Aspergillus oryzae	0/13	C.B.S.	de Vogel *et al.*, 1965
Aspergillus parasiticus	1/3	C.B.S.	de Vogel *et al.*, 1965
Aspergillus parasiticus	2/2	Hawaiian beetle	de Vogel *et al.*, 1965
Aspergillus parasiticus	4/4	Q.M.C.C.	Parrish *et al.*, 1966
Aspergillus gymnosardae	0/1	C.B.S.	de Vogel *et al.*, 1965
Aspergillus effusus	0/2	C.B.S.	de Vogel *et al.*, 1965
Aspergillus micro-virido-citrinus	1/1[c]	Brazil peanuts	de Vogel *et al.*, 1965
Aspergillus sp.	1/2	Brazil peanuts	de Vogel *et al.*, 1965
Penicillium puberulum	1/1[d]	Peanuts	Hodges *et al.*, 1964

[a] Centraalbureau voor Schimmelcultures, Baarn, The Netherlands.
[b] Quartermaster Culture Collection, U.S. Army Laboratories, Natick, Mass., U.S.A
[c] Uncertain that aflatoxin is produced.
[d] Uncomfirmed report.

Considering the wide use made of *Aspergillus oryzae* and related organisms for the production of fermented foods in the Orient, the finding that pure koji starters used industrially were not a source of aflatoxin-producing strains (Hesseltine *et al.*, 1966) is of interest. These authors expressed the opinion that starters used in home-grown koji production frequently contaminated with unwanted fungi might contain aflatoxin-producing organisms as contaminants.

The various strains which produce aflatoxins synthesize these compounds in amounts which vary both with the strain and with the medium and culture conditions (see Table III), as is the case for most secondary metabolites of fungi.

TABLE III. Effects of Strain and Culture Conditions on the Relative Amounts of Aflatoxins Formed

Strain	Medium	Total aflatoxins produced (mg./l. or mg./kg.)	Relative amounts B_1	B_2 (% total)	G_1	G_2	References
Aspergillus flavus (ATCC 15517)	Synthetic medium, resting culture	45	87	4	9	<1	Mateles and Adye, 1965
Aspergillus flavus (5 strains)	Peanuts	25–265	44	1	54	1	Codner *et al.*, 1963
Aspergillus flavus (MRE 1)	Peanuts	14	98	2	0	0	Codner *et al.*, 1963
Aspergillus flavus (NRRL 2999)	Wheat	870	35	9	48	7	Hesseltine *et al.*, 1966
Aspergillus flavus (NRRL 2999)	Wheat + methionine	1700	44	11	38	7	Hesseltine *et al.*, 1966
Aspergillus flavus (NRRL 3000)	Sucrose-amino acids, submerged (Mateles and Adye, 1965) 72 hr., 20°	86	26	0	74	0	Ciegler *et al.*, 1966
Aspergillus flavus (NRRL 3000)	ibid, 25°	154	70	0	30	0	Ciegler *et al.*, 1966

B. Media and Culture Conditions

The conditions which favour aflatoxin production on field crops have been described in a number of publications (Bampton, 1963; McDonald and Harkness, 1963, 1964; McDonald and A'Brook, 1963; Burrell *et al.*, 1964; McDonald *et al.*, 1964; Schroeder and Ashworth, 1965; Ashworth *et al.*, 1965). The production of aflatoxins under laboratory conditions has been investigated primarily with the aim of securing aflatoxins for chemical, physiological, and toxicological purposes, and to a lesser extent in order to elucidate the pathway of aflatoxin biosynthesis.

Many workers have had considerable success using moistened, crushed or whole peanuts (Sargeant *et al.*, 1963; Nesbitt *et al.*, 1962; Diener *et al.*, 1963; van der Zijden *et al.*, 1962; de Iongh *et al.*, 1962; Codner *et al.*, 1963; Newberne *et al.*, 1964) or whole, crushed, or shredded moistened wheat (Schumaier *et al.*, 1961; Asao *et al.*, 1963, 1965; Chang *et al.*, 1963; Hesseltine *et al.*, 1966) inoculated with various strains of *A. flavus* or *A. parasiticus* and incubated at 30° for 5–10 days. Yields ranging up to 1·5 g. of mixed aflatoxin per kg. of substrate were obtained (Hesseltine *et al.*, 1966). Other grains such as corn, oats, rye, buckwheat, rice, soybeans, and sorghum support considerable aflatoxin production and have been used as substrates (van der Merwe *et al.*, 1963; Armbrecht *et al.*, 1963; Hesseltine *et al.*, 1966).

Although solid substrates may be used for aflatoxin production, they are rather inconvenient for use in preparing large amounts of aflatoxins and are also unsuited to studies on the biosynthesis of these compounds. For these, some workers have studied various synthetic and defined liquid media. De Iongh *et al.* (1962), van der Zijden *et al.* (1962), and Nesbitt *et al.* (1962) mention briefly the production of small quantities of aflatoxins in a glucose–ammonium nitrate medium, and Nesbitt and his colleagues (1962) suggest the desirability of adding zinc sulphate to the basic medium. P. de Vogel (personal communication) reported obtaining 3 mg. of mixed aflatoxins per litre with a glucose–ammonium nitrate-salts medium (Brian *et al.*, 1961) in agitated bottles incubated for 7 days at 30°. He also found that the addition to Czapek–Dox medium of an aqueous extract of peanuts increased the yield from 0·3 mg./l. to about 72 mg./l. Codner *et al.* (1963) found that a Czapek–Dox medium supplemented with corn-steep liquor gave yields of 100–200 mg./l. in 3 days in shaken flasks at 27°, although they were unable to secure adequate yields in stirred fermentors. Adye and Mateles (1964) described an entirely synthetic glucose–ammonia medium, containing Zn^{2+}, which supported aflatoxin production by growing or resting cultures, and further demonstrated that such a medium could give yields of 50–80 mg./l. in shaken flasks at 30° (Mateles and Adye, 1965).

As is frequently found in the production of secondary fungal metabolites, the yield of product was greatly influenced by the nature of the carbon and nitrogen sources, although it was not possible to interpret the results meaningfully. Sucrose, glucose, and fructose were almost equally good as carbon sources, while an amino-acid hydrolysate was the preferred nitrogen source. At least 0·4 p.p.m. of Zn^{2+} was required for maximum production of aflatoxins. Lee *et al.* (1966) confirmed the requirement for Zn^{2+}, and found that, with a different strain of *A. flavus*, about 0·8 p.p.m. Zn^{2+} was required for maximum production of aflatoxins. These workers further reported that the addition of Cd^{2+} (4 μM) increased the yield of aflatoxins about 50% above a control, although the concentration thus achieved was only about 4–5 mg./l.

III. Biosynthesis of Aflatoxins

Relatively little has been published concerning the biosynthetic pathway of the aflatoxins. Virtually the only experimental results available in the literature are those of Adye and Mateles (1964), who found that a variety of compounds could be incorporated into aflatoxins. This study was carried out primarily to determine which compounds would be suitable to use in preparing ^{14}C-labelled aflatoxins. In order to obtain aflatoxin of a high specific activity, the suspected precursor was incorporated into the medium of a resting culture at a concentration of 0·5–1·0 mM. Excellent incorporation of L-[Me-^{14}C]methionine, DL-[3-^{14}C-alanine]phenylalanine, DL-[2-^{14}C alanine]tyrosine, DL-[3-^{14}C alanine]-tryptophan and [1-^{14}C]acetate were obtained. It was suspected that the incorporation of the label from tryptophan was an artifact caused by the degradation of tryptophan, yielding [3-^{14}C]alanine, which could give rise to [3-^{14}C]pyruvate and [2-^{14}C]acetate. That this was so was later demonstrated by the finding that the addition of unlabelled alanine or acetate drastically decreased the amount of label incorporated from tryptophan (J. Adye and R. I. Mateles, unpublished observations).

A possibility of degradation by enzymes induced by the relatively high concentrations of precursor added to the medium, followed by incorporation of the degradation products, e.g. acetate, was suspected when it was found that the amount of label incorporated from carboxyl-labelled phenylalanine or tyrosine was very much less than from these amino acids labelled in the side-chain. A study of the incorporation of label from phenylalanine and tyrosine labelled in various positions indicated that the incorporation of label fits the pattern calculated from the assumption that these aromatic acids are being degraded according to the pathway shown in Fig. 2 (J. Adye and R. I. Mateles, unpublished observations).

Recent studies in our laboratories indicate that, when ^{14}C-labelled tyrosine or phenylalanine is present at tracer concentrations in the medium, so that degradation is unlikely to occur, the extent of label incorporated into aflatoxin is unaffected by the position of the label in the precursor molecule. Thus, at this time, we cannot exclude the

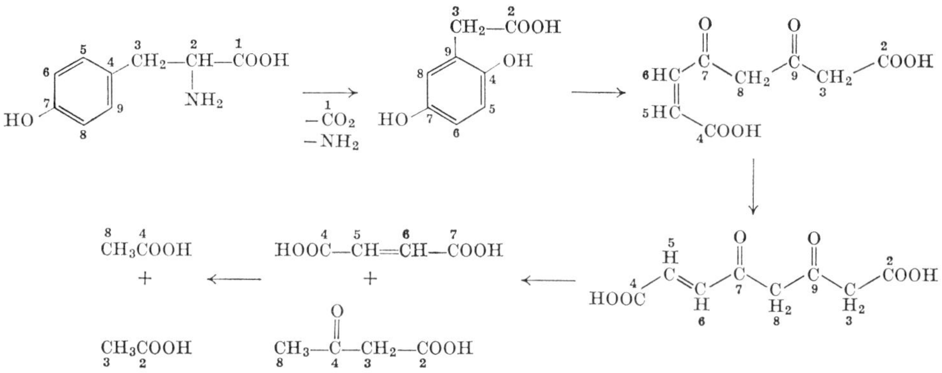

FIG. 2. Probable pathway for the degradation of tyrosine to acetate.

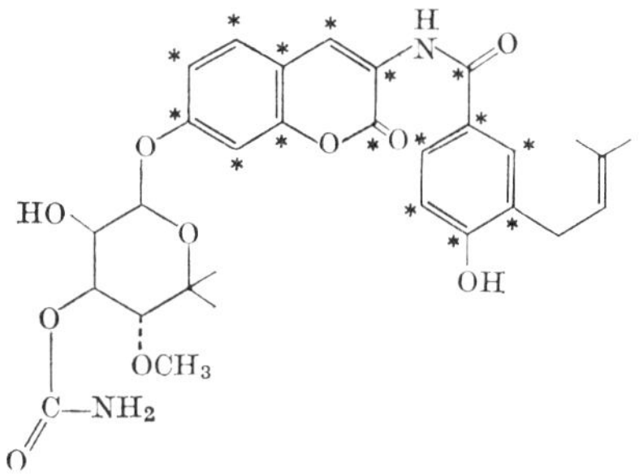

FIG. 3. The structure of novobiocin. The carbon atoms derived from tyrosine are marked *.

possibility that the coumarin portion of the aflatoxin molecule is derived from tyrosine or phenylalanine as has been found to be so with the streptomycete antibiotic, novobiocin (Fig. 3; Chambers *et al.*, 1960). The activity incorporated into aflatoxin from L-[Me-^{14}C]methionine has been shown to be exclusively in the O-methyl group (S. A. Brechbuehler and G. H. Buchi, unpublished observations), as has previously been demonstrated for the O-methyl group of novobiocin (Birch *et al.*, 1960) It is clear that the remainder of the molecule may be derived at least in part from acetate units.

Holker and Underwood (1964) suggested that aflatoxins may be derived from the *Aspergillus versicolor* metabolite, sterigmatocystin, by the pathway shown in Fig. 4. Furthermore, Thomas (1965, and personal communication) proposed a pathway by which sterigmatocystin may be derived from the anthraquinone pigment, averufin, also a metabolite of *A. versicolor*, and another pathway for the conversion of sterigmatocystin into aflatoxin (Fig. 5).

Holker and Underwood (1964) failed to find any incorporation of ^{14}C-sterigmatocystin (produced by *A. versicolor* on media containing [1-^{14}C]acetate) into aflatoxin by either *A. flavus* or *A. parasiticus*.

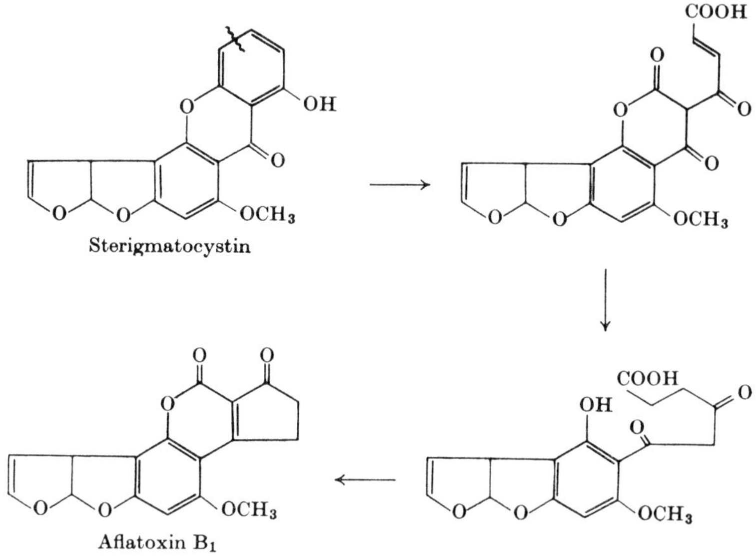

FIG. 4. Possible pathway for the biosynthesis of aflatoxin from sterigmatocystin as suggested by Holker and Underwood (1964).

They also found that these organisms did not produce detectable quantities of sterigmatocystin, while *A. versicolor* did not produce detectable quantities of aflatoxins. This evidence, however, does not rule out their hypothesis.

Other hypotheses, without experimental evidence, are those of Moody (1964) who suggested that mevalonate is a precursor of aflatoxin, and Heathcote *et al.* (1965) who suggested that kojic acid may be a precursor of the non-furan portions of the sterigmatocystin and/or aflatoxin molecules.

It is likely that, in the near future, various groups working on the problem of aflatoxin biosynthesis will be able to shed further light on the question of the involvement of aromatic amino acids in its biosynthesis.

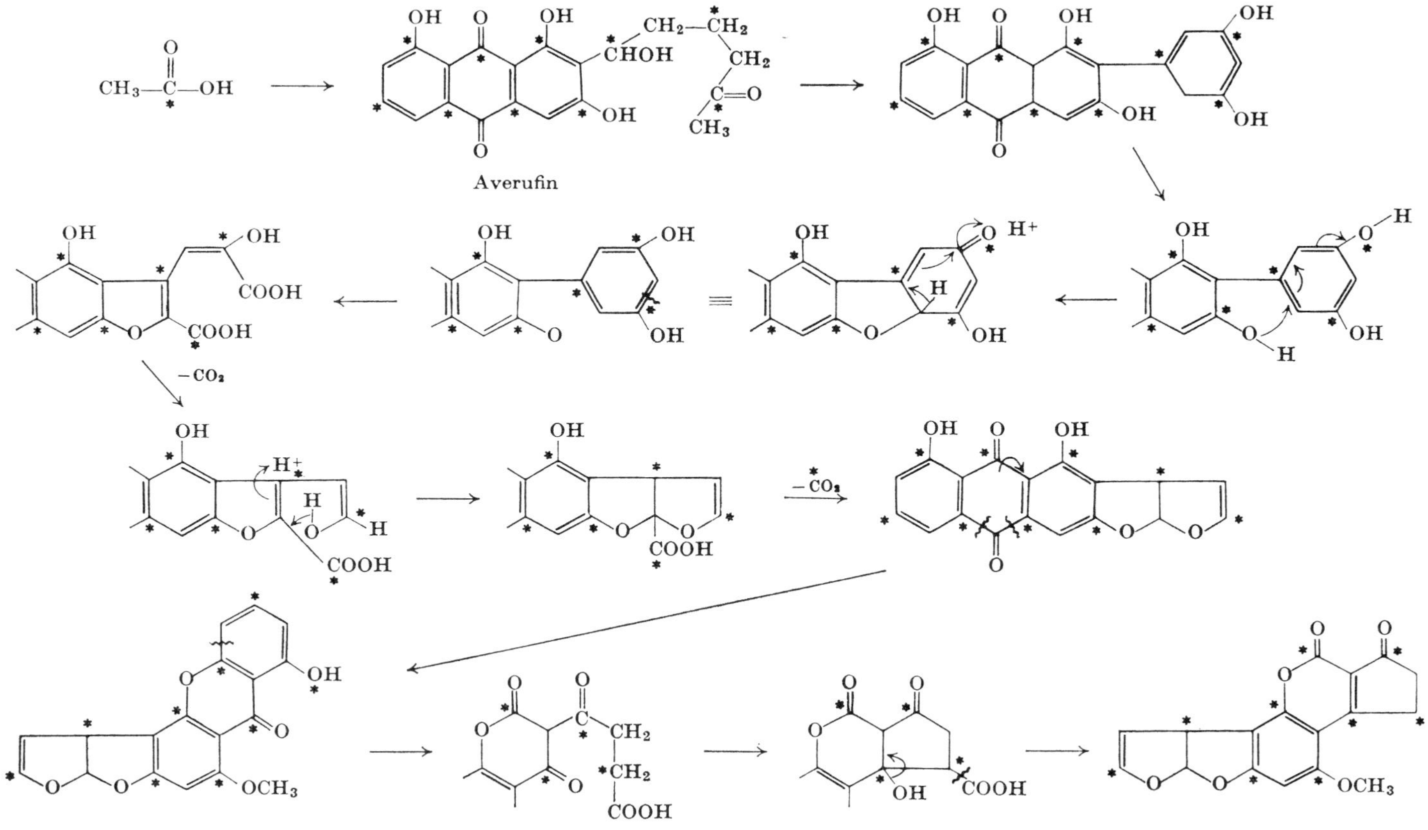

Sterigmatocystin

Aflatoxin B_1

FIG. 5. Possible pathway for the biosynthesis of aflatoxins from acetate, involving sterigmatocystin as an intermediate. The carbon atoms, representing the carboxyl carbons of acetate groups, are marked *. The pathway was suggested by Thomas (1965).

IV. Acknowledgements

The research in the authors' laboratories described in this report was supported in part by grant EF-00694 from the Division of Environmental Engineering and Food Protection, U.S. Public Health Service, and from contract PH 43-62-468 of the National Cancer Institute, National Institutes of Health. The authors appreciate the continued interest and collaboration of Dr. G. H. Buchi, Department of Chemistry, Massachusetts Institute of Technology. Dr. Ronald Bentley, University of Pittsburgh, and Dr. Robert Thomas, Imperial College, University of London, participated in useful discussions concerning the possible pathways of biosynthesis of aflatoxins and related compounds.

References

Adye, J. and Mateles, R. I. (1964). *Biochim. biophys. Acta* **86**, 418.

Allcroft, R. and Carnaghan, R. B. A. (1963). *Chem. Ind.* (*Lond.*), p. 50.

Armbrecht, B. H., Hodges, F. A., Smith, H. R. and Nelson, A. A. (1963). *J. Assoc. Offic. Agr. Chem.* **46**, 805.

Asao, T., Buchi, G., Abdel-Kader, M. M., Chang, S. B., Wick, E. L. and Wogan, G. N. (1963). *J. Amer. chem. Soc.* **85**, 1706.

Asao, T., Buchi, G., Abdel-Kader, M. M., Chang, S. B., Wick, E. L. and Wogan, G. N. (1965). *J. Amer. chem. Soc.* **87**, 882.

Ashley, L. M., Halver, J. E., Gardner, W. K. Jr. and Wogan, G. N. (1965). *Fed. Proc.* **24**, 627.

Ashworth, L. J. Jr., Schroeder, H. W. and Langley, B. C. (1965). *Science* **148**, 1228.

Austwick, P. K. and Ayerst, G. (1963). *Chem. Ind.* (*Lond.*), p. 55.

Bampton, S. S. (1963). *Trop. Sci.* **5**, 74.

Barnes, J. M. and Butler, W. H. (1964). *Nature, Lond.* **202**, 1016.

Birch, A. J., Cameron, D. W., Holloway, P. W. and Rickards, R. W. (1960). *Tetrahedron Letters* **25**, 26.

Brian, P. W., Dawkins, A. W., Grove, J. F., Hemming, H. G., Lowe, D. and Norris, G. L. F. (1961). *J. exp. Bot.* **12**, 1.

Burrell, N. J., Grundey, J. K. and Harkness, C. (1964). *Trop. Sci.* **6**, 74.

Butler, W. H. and Barnes, J. M. (1964). *Brit. J. Cancer* **17**, 699.

Carnaghan, R. B. A. (1965). *Nature, Lond.* **208**, 308.

Chambers, K., Kenner, G. W., Temple Robinson, M. J. and Webster, B. R. (1960). *Proc. chem. Soc.* p. 291.

Chang, S. B., Abdel-Kader, M. M., Wick, E. L. and Wogan, G. N. (1963). *Science* **142**, 1191.

Cheung, K. K. and Sim, G. A. (1964). *Nature, Lond.* **201**, 1185.

Ciegler, A., Peterson, R. E., Lagoda, A. A. and Hall, H. H. (1966). *Appl. Microbiol.* **14**, 826.

Codner, R. C., Sargeant, K., and Yeo, R. (1963). *Biotechnol. Bioeng.* **5**, 185.

Daniels, M. R. (1965). *Brit. J. exp. Pathol.* **46**, 183.

de Iongh, H., Beerthuis, R. K., Vles, R. O., Barrett, C. B. and Ord, W. O. (1962). *Biochim. biophys. Acta* **65**, 548.

de Iongh, H., Vles, R. O. and van Pelt, J. G. (1964). *Nature, Lond.* **202**, 466.

de Vogel, P., van Rhee, R. and Blanche Koelensmid, W. A. A. (1965). *J. appl. Bact.* **28**, 213.

Dickens, F. and Jones, H. E. H. (1964). *Brit. J. Cancer* **17**, 691.
Diener, U. L., Davis, N. D., Salmon, W. D. and Prickett, C. O. (1963). *Science* **142** 1491.
Gabliks, J. Z., Schaeffer, W., Friedman, L. and Wogan, G. N. (1965). *J. Bact.* **90**, 720.
Heathcote, J. G., Child, J. J. and Dutton, M. F. (1965). *Biochem. J.* **95**, 23P.
Hesseltine, C. W., Shotwell, O. L., Ellis, J. J. and Stubblefield, R. D. (1966). *Bact. Rev.* **30**, (in Press).
Hodges, F. A., Zust, J. R., Smith, H. R., Nelson, A. A., Armbrecht, B. H. and Campbell, A. D. (1964). *Science* **145**, 1439.
Holker, J. S. E. and Underwood, J. G. (1964). *Chem. Ind.* (*Lond.*), p. 1865.
Lancaster, M. C., Jenkins, F. P. and Philp, J. McL. (1961). *Nature, Lond.* **192**, 1095.
Lee, E. G. H., Townsley, P. M. and Walden, C. C. (1966). *J. Food Sci.* **31**, 432.
Legator, M. S. and Withrow, A. (1964). *J. Assoc. Offic. Agr. Chem.* **47**, 1007.
Mateles, R. I. and Adye, J. (1965). *Appl. Microbiol.* **13**, 208.
McDonald, D. and A'Brook, J. (1963). *Trop. Sci.* **5**, 208.
McDonald, D. and Harkness, C. (1963). *Trop. Sci.* **5**, 143.
McDonald, D. and Harkness, C. (1964). *Trop. Sci.* **6**, 12.
McDonald, D., Harkness, C. and Stonebridge, W. C. (1964). *Trop. Sci.* **6**, 131.
Moody, D. P. (1964). *Nature, Lond.* **202**, 188.
Nesbitt, B. F., O'Kelly, J., Sargeant, K. and Sheridan, A. (1962). *Nature, Lond.* **195**, 1062.
Newberne, P. M., Wogan, G. N., Carlton, W. W. and Abdel-Kader, M. M. (1964). *Toxicol. appl. Pharmacol.* **6**, 542.
Parrish, F. W., Wiley, B. J., Simmons, E. G. and Long, L. Jr. (1966). *Appl. Microbiol.* **14**, 139.
Raper, K. B. and Fennell, D. I. (1965). "The Genus Aspergillus", pp. 106–109, Williams and Wilkins Co., Baltimore.
Sargeant, K., Sheridan, A., O'Kelly, J. and Carnaghan, R. B. A. (1961). *Nature, Lond.* **192**, 1096.
Sargeant, K., Carnaghan, R. B. A. and Allcroft, R. (1963). *Chem. Ind.* (*Lond.*), p. 53.
Schroeder, H. M. and Ashworth, L. J. Jr. (1965). *Phytopathol.* **55**, 464.
Schumaier, G., Panda, B., de Volt, H. M., Laffer, N. C. and Creek, R. D. (1961). *Poultry Science* **40**, 1132.
Thomas, R. (1965). *In* "Biogenesis of Antibiotic Substances", (Z. Vanek and Z. Hostalek, eds.), pp. 160–161, Academic Press, New York.
van der Merwe, K. J., Fourie, L. and Scott, de B. (1963). *Chem. Ind.* (Lond.), p. 1660.
van der Zijden, A. S. M., Koelensmid, W. A. A., Bolding, J., Barrett, C. B., Ord, W. O. and Philp, J. (1962). *Nature, Lond.* **195**, 1060.
van Dorp, D. A., van der Zijden, A. S. M., Beerthuis, R. K., Sparreboom, S., Ord, W. O., de Jong, K. and Keuning, R. (1963). *Rec. Trav. Chim. Pays-Bas* **82**, 587.
Wogan, G. N. (ed.) (1965). "Mycotoxins in Foodstuffs", M.I.T. Press, Cambridge.
Wogan, G. N. (1966). *Bact. Rev.* **30**, 460.

Regulation of Nucleic Acid and Protein Formation in Bacteria

NIELS OLE KJELDGAARD

University Institute of Microbiology, Copenhagen, Denmark

Lorsqu'il n'est pas en notre pouvoir de discerner les plus vraies opinions, nous devons suivre les plus probables.

RENÉ DESCARTES

I. Introduction 39
II. Regulation and Bacterial Growth 40
A. DNA 42
B. RNA 43
C. Proteins 46
D. States of Transition 47
III. Regulation of DNA Formation 50
A. Molecular Weight of Bacterial DNA 50
B. The Replication Cycle 52
C. Induction of DNA Replication 55
D. Genetic Studies of DNA Replication 60
IV. Regulation of RNA Formation 62
A. Synthesis of RNA 62
B. Modes of Regulation 63
C. Role of Amino Acids 64
V. Regulation of Enzyme Synthesis 76
A. The Operator Model 76
B. Regulatory Genes 83
C. Models of Regulation 88
References 90

I. Introduction

It is only during the last 10 years that the concept of regulation of the formation of nucleic acids and proteins has been recognized as a function of great physiological significance to cells. These notions have partly been formulated into theories, and are now so widely acclaimed in all systems that it frequently seems difficult to separate hypothesis from facts, reality from extrapolations.

Numerous and detailed reviews of all facets of the present subject have appeared (Lark, 1966a, b; Neidhardt, 1964; Maaløe and Kjeldgaard, 1966; Jacob and Monod, 1963), and it will not, and indeed cannot, be my aim in the space available to present a complete coverage of the field to be discussed. By discussing a rather limited number of findings, I hope to convey some personal feelings towards these subjects which for several years have been the focus of interest of this laboratory.

We shall here define regulation as a process which, under the influence of environmental conditions, leads to variations in the quantity per cell of the macromolecule in question.

A regulation can be *general*, simultaneously affecting all species within a class of macromolecules, or *specific*, when it affects only a single species. It is obvious that the latter type of regulation, up till now, has mostly been recognized as affecting proteins but it is likely, as our methods improve for scoring specificities among transfer (t-RNA) and ribosomal (r-RNA) ribonucleic acid molecules, that specific regulation will be shown also in these groups of macromolecules.

It is known that regulation can occur as a result of changes in the overall rate of synthesis of a macromolecule. With metabolically unstable molecules, however, it should not be overlooked that an observed regulation might manifest variations in the rate of breakdown of these molecules.

II. Regulation and Bacterial Growth

The apparently rather trivial fact that a bacterial strain can grow at widely different rates, depending on the growth media, has given valuable information about the general regulation of macromolecular synthesis. Such studies obviously require that growth conditions can be reproduced, and it might be pertinent here again to stress the importance of this point. It is essential for reproducibility that a condition of balanced growth be established (Campbell, 1957) and maintained in the culture. In most standard media, with good aeration, this requires that the cell density be maintained at least below 0.2 mg. dry weight per ml. (see Maaløe and Kjeldgaard, 1966).

In bacterial cultures, growing under such steady-state conditions at a given temperature, the composition of the cells with respect to DNA, RNA, and proteins is characteristic of the rate of growth supported by the medium (Schaechter *et al.*, 1958). Comparing the amount of cell mass, DNA and RNA per cell of *Salmonella typhimurium*, we found that these values vary in exponential fashion with the growth rate. On a semilogarithmic plot, this gives a linear relationship with different slopes for the various macromolecules (Fig. 1). Rather similar results have been obtained with other organisms (Neidhardt and Magasanik, 1960).

The large variations clearly indicate that regulation occurs, and the regularity of the changes suggests that, with each component, the same mechanisms of regulation are active at all growth rates. The low-molecular weight components of the media, in a unique sense, determine a series of induction and repression mechanisms which again, through their action, set the growth rate and the overall composition of the cells. We shall now discuss in more detail some of the observed variations in the synthesis of macromolecular cell constituents.

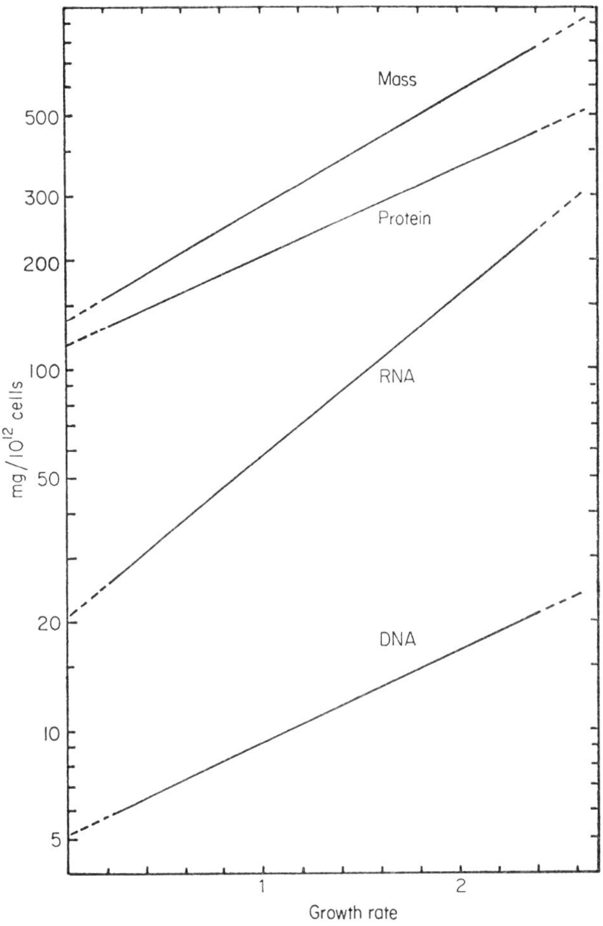

Fig. 1. Relationship in *Salmonella typhimurium* between cell composition and growth rate, measured in doublings per hour (after Maaløe and Kjeldgaard, 1966).

A. DNA

In cytological studies, the average number of nucleoids in *Salmonella* was found to vary from about 1 in slow-growing cells, through about 1·5 in cells grown in glucose-minimal medium (generation time 50 min.) to about 3 in cells growing in a rich medium at a generation time of 25 min. This variation corresponds roughly to the changes in the DNA content of the cells, indicating that the observed cytological units each contain an approximately fixed amount of DNA. Furthermore, it can be concluded that there is no obligatory linkage between nuclear division and cellular division.

It is interesting to note that, although the amount of DNA per cell increases with the growth rate, there is a decrease per unit mass under the same conditions. This might easily correspond to a constant amount of DNA per unit of bacterial surface since, in *Salmonella typhimurium*, a decrease in the surface area : mass ratio with increasing growth rate is to be expected. This indeed has been observed with *Bacillus megaterium* by measurements of membrane lipid phosphorus and cell wall hexosamine (Sud and Schaechter, 1964). It is known from the work of Schaechter *et al.* (1959) that, in glucose-grown cells, DNA is synthesized during at least 80% of the generation time. If that holds true for all growth rates, it is obvious that the rate of replication of the DNA must show large variations. It is known, as will be discussed later, that, in cells grown in minimal medium, each DNA molecule has one growing point only. The step time for the addition of a deoxynucleotide to a polynucleotide strand therefore might vary dramatically. However, Yoshikawa *et al.* (1964) found that, in rapidly growing cells of *B. subtilis*, the number of replication points per NDA molecule had increased to three. Such an increase would allow for an augmentation in the overall rate of replication, with the simultaneous conservation of the step time. Preliminary experiments with slow-growing bacteria made us suggest that the step time is essentially constant at all growth rates, and that the variations in DNA content of the cells are the effects of intermissions between successive rounds of DNA replication (Maaløe and Kjeldgaard, 1966). This is only partly confirmed by the experiments of Lark (1966a, b, c) in which, at low growth rates, autoradiographic experiments showed that short time labelling with thymine resulted in a relatively large unlabelled fraction. From their experiments however, it is concluded that, in slow-growing cells, there is a decrease in the rate of DNA replication and in the duration of the intermission between successive rounds of replication. If this is so, the rate of multiplication of the genomes can be affected both by the rate of replication and by varying the duration of the intermission. At intermediate growth rates, the cells are thought to contain two genomes, undergoing alternate replications; in bacteria growing in a glucose-minimal medium, the two genomes would appear to replicate simultaneously (Lark, 1966a, b).

B. RNA

The metabolically stable species of RNA, r-RNA and t-RNA, account for the major part of the RNA, and only a few percent is found in the messenger RNA (m-RNA) fraction. As can be seen from Fig. 1, the RNA content of the cells varies in amount from about 10% to 30% of the total mass.

By measuring the relative amounts of ribosomes, using the analytical ultracentrifuge, Ecker and Schaechter (1963) found a steady decrease extrapolating virtually to zero at extremely low growth rates. These large variations in ribosomal content are clearly indicated in the electron micrographs of Fig. 2 which show sections of rapidly growing and slow-growing cells of *Salmonella typhimurium*. It is evident that, in the cells from the rapidly growing culture, the ribosomes are rather uniformly distributed throughout the cytoplasm. In the slow-growing cells, the cytoplasm is almost devoid of ribosomes, which seem to exist mostly in contact with the nuclear region.

To obtain more accurate data, Kjeldgaard and Kurland (1963) measured the amount of r-RNA and t-RNA by sucrose gradient centrifugation of the RNA isolated from cells grown in different media. Large variations were observed in the relative amounts of these two fractions, ranging from about 18% t-RNA in broth-grown cells to about 65% t-RNA in cells grown in a glutamate-minimal medium with a generation time of 5 hr. The total RNA was usually extracted from broth-grown carrier cells mixed with $^{32}PO_4$-labelled cells grown in a different medium. The cells were lysed with sodium dodecyl sulphate (SDS) and purified by phenol treatment and ethanol precipitations. All purification steps were controlled for differential losses of the RNA fractions and, to control the extraction process, broth-grown cells were lysed either by SDS or disrupted by ultrasonic treatment prior to the SDS addition. The results obtained gave reason for confidence in both the purification and the extraction procedures. Later observations, however, have shown that our extraction control, although fully valid, was an unfortunate choice, since it is only with broth-grown cells that identical results are obtained with and without ultrasonic treatment. At other growth rates, cell lysis by SDS, for unknown reasons, gives rise to a selective loss of r-RNA. The results reported, therefore, are not entirely correct. Repetition of the measurements using ultrasonic treatment before SDS addition has given

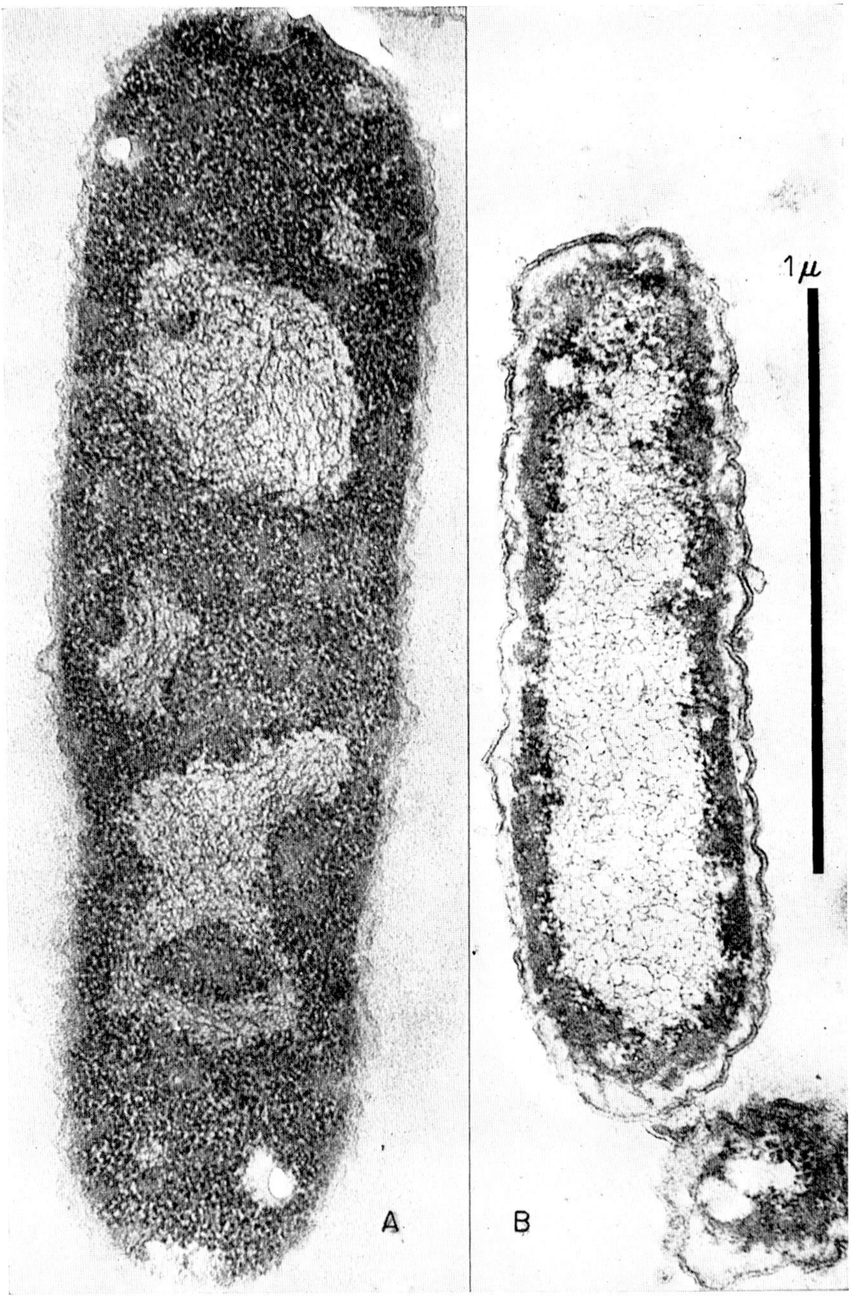

Fig. 2. Electron micrographs of thin sections of *Salmonella typhimurium*. The cells were grown in a broth medium (growth rate, $\mu = 2{\cdot}4$; A), or in a glutamate-minimal medium ($\mu = 0{\cdot}2$; B). The ribosomes of the rapidly growing cells are packed throughout the cytoplasm, whereas the ribosomes of the slow-growing cells are found clustered around the nuclear region. These electron micrographs were prepared by Mrs. Annelise Fiil and are reproduced with her kind permission.

evidence for a t-RNA content of about 25% in glucose-grown cells, and about 45% in bacteria grown in the glutamate medium.

Even less pronounced variations have been reported for *Escherichia coli* by Rosset *et al.* (1964, 1966). According to R. Lavallé (personal

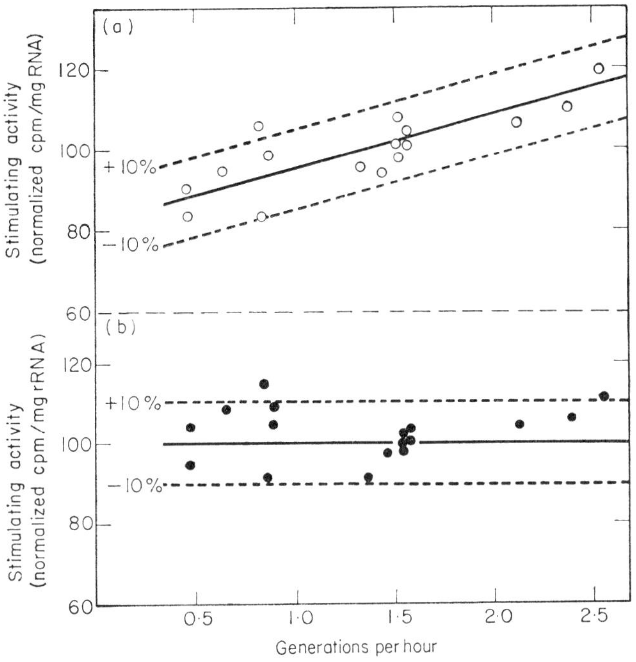

FIG. 3. The stimulating activity of RNA, isolated from *Escherichia coli* grown at different rates, on the extent of amino acid incorporation *in vitro*. RNA was isolated from cells growing exponentially in media which supported different growth rates. In each experiment, the stimulation of the incorporation of [6-^{14}C]leucine by a subcellular protein-forming system was followed as a function of the amount of added RNA. The normalized responses measured as counts/min./mg. RNA are shown as a function of the growth rate of the culture (a). For all RNA preparations the proportion of r-RNA was determined by sucrose gradient centrifugation. In (b) the normalized counts/min./mg. r-RNA are plotted against the growth rates of the cultures.

communication), the relative amounts of t-RNA found in the RNA preparations are extremely dependent upon the cell concentration used for the isolation of RNA. At relatively high cell concentrations, the result of Rosset *et al.* (1964) are reproduced. In all cases, however, the trend is the same, namely that slow-growing cells contain a higher proportion of t-RNA than rapidly growing cells. This phenomenon has also been observed with *Bacillus subtilis* (Doi and Igarashi, 1964). It is therefore clear that, besides a general regulation of the total RNA of the cells, the formation of r-RNA and of t-RNA can be regulated independently. Attempts to measure the amount of m-RNA under different growth conditions have encountered extreme difficulties.

The development of cell-free protein-forming systems by Nirenberg and Matthaei (1961) gave a possible way of evaluating the total amount of m-RNA in RNA preparations. The messenger fraction seems to be the only normally occurring RNA which stimulates the incorporation *in vitro* of amino acids in a cell-free protein-forming system (Wilson and Gros, 1964). Recently, we have used this method to determine the m-RNA concentration in RNA isolated from bacteria grown under conditions of balanced growth in different media (J. Forchhammer and N. O. Kjeldgaard, unpublished observations). Compared to the total RNA content the stimulating activity shows an increase with increasing growth rate (Fig. 3a). If, however, the values are corrected for the increase in r-RNA with the growth rate, the stimulating activity is constant at all rates of growth (Fig. 3b). This proportionality between m-RNA and r-RNA might imply an important regulatory function of the ribosomes in at least messenger synthesis, as will be discussed later.

C. Proteins

Proteins account for 70–80% of the bacterial mass. There is a slight decrease in the amount per cell with increasing growth rate. A comparison of the number of ribosomes per cell with the rate of protein synthesis gave indications for a relatively constant (within a factor of two) rate of protein synthesis per ribosome at all growth rates (Kjeldgaard and Kurland, 1963). A similar conclusion was reached by Ecker and Schaechter (1963) in their ultracentrifugal studies of the ribosomal content of cells grown in different media. The idea is further supported by the work of McCarthy (1962) and Kennell and Magasanik (1962) on magnesium starvation of *Escherichia coli*. The starvation leads to a breakdown of the ribosomes. Re-addition of Mg^{2+} to starved cultures results in a renewed ribosomal synthesis proceeding in an exponential fashion, and a resumption of protein synthesis at a rate which at all stages is proportional to the amount of ribosomes present in the bacteria.

However, it is extremely difficult to obtain a clear verification of the notion that the ribosomes show a constant efficiency in protein synthesis. Besides determinations of the total number of ribosomes and the amount of protein synthesized, neither of which can be determined with great precision, a rigorous proof requires an estimate of the fraction of ribosomes actually engaged in protein formation.

An estimate of this value was attempted by Schaechter *et al.* (1965), by measuring the proportion of 70S ribosomes bound in polysomal structures. In *Escherichia coli* grown in glucose-minimal medium, Casamino acid medium, or in Penassay broth, they found that 75–80% of the 70S ribosomes were bound as polysomes, and therefore presumably active in protein synthesis. On the other hand, in experiments in which protein synthesis in bacteria was studied at 0°, Goldstein *et al.* (1964) found that the number of polypeptide chains being synthesized was about one-third of the number of ribosomes present. If one assumes that the ribosomes are involved in regulatory steps in addition to their function as units of protein synthesis, it is not unlikely that the distribution of the ribosomes over these functions might vary as a result of different growth conditions.

D. States of Transition

To observe the immediate effects of the regulatory functions which have been revealed by the steady-state measurements, two types of experiments were employed both involving abrupt changes in the growth conditions of bacterial cultures. In the "upshift experiments", cultures growing under conditions of balanced growth were diluted into another medium supporting faster growth. In the "downshift experiments", the bacteria were transferred by a filtration procedure into a medium supporting a lower rate of growth (Kjeldgaard *et al.*, 1958). The results of a typical upshift of a *Salmonella typhimurium* culture from a glucose-minimal medium into nutrient broth are given in Fig. 4, and show a characteristic pattern of changes in the rate of synthesis of the different macromolecules. The observed pattern seems to be universal in character, and is exactly reproduced by *Escherichia coli* strains both of the stringent and the relaxed type (see below), as well as by a Gram-positive organism like *Bacillus megaterium* (Sud and Schaechter, 1964).

Let us now discuss briefly some of the main features of this upshift pattern. Immediately after the shift, the synthesis of RNA shows a remarkable change in rate. By using pulse labelling for 20 sec. with [7,8-^{14}C]guanine as a measure of the rate of RNA synthesis, Koch (1965) showed that the rate starts to increase in less than 5 sec. In about 100 sec. the eventual rate in the new medium was established. In the shift employed, this change corresponded to an increase in rate by a factor of about 10. It is obvious from this experiment that the low rate of synthesis before the shift cannot be caused by a limitation of the amount of RNA polymerase present.

The RNA formed immediately after the upshift seems to be preferentially r-RNA (Mitsui *et al.*, 1963), and, during the first 30 min., the ribosomal concentration becomes adjusted to that characteristic of rapidly growing cells (Fig. 5; Kjeldgaard, 1963). It is clear that ribosome synthesis also requires the formation of ribosomal proteins. During the first few minutes after an upshift, this process must occupy the rather restricted protein-forming machinery very heavily, since it can be

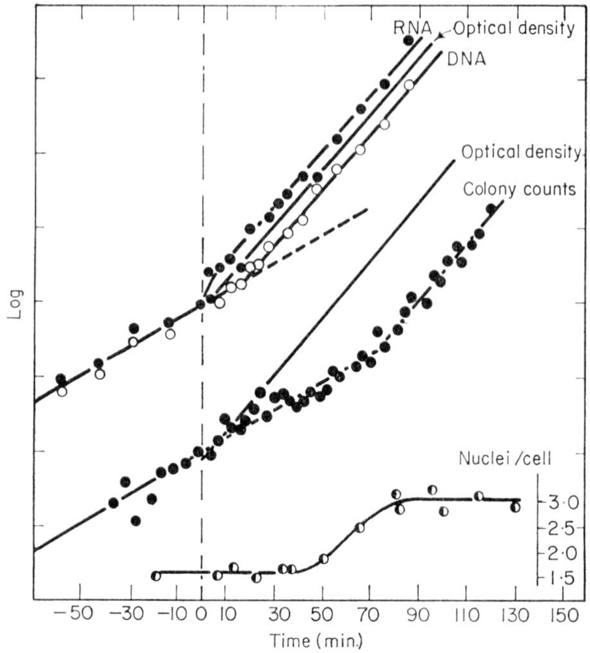

FIG. 4. Results of a shift-up experiment with *Salmonella typhimurium*. A culture was shifted, at 37°, from glucose–salts-minimal medium to broth at time 0. Optical density, viable counts, and RNA and DNA contents were then determined at frequent intervals. For clarity, two separate sets of curves are presented. The top curves show changes in optical density and RNA and DNA contents. The middle curves show the relation between optical density and viable counts. With each set of data, the logarithms of the measured values are plotted against time, and all values are transposed so as to make the curves representing balanced growth in minimal medium co-incide. The distance between horizontal lines corresponds to one doubling. In the curve in the lower right-hand corner, the average numbers of nuclei per cell that were estimated from direct counts on stained preparations are plotted against time (after Kjeldgaard *et al.*, 1958).

calculated that about 30% of the proteins synthesized during this period are ribosomal proteins. In glucose-grown cells, the ribosomal proteins account for about 12% of the total proteins.

It has been suggested (Neidhardt and Fraenkel, 1961) that the rapid change in RNA synthesis could be caused by changes in the concentration of the pool of triphosphates within the cells. Analysis of this pool using a very sensitive thin-layer chromatographic technique (Neuhard and Munch-Petersen, 1966) has not yielded support for this assumption, but shows that the pool of triphosphates drops immediately after an upshift and increases again after 10-15 min. (J. L. Ingraham, personal communication).

The rate of protein synthesis, as measured by the uptake of radio-active amino acids, shows an increase after the shift corresponding to the

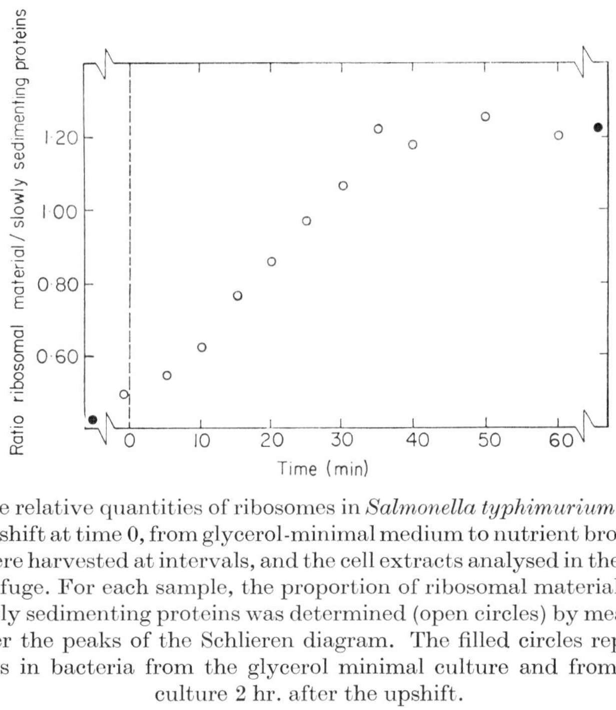

FIG. 5. The relative quantities of ribosomes in *Salmonella typhimurium* before and after an upshift at time 0, from glycerol-minimal medium to nutrient broth. Culture samples were harvested at intervals, and the cell extracts analysed in the analytical ultracentrifuge. For each sample, the proportion of ribosomal material compared to the slowly sedimenting proteins was determined (open circles) by measuring the areas under the peaks of the Schlieren diagram. The filled circles represent the proportions in bacteria from the glycerol minimal culture and from the broth culture 2 hr. after the upshift.

increase in the number of ribosomes (Kjeldgaard, 1961). Under upshift conditions, a constant efficiency of the ribosomes in protein formation seems to hold true.

These findings therefore put strong emphasis on the regulation of ribosome formation as a central function in the adjustment of the cells to the environment.

After upshift, the synthesis of DNA continues at the old rate for about 20 min., and then changes rapidly to the rate characteristic of the shift medium. As mentioned above, Yoshikawa *et al.* (1964) obtained evidence for the existence in broth-grown *Bacillus subtilis* cells of three growing points per genome. Lark (1966b) suggests that bacteria grown in glucose-minimal medium contain two genomes each having a single point of replication. In succinate-grown cells, the two genomes show alternating replications, i.e. at any time only one growing point is acting. At slower rates of growth, the cells contain only one genome with one growing point, and intermissions are observed between successive rounds of replication (Lark, 1966c).

During the transition period in an upshift experiment, it appears that, once the replication has started, it goes to completion according to the mode prevalent at the start. It is only at the end of this replication that the mode of replication of the new medium is imposed onto the replication mechanism (Lark, 1966b). By combining the delayed change with the variations in the mode of replication, it seems possible, at least with some organisms, to explain the course of DNA synthesis (O. Maaløe, unpublished observations). After the increase in rate of DNA synthesis has become manifest, and probably as a consequence thereof, the average number of nuclei per cell gradually increases to the number characteristic of broth-grown cells.

It is remarkable that the rate of cell division stays at the preshift rate for about 60 min. after the shift. As is the case for the other functions, the time of rate shift is found to be independent of the actual rates before and after the shift. It is still not understood what functions are responsible for the delay, but it might reflect a subtle relationship between the DNA replication and the act of cell division.

The results of downshift experiments, e.g. a shift of broth-grown organisms into a glucose-minimal medium, are in many respects reverse copies of the results of upshift experiments. The most remarkable feature is the immediate halt of RNA synthesis, which in fact corresponds to a situation of multiple amino acid starvation since the enzymes necessary for the synthesis of amino acids have been repressed during growth in the rich medium. It is therefore not surprising that a downshift using a mutant in which RNA synthesis is under relaxed control (Stent and Brenner, 1961) leads to a continued synthesis of RNA (Neidhardt, 1963). A more detailed discussion of the behaviour of relaxed mutants will be presented later in this chapter.

III. Regulation of DNA Formation

A. Molecular Weight of Bacterial DNA

From the steady-state measurements mentioned above, one can calculate that, for *Salmonella typhimurium*, one "nuclear equivalent" contains $4–5 \times 10^{-12}$ mg. of DNA corresponding to about 3×10^{9} or about $4{\cdot}5 \times 10^{6}$ base pairs. In *Bacillus subtilis*, the amount of DNA in outgrown cells was found to be between $6{\cdot}4 \times 10^{-12}$ and $6{\cdot}9 \times 10^{-12}$ mg. per cell (Massie and Zimm, 1965). Since these cells are thought to contain one or two nuclear bodies, the results are in fair accordance with the *Salmonella* data. In stationary cultures of *Escherichia coli*, the DNA content has been found to be 7×10^{-12} mg. per cell. It seems reasonable, therefore, to assume that the amount of DNA per nucleus is fairly constant.

In the experiments of Cairns (1963), the high-power resolution of autoradiography of [^{3}H]-labelled molecules was employed to trace the linear dimensions of radio-active DNA from *Escherichia coli*. The bacteria labelled with [^{3}H]thymidine were cautiously lysed by detergents and the DNA allowed to extrude slowly from the cells to prevent strand breakage by shearing forces. The DNA was then trapped on a membrane filter which, after drying, was overlaid with photo-emulsion. The grains produced by the decay of the tritium give a photographic image of the DNA molecules and among those, circular DNA molecules with a length of about 1200 μ are seen. If we now assume that the DNA when dried on the membranes exists in the A-configuration (Fuller *et al.*, 1965), a distance of 1200 μ corresponds to about $4{\cdot}5 \times 10^6$ nucleotide pairs. Therefore, what we observe in stained preparations as nuclear bodies are most likely single closely packed DNA molecules.

From these autoradiographic experiments it is, however, not possible to exclude the existence of protein linkers connecting smaller stretches of DNA. Measurements of the molecular weight of DNA by sedimentation analysis or viscosity measurements naturally require an extensive handling of the DNA preparations and thereby exposure to shearing forces (Levinthal and Davison, 1961). During the last few years, increased precautions during the isolation have yielded continuously increasing estimates of the molecular weight of bacterial DNA. Thus Berns and Thomas (1965) obtained DNA preparations from *Haemophilus influenzae* with a molecular weight of 4×10^8. Massie and Zimm (1965) found the molecular weight of DNA from both *Escherichia coli* and *Bacillus subtilis* to about 250×10^6, when cells were treated in a dialysis bag with lysozyme, pronase, and RNase, and the lysate exposed to phenol without shaking at 65°. If protein linkers do exist, they cannot therefore be present in numbers greater than 8–10.

The fact that the DNA molecules in Cairns' experiments appear as circular structures is in beautiful agreement with the continuous genetic map of the *Escherichia coli* genome (Jacob and Wollman, 1957). This again raises the question of a protein linker, as the unwinding of the DNA strands during replication must require at least one bond per chain allowing for the rotational movement not permitted by the rigid polynucleotide chain itself.

B. The Replication Cycle

The experiments of Meselson and Stahl (1958), demonstrating the semi-conservative replication of DNA, simultaneously showed that after one generation of growth at least 90% of the DNA had replicated only once. The synthesis of a new molecule of DNA therefore does not seem to start until the replication of the entire old molecule is finished. The experiments of Meselson and Stahl, however, gave no information about the number of points of replication on the DNA. Using short-term labelling of DNA with 5-bromo-[2-^{14}C]uracil, Bonhoeffer and Gierer (1963) were able to show that probably only one, but at most two, regions of each DNA molecule became labelled during exposure of a bacterium to the "heavy" thymine analogue. That the number of growing points is indeed one was clearly demonstrated in the experiments of Cairns (1963). If cells were labelled with [^{3}H]thymidine for more than one generation, the course of the second round of replication in the radio-active medium could be observed. The radioautographs revealed the existence of DNA molecules about 1000 μ long in the process of replication, with a loop of DNA containing one branch with the double grain density (two labelled DNA strands) of the rest of the molecule. Taken together with the results of Meselson and Stahl (1958) this suggests that, in *Escherichia coli*, the DNA replication starts from a fixed point on the chromosome and proceeds to the end of the DNA molecule.

During experiments on thymine-less death of *Escherichia coli* 15 TAU$^-$ (*thy*$^-$, *arur*$^-$)*, Hanawalt and Maaløe (1961) observed that, when this strain was starved of the required amino acid, DNA synthesis continued for about 90 min. at 37°, until the total amount of DNA had increased by about 40%. At the same time, the cells became refractive to thymine-less death.

This was interpreted to mean that the cells can finish an already started replication, but are unable to initiate a second round of replication in the absence of the amino acid. Proof of this hypothesis was given in a series of elegant experiments by Lark and his associates. An amino acid auxotrophic substrain of *E. coli* 15 *thy*$^-$ was starved of a required amino acid for 80 min. Growth was re-initiated, and radio-active thymine added

* To designate the genetic characters of the various bacterial mutants described in the text, the following abbreviations have been used: *ac*, *ara*, *gal*, and *lac* indicate utilization of acetate, arabinose, galactose and lactose respectively; *arg*, *his*, *leu*, *met*, *pro*, *pyr*, *thr*, *thy*, *try*, and *ura* indicate requirements for arginine, histidine, leucine, methionine, proline, pyrimidine, threonine, thymine, tryptophan, and uracil respectively; *arur* and *ilv* indicate one-step mutations resulting in simultaneous requirements for arginine and uracil and for isoleucine and valine; *pho* indicates production of alkaline phosphatase; *str*^{r} indicates resistance to streptomycin and *str*^{s} sensitivity to streptomycin. Other terms are defined in the text.

during the period, when DNA synthesis started and continued until an increase of about 10% in DNA content was observed. According to the hypothesis, the starvation would allow DNA replication to go to completion, and when replication was re-initiated the beginning of all DNA molecules will be labelled. The culture was then allowed to grow in a non-radio-active medium for several generations and again starved of a required amino acid for 80 min. Growth was once more re-initiated by amino acids, but this time 5-bromouracil (BU) was added in place of thymine. At different times after BU addition, samples were withdrawn and the DNA liberated by lysis of the cells using SDS and digestion with papain. The DNA was centrifuged through a CsCl gradient and bands representing fully "light", "half heavy" (one "light" and one BU strand), and "all heavy" DNA were collected. For each time of sampling, the amount of radio-activity associated with the "half heavy" band was determined and compared to the total amount of BU-labelled DNA.

If the amino acid starvation in fact permitted the completion of DNA replication, it would be expected that the radio-active DNA would be found in the half heavy fraction very early after the start of replication. This indeed was observed (Fig. 6). Furthermore, the early transfer was found to be rather independent of the number of generations of growth between the two periods of starvation. This clearly shows that DNA replication not only starts from one point and proceeds through a cycle of replication, but that the starting point is strictly maintained. However, it does not tell us whether all bacteria in the culture have the starting point at the very same genetic location (Lark *et al.*, 1963).

In *Bacillus subtilis*, this problem was elegantly approached by Yoshikawa and Sueoka (1963a, b). They assumed that, by analogy to amino acid-starved cells, both "resting" cells and spores contain a "finished" genome. Every genetic marker, therefore, should be represented by the same number of copies. In exponentially growing cultures, on the other hand, replication from a fixed point on the chromosome implies that the markers close to the starting point are present in twice as many copies as markers close to the end. The frequencies of transformation of eleven genetic characters were measured for DNA isolated from cells growing exponentially in a glucose-minimal medium, and compared to the values obtained for DNA from "resting" cells. For two wild-type strains of *Bacillus subtilis*, the ratio varied from 1 to 2 and the same genetic marker, *ade*, was found close to the starting point, and the *met* locus close to the terminus (Yoshikawa and Sueoka, 1963a, b; Yoshikawa *et al.*, 1964).

In the experiments of Nagata (1963), an attempt was made to locate the region of the starting point on the *Escherichia coli* genome by quantitative measurements of gene dosage. The "genes" selected were two prophages, λ and 424, which are located far apart on the genetic map. It was assumed that, after induction by u.v. radiation, the number of phages produced would be related to the number of prophages present in the cells. Stationary-phase cultures of two *E. coli* Hfr strains and one F^- strain, all double lysogenic for λ and 424, were synchronized by filtration through a pile of filter paper according to the technique of Maruyama and Yanagita (1956), and the small cells that passed through the paper

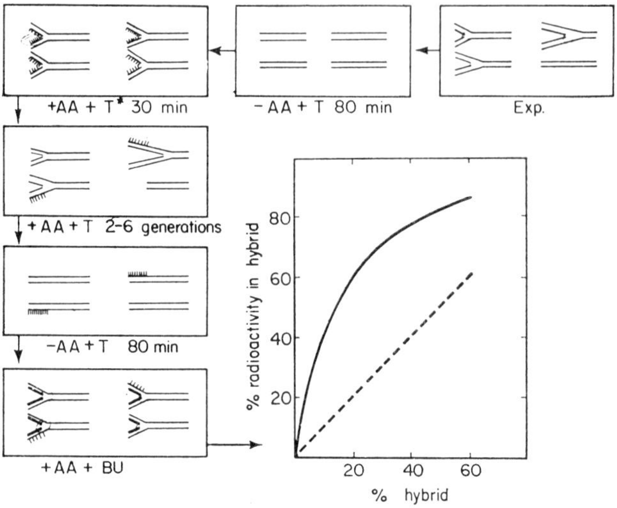

FIG. 6. Schematic representation of the course of a double labelling experiment of DNA according to Lark *et al.* (1963). The DNA from the bromouracil-grown *Escherichia coli* was analysed by CsCl gradient centrifugation, and the quantity of hybrid density material and the amount of radio-activity associated with this fraction were determined. The full line on the graph shows the results from the experimental culture, and the broken line those from a control culture not submitted to amino acid starvation. AA denotes amino acid; T, thymine; T* [^{3}H]-thymine; BU, bromouracil. See text for explanation. Redrawn from Lark *et al.* (1963).

were inoculated into fresh medium. At different times during the subsequent growth, cells were induced and the titres of λ and 424 prophages determined in the lysate. Variations in the proportion of the two phages were observed in the two Hfr strains, but no changes were seen with the F^- strain.

The results were interpreted to mean that, in Hfr strains, the replication of DNA has a definite polarity, starting at the point of attachment of the F factor. In the two strains tested, this means that replication proceeds in opposite directions. In the F^- strain, no polarity was observed and it was suggested that in such strains the starting point of replication has a random location.

The interpretation of the results, however, is based on the assumption that synchronous growth was established in the cultures, and that synchronous cell division was accompanied by synchronous DNA replication. However, by re-examining the data of Nagata (1963), the viable counts that were taken to indicate synchronous growth equally well fit a conventional exponential growth curve. The same is true for the DNA curve. Although the variations in the ratio of λ to 424 prophages are reported to be statistically significant, the synchronization data seem not equally significant and the interpretations, although they might be true, are not convincingly supported by the experimental evidence.

C. Induction of DNA Replication

1. *Induction During Mating*

During recombination between an Hfr and an F^- strain of *Escherichia coli*, the genetic transfer is polarized. The Hfr genome has an origin, *O*, which is the first to be transferred to the recipient bacteria, and a terminus which at 37° is transferred to the F^- cell some 90–100 min. later. The origin and the direction of transfer vary from strain to strain.

It is obvious that, during transfer, the genome physically exists as a linear structure. Two theories have been proposed to explain the mechanism of transfer of DNA which seem to be the only macromolecules to enter the F^- cells (Silver, 1963). Bouck and Adelberg (1963) proposed that, when replication of the Hfr DNA is finished, it exists in a linear state, permitting the transfer of one daughter chromosome to the recipient cell. In accordance with the results of Nagata (1963), they suggested that the transfer starts from the end of the DNA that is replicated last. According to the hypothesis of Jacob *et al.* (1963), transfer and replication are intimately coupled. By contact with an F^-cell, replication of the Hfr genome is induced starting from the origin and injecting one parent and one newly synthesized DNA strand into the recipient cell. The direction of this replication is not necessarily the same as the normal chromosomal replication, although it is not easy to visualize what happens if two points of replication meet. Without going into details in arguing for or against these two theories, a discussion which can be found in the paper by Gross and Caro (1965), it is clear that the evidence is mostly in favour of the Jacob *et al.* (1963) model. Only two papers in support will be mentioned.

Blinkova *et al.* (1965) pulse labelled conjugating pairs of *Escherichia coli* with ^{32}P and followed the survival during decay of the radio-active isotope in recombinants and in the F^- population as a whole. When a mixture of Hfr H, thr^+, leu^+, try^+, str^s, and F^-, thr^-, leu^-, try^- str^s cells were labelled for 20 min., the rate of decay of the leading *thr*, *leu* locus, as measured by the number of thr^+, leu^+, str_r recombinants, was found to be eight times greater than the suicide rate of the F^- cells. In another experiment, a mating culture was labelled for 20 min. and subsequently chased with non-radio-active phosphate, and the rate of decay of the *try* marker, which is injected after 33 min., was followed. The decay was found to be similar to that in the F^- population, indicating that no radio-activity was associated specifically with the tryptophan locus. The replication of DNA therefore must occur simultaneously with transfer.

In the experiments of Gross and Caro (1966), Hfr *Escherichia coli* cells were mixed with cells of an adenine-requiring F^- strain and starved of adenine to stop DNA synthesis. Control experiments had shown that, in the experiments, all F^- cells were engaged in mating 20 min. after mixing. Tritium-labelled thymine was therefore added after 20 min., and at different times the recombinants were separated by vigorous agitation, and the amount of label transferred to the F^- cells was measured by autoradiography. The F^- cells which could be morphologically distinguished from the Hfr cells became progressively more labelled with increasing time of contact. Dilution of the radio-active thymine by non-radio-active thymine resulted in an immediate cessation of the increased labelling of the F^- cells. To rule out the possibility that replication of DNA took place in the F^- cells in spite of the adenine starvation, the experiments were repeated with heavily u.v.-irradiated F^- cells. In these cells, DNA synthesis was completely arrested, but the results obtained were similar to those using adenine-starved cells.

These experiments were taken to prove that DNA synthesis, which occurs during mating, takes place in the Hfr cells. However they do not entirely preclude the possibility that DNA synthesis took place in the F^- cells. This could be the case if sufficient amounts of adenine were transferred from the Hfr to the F^- cells. In the case of the u.v.-irradiated cells it is possible that, although the DNA templates of the F^- cells are destroyed by the irradiation, the DNA-synthesizing mechanism is perfectly able to function with an intact template transferred to the cells.

Recent experiments by Bonhoeffer (1966) seem to provide rather convincing support for the notion that DNA replication during transfer takes place in the F^- cells. Two strains, an Hfr and an F^- strain, both with temperature-sensitive (*ts*) DNA synthesis, were mated with their normal counterparts at 37° and 43°. They behaved normally when mating took place at 37°, but, when the temperature was increased to 43°, the frequency of recombination decreased dramatically when the

F^-_{ts} strain was used as recipient. No effect was observed by mating Hfr_{ts} at 43°.

In all bacteria we must look upon DNA transfer during mating as an induced replication proceeding from the origin, but at a rate lower than the normal DNA replication. Since such a replication exists, does this mean that the DNA of Hfr cells is invariably present in a linear form? Apparently this is not so. When Hfr cells are grown to a saturation density in broth, and by analogy with *Bacillus subtilis* probably exist mostly with finished DNA strands, the cells behave as F^- phenocopies and are able to receive genetic material from other Hfr species. In such crosses, the same linkage exists between early and distal markers as found in a cross with F^- cells (Taylor and Adelberg, 1961). That a circular chromosome also exists in growing Hfr cells is indicated by the results of Fulton (1965), who showed that more than an entire genome can be transferred to the F^- cells.

When mating pairs are immobilized on the surface of a membrane filter, the spontaneous interruption of chromosome transfer is greatly decreased (Matney and Achenbach, 1962). By this technique, it has been possible to show that, after very long times of mating, not only the terminal markers including the Hfr character were transferred to the recipient, but the early markers were injected a second time linked to the distal markers (Fulton, 1965). Accordingly there is little doubt that the chromosome within the Hfr cells is also present in a circularized form during mating.

2. *Induction During Thymine Limitation*

During amino acid starvation of a bacterium, the DNA replication goes to completion in all cells, and a new round of replication is only started when the required amino acid is added once more to the culture. That the synthesis of a protein is responsible for the resumption of synthesis is supported by the observation that chloramphenicol mimics the amino acid starvation when added to the culture at a concentration which completely inhibits protein synthesis.

In thymine-requiring *Escherichia coli* 15 strains, a lowering of the thymine concentration to 0.05 μg./ml. decreases the rate of DNA synthesis to about 30% of the control but, contrary to the situation when thymine is completely removed from the culture, the cells remain viable for several hours at 37°. Neither RNA synthesis, nor the overall growth rate as measured by optical density determinations is affected by the low thymine concentration for about 60 min.

To follow the residual DNA synthesis during amino acid starvation in such a culture, an exponentially growing culture of *Escherichia coli* 15 *thy*$^-$, *arur*$^-$ was transferred to a medium containing 0·05 μg. thymine/ml. and, after different times of incubation, samples were withdrawn and transferred to a medium lacking a required amino acid but containing 1 μg. thymine/ml. A parallel culture in a medium containing 1 μg. thymine/ml. was starved of the amino acid as a control. When DNA synthesis was followed, it was found that, during the first

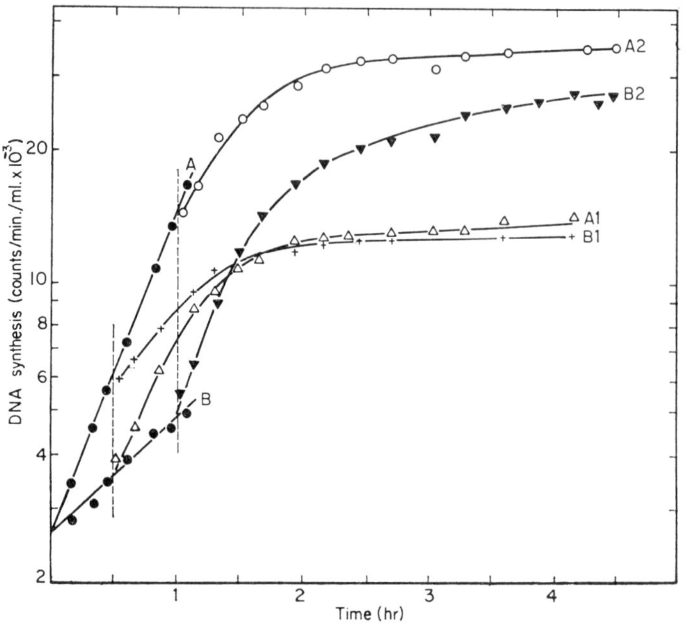

FIG. 7. Synthesis of DNA in *Escherichia coli* 15 *thy*$^-$, *arur*$^-$ in the absence of arginine after growth in a medium containing a low concentration of thymine. A culture was grown in a medium containing radio-active thymine (1 μg./ml.; A). At time 0, half the culture was transferred to a medium containing 0.05 μg. thymine/ml. (B) and, after 0.5 and 1 hr., aliquots of this culture (B1 and B2) and of a control culture (A1, A2) were starved of arginine in the presence of 1 μg. thymine/ml. DNA synthesis was followed by measuring the incorporation of radio-activity into the DNA. The specific activity of thymine was the same in all cultures.

60 min. of incubation in the presence of a low concentration of thymine, the amount of DNA formed in the absence of the amino acid made up for the deficit due to the difference in the rate of DNA synthesis in bacteria grown in the high and the low thymine media (Fig. 7). The rate of DNA synthesis during amino acid starvation was also observed to be higher in bacteria grown in the low thymine medium than in the control (Maaløe and Rasmussen, 1963). After 60 min., DNA synthesis in the bacteria

grown in the low thymine medium did not catch up with that in the control culture, and it might be significant that this changed behaviour coincides with the time that the growth curves of the two cultures started to deviate. Under these conditions, new rounds of replications are induced, and it seems that this induction is not linked to DNA replication as such but rather to cellular growth.

During thymine starvation, very similar events are observed. Using the BU-labelling technique described above, Pritchard and Lark (1964) showed an induction of new growing points during thymine starvation in the presence of amino acids. These growing points were structurally located at the starting point for the normal replication. They found that the increase in DNA content during amino acid starvation subsequent to thymine starvation reached a maximum value of 110%. This is compatible with the idea that DNA synthesis, after re-addition of thymine, proceeds from the original growing point where it had stopped during thymine starvation, and that a new round of replication is initiated in only one of the two daughter chromosomes. It is clear that the same quantitative results would be obtained if the premature replication was started in both daughter chromosomes in only half the bacteria. Autoradiographic analysis of microcolonies containing the progeny of cells having premature replication of DNA is taken to support the first possibility (Lark and Bird, 1965a, b).

In cultures supplemented with either fluorouracil or chloramphenicol in concentrations not sufficient completely to inhibit protein synthesis, Lark and Lark (1964) demonstrated that normal and premature initiation of replication show differences in sensitivity to the two inhibitors. The premature initiation appears to be much more sensitive to the two compounds than the normal initiation. It is, however, not excluded in their experiments that the effectiveness of the inhibitors varies under the different experimental conditions used to analyse the two cases, e.g. by differences in the extent of penetration.

3. *Induction During Growth*

We may now ask at what stage during the normal division cycle the onset of a new round of replication is triggered. This problem has been approached by D. J. Clark (see Maaløe and Kjeldgaard, 1966) using cultures of *Escherichia coli* B/r synchronized according to the method of Helmstetter and Cummings (1963). Exponentially growing cells are allowed to adsorb to a Millipore filter and, by washing with fresh pre-warmed medium, newly divided cells can constantly be eluted over several hours at a density of about 10^7 cells per ml. The degree of synchronization of such a physiologically sound culture was followed in the

Coulter counter by counting about 10,000–30,000 cells per sample. At different times during growth of a synchronized culture, radio-active thymidine and 150 μg. chloramphenicol per ml. were added to the culture, and the incorporation of label measured over 60 min. In cultures grown in glucose-minimal medium with a generation time of 45 min., and in succinate-minimal medium with a generation time of 75 min., it was observed that initiation occurred about half-way between cell divisions. Similar data were reported for synchronously germinating spores of *Bacillus subtilis*. The initiation of replication is claimed to occur early during the division cycle (Yoshikawa, 1965). Several proteins must be involved in the initiation and in the replication of DNA, but only the DNA polymerase, which is believed to be responsible for the normal replication, has been isolated and purified (cf. Richardson *et al.*,1964).

D. Genetic Studies of DNA Replication

Mutational studies of defects in the DNA replication system have to be done with conditional mutations such as those present in temperature sensitive(*ts*) mutants. There are only few reports concerning such mutational alterations of the normal DNA replication. Bonhoeffer and Schaller (1965) isolated a few *ts* mutants of *Escherichia coli* selected by the increased sensitivity to u.v. radiation at 313 mμ of bacteria having incorporated 5-bromouracil into their DNA. The properties of these bacteria, however, have not been published. Other *ts* mutants have been isolated by Kohiyama *et al.* (1963, 1966), some of which were shown, at 42°, to be blocked at the step of re-initiation of DNA replication. The DNA polymerase was found to be unaffected by the higher these mutants temperature.

A number of mutant strains have been isolated from Hfr strain AB 311 showing no transfer of genetic markers to F^- cells at high temperature. Since Flac particles can be transferred to, and multiply autonomously in, this Hfr strain (Cuzin, 1962) a dominance test can be performed. If a normal wild-type Flac is transferred to the *ts* strains, the property of transfer at 42° is restored.

Temperature-sensitive mutants have also been isolated in strains carrying a Flac episome, and during growth at high temperature these strains lose their episome (Jacob *et al.*, 1963). A number of these mutations have defects in the system for replicating the Flac particle. When these Flac mutants are transferred into the *ts* Hfr mutants, complementation is observed in a number of cases leading to the re-establishment of Flac multiplication and of chromosome transfer to F^- cells (Cuzin and Jacob, 1965a).

It is clear from these experiments that one or more diffusible products are involved in the replication of DNA, at least the replication which occurs during episomal multiplication. However, this might be different from normal chromosome replication.

To account for these and many other facts, Jacob *et al.* (1963) advanced the hypothesis that the multiplication of the different units of replication, termed replicons, is governed by two determinants, the replicator and the initiator. The initiator is supposed to be a diffusible element specified by a genetic locus, which activates the initiation of replication. The replicator is the starting point of replication and has a structural specificity towards the initiator. The specificity of the interaction between the initiator and the replicator accounts for the independent replication of different replicons within the same cell.

Episomes are autonomously replicating DNA molecules, which therefore must be controlled independently of the chromosome. When an episome is integrated into the chromosomal structure, the episomal replicator and initiator become non-functional and the episomal genetic markers are replicated in pace with the chromosome. Furthermore, DNA molecules devoid of a replicator, when introduced into a bacterial cell, are unable to replicate independently.

The simple model of regulation based on the interaction of a replicator and an initiator has to be somewhat modified to account for the re-initiation of DNA replication in one of the two daughter strands only during thymine starvation. Lark and Lark (1965) obtained evidence for the alternate replication of two chromosomes in cells grown in succinate medium. Accordingly, the production of an initiator seems not to be sufficient for the start of DNA replication. It is likely that the replicator must be in a specific structural conformation to be able to respond to the initiator. The suggested localization of the replicator on the cell membrane could possibly give rise to such changes in conformation. On the other hand, selective replication might be caused by an extreme limitation in the quantity of initiator produced as suggested by Lark and Lark (1965). The importance of the cell membrane as a site of attachment for the genetic elements of a bacterial cell was first hypothetically emphasized by Jacob *et al.* (1963).

It is evident that DNA replication during mating is triggered by cellular contact between the male and female cells. It is also evident that a mechanism must exist to ensure an even distribution of chromosomes and episomes, e.g. F particles among sister cells, since both of these replicons are present in the bacterial cells in only very few copies. The theoretical arguments for the attachment of the nucleoid to the cellular membrane were further substantiated by the electron microscopic studies of Ryter and Jacob (1963). In serial sections of *Bacillus subtilis,* they observed that the DNA of the cells always seemed to be connected to one or more invaginations of the membrane, the mesosomes. In similar electron microscope studies, Fuhs (1965) found only a few nucleoids attached to more than one mesosome.

Although the F particles can be transferred to F^- cells independently of the chromosome and can be lost selectively from the cells by treatment with acridine orange, the particles seem to be associated with the chromosome possibly through attachment to the membrane. Two experiments are relevant to this point. Among the *ts* diploid mutants that lose the Flac particle at high temperatures, there exists a group in which the lesion is not located at the episome but is a chromosomal mutation (Jacob *et al.*, 1963). This *ts* mutation might lead to a loss of the site for F attachment to the membrane at high temperatures. In another type of experiment (Cuzin and Jacob, 1965b), diploid cells carrying a *ts* $F_t$lac and having a *lac* deletion on the chromosome were labelled with $^{32}PO_4^{3-}$ for several generations at 25°. The culture was then transferred to a non-radio-active medium at 42°, at which temperature the $F_t$lac particle does not multiply. Samples of the culture were frozen in liquid nitrogen after 0, 0·8, and 3·6 generations of growth. The rate of suicide of the Flac-carrying bacteria and of the entire population was followed and found to be similar after 0 and 0·8 generations. At 3·6 generations, however, the bacterial population was only slightly affected whereas the rate of killing of cells carrying the episome was the same as after 0·8 generations. Since the Flac particles form only a very small fraction of the total DNA, the suicide must be due to the radio-active decay in the chromosome. An unchanged rate of suicide of the diploid cells can only mean that the episomes are associated with radio-active chromosomal DNA.

It is possible that the common site of association of the two replicons is an organelle in the cell membrane. Similarly it seems possible to explain the premature induction of chromosomal replication during growth in media containing a low concentration of, or lacking, thymine, as an expression of an unbalance between membrane growth and DNA replication. The argument might also be valid during an upshift experiment, when the rapid mass increase, which certainly must be followed by a rapid membrane synthesis, is coupled to an increase in the number of growing points.

IV. Regulation of RNA Formation

A. Synthesis of RNA

All species of RNA are thought to be copies of one strand of appropriate regions of the DNA. This is not only true for the m-RNA molecules, which by definition are transcripts of structural genes, but annealing experiments have indicated that r-RNA and t-RNA are able to form hybrids with 0·2% and 0·02% of the DNA respectively (Yankofsky and

Spiegelman, 1962a, b; Giacomoni and Spiegelman, 1962; Goodman and Rich, 1962). In *Bacillus megaterium*, two separate regions were found for 23S and 16S r-RNA (Yankofsky and Spiegelman, 1963). For all RNA species, the transcription is believed to be catalysed by the same enzyme, the DNA–RNA polymerase. This enzyme is known *in vitro* to synthesize RNA molecules with a base sequence corresponding to that of the DNA primer (Chamberlin and Berg, 1962; Geiduschek *et al.*, 1961). The asymmetric synthesis of RNA *in vivo*, in which only one of the strands of DNA is copied (cf. Hall *et al.*, 1963; Tocchini-Valentini *et al.*, 1963) has also been demonstrated *in vitro* using intact circular ϕX 174 DNA as the primer for the polymerase (Hayashi *et al.*, 1964). Similarly, Geiduschek *et al.* (1964) demonstrated an asymmetric synthesis of RNA using a crude extract of *B. megaterium* as a source of enzyme and phage α DNA as template, in its native but not circular form. Actinomycin D acts as an inhibitor of the polymerase *in vitro* (Hurwitz *et al.*, 1962), just as this antibiotic *in vivo* inhibits all DNA-dependent RNA synthesis (Reich *et al.*, 1961; Goldberg and Rabinowitz, 1962). The inhibition does not involve the enzyme itself, but results from the binding of actinomycin to GC pairs in the DNA (Goldberg *et al.*, 1962; Kahan *et al.*, 1963; Gellert *et al.*, 1965).

It is known from the work of Bremer *et al.* (1965) that the direction of RNA synthesis by the polymerase is from the 5′ to the 3′ end. The direction of *in vivo* synthesis of RNA is similarly found to proceed from the 5′ to the 3′ end (Goldstein *et al.*, 1965).

Although it is not a proven fact, it seems likely that the DNA–RNA polymerase observed *in vitro* is the very same enzyme which is responsible for the *in vivo* transcription of DNA. This assumption is substantiated by the findings of Hurwitz *et al.* (1963) that the product of an *in vitro* copying of DNA can be methylated by an enzyme, which is specific for the methylation of t-RNA.

B. Modes of Regulation

Departing from these facts, we can attempt to analyse the regulatory functions which might interact:

(1) At the level of DNA, regulating the frequency of transcription.
(2) At the level of the polymerase, regulating the rate of the copying process. Two situations must be analysed where the rate limitations are instituted by (a) the activity of the polymerase, and (b) the concentration of the polymerase.
(3) At the level of the substrate.
(4) At the level of RNA breakdown.

Two of these possible modes of regulation can immediately be eliminated. Concerning possibility 2b, it is clear from upshift experiments, as well as from experiments involving addition of chloramphenicol (Fraenkel and Neidhardt, 1961), that the rate of RNA formation can strongly and suddenly be increased without a concomitant enzyme synthesis. It seems obvious therefore that the regulation does not primarily involve enzyme concentrations.

It has already been pointed out that the variations in the concentration of triphosphates in cells during up- and downshifts show the transient changes that would be expected from the changes in the overall metabolism (J. L. Ingraham, personal communication). During amino acid starvation of a stringent strain, it was found that the pool of triphosphates shows a gradual decrease to about 50% over a period of 30–60 min. at 37° (G. Edlin, personal communication). Furthermore, it has been found that the pools of triphosphates vary in size by a factor of only about two in cells growing at widely different rates (Smith and Maaløe, 1964; Franzen and Binkley, 1961). The general metabolic role of all four ribosetriphosphates also seems to militate against a sensitive regulatory role for these compounds in RNA synthesis. We shall therefore, in this discussion, restrict ourselves to a closer examination of the possibilities 1, 2a, and 4.

C. Role of Amino Acids

1. *The Effect of Starvation*

In most bacterial strains, RNA formation is dependent upon an adequate supply of amino acids. Starvation of an auxotrophic strain of a required amino acid results not only in a cessation of net protein synthesis but the rate of RNA formation is decreased to around 10% of the control (Fig. 8). Analysis of the RNA isolated from a methionine-starved culture that had been pulse-labelled with [2-^{14}C]uracil showed a normal pattern of labelling, very similar to that obtained for RNA isolated from a control culture labelled for one-fifteenth of the time (Fig. 9). Consequently, the effect of amino acids is more quantitative than qualitative.

If protein synthesis is stopped in a wild-type strain by the addition of either chloramphenicol (CM) or puromycin, RNA synthesis continues at an unchanged or even increased rate (Fraenkel and Neidhardt, 1961). Addition of CM to an amino acid-starved culture results in two different responses depending upon the concentration of the antibiotic. At concentrations around 10 μg./ml., no effect on RNA synthesis is observed. Addition of the required amino acid in trace amounts immediately releases the inhibition of RNA synthesis. Addition of the same small amount of amino acid to a starved culture, without chloramphenicol, only results in an infinitesimally small increase in the RNA content (Pardee and Prestidge, 1956; Gros and Gros, 1958). This reaction is common to all amino acids, and some analogues; e.g. *p*-fluorophenylalanine and ethionine can replace the corresponding amino acids.

When CM is added to amino acid-starved cells at a concentration of 100 μg./ml., RNA synthesis is resumed to the same extent as in non-

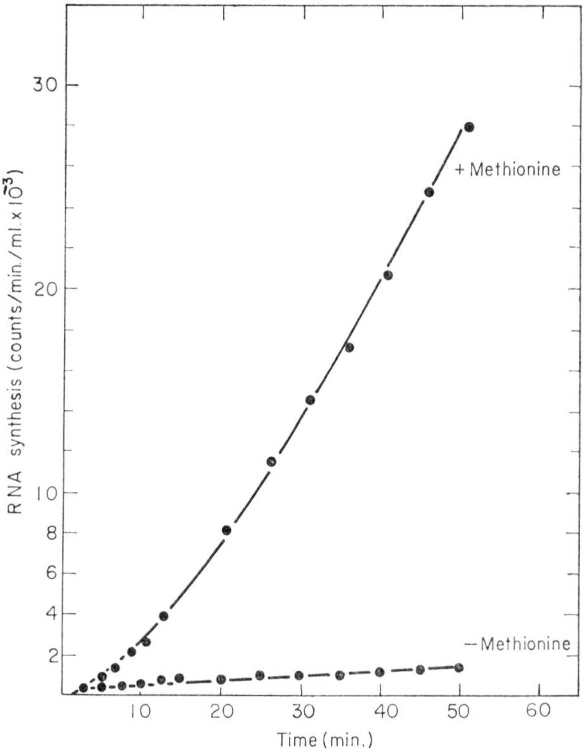

FIG. 8. RNA synthesis by the stringent strain *Escherichia coli* 15 *thy*$^-$, *ura*$^-$, *arg*$^-$, *pro*$^-$, *try*$^-$, *met*$^-$ in the presence and absence of methionine. RNA synthesis was followed by the incorporation of [2-^{14}C]uracil.

starved cultures (Aronson and Spiegelman, 1961). In their detailed study of this concentration effect, Kurland and Maaløe (1962) found that RNA synthesis became independent of added amino acids when the CM concentration was sufficient (above 50 μg./ml.) completely to block protein synthesis. They suggested that, at high concentrations of the antibiotic, the endogenous breakdown of proteins is sufficient to supply enough of the required amino acid to allow RNA synthesis to proceed.

The independence for this phenomenon of the amino acid in question led Kurland and Maaløe (1962) to the hypothesis that the t-RNA molecules, as a structurally common denominator, were the effectors in the regulatory functions. During treatment with CM, RNA-containing

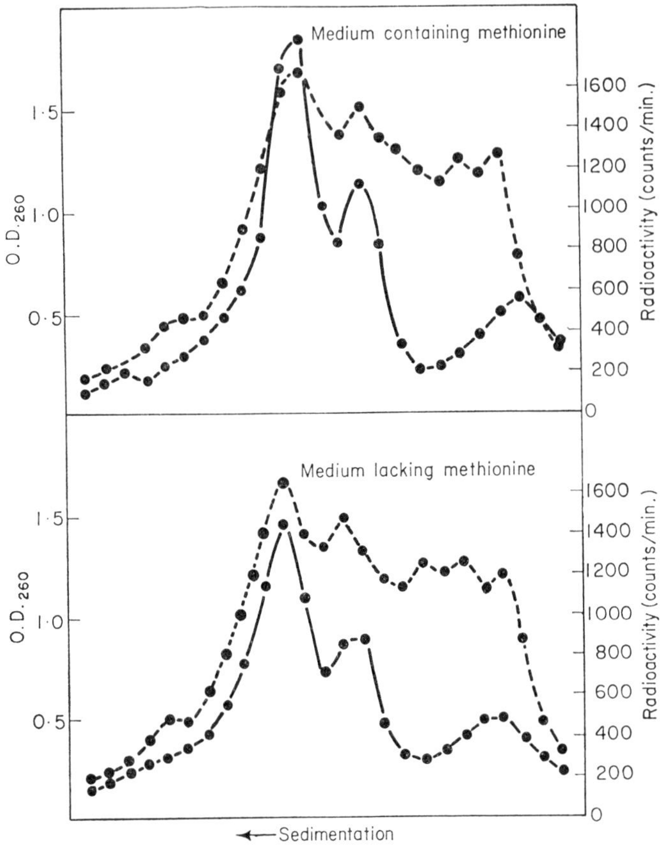

FIG. 9. Sucrose density gradient pattern of [2-^{14}C]uracil pulse-labelled RNA from the stringent strain, *Escherichia coli* 15 *thy*$^-$, *ura*$^-$, *arg*$^-$, *pro*$^-$, *try*$^-$, *met*$^-$ incubated in the presence or absence of methionine. The control culture was pulse-labelled for 1 min., the starved culture for 15 min. The same quantity of RNA was analysed in the two cases. ----- indicates radio-activity; ——— indicates optical density at 260 mμ. The direction of sedimentation is from right to left. The sedimentation coefficients of the two middle peaks of r-RNA are 23S and 16S respectively. The peak near the top of the gradient has a sedimentation coefficient of about 4S.

particles are formed, having sedimentation coefficients of 25S and 18S and containing 23S and 16S ribosomal RNA respectively (Nomura and Watson, 1959; Kurland *et al.*, 1962). The protein content of the preribosomal particles was 25% which, however, might be an overestimate due to the difficulties of separating the particles from soluble proteins. The protein associated with the particles naturally is formed in the cells before addition of CM. The kinetics of the formation of the RNA fractions were followed by Kurland and Maaløe (1962) during the first 2 hr. after CM addition at 25°. Analysis of the [^{32}P]labelled RNA in sucrose density gradients showed a synthesis of relatively large amounts of 4S RNA compared to ribosomal RNA. This was especially accentuated by labelling after longer periods of CM treatment. They suggested that the production of r-RNA is limited by the supply of particle proteins.

The accumulation of RNA during CM treatment usually levels off after an increase of about 60%. Assuming that about two-thirds of this is in CM-particles, we can calculate that, before addition of the antibiotic, the cells contain the particle proteins in a concentration of around 2% of the total proteins. Treatment of cultures with puromycin leads to the accumulation of similar preribosomal particles (Dagley *et al.*, 1962).

Contrary to the stringent behaviour of most strains, the classical *Escherichia coli* K 12 F^+ strain W6 continues to form RNA after deprival of the required amino acid, namely methionine (Borek and Ryan, 1958). This "relaxed" character was further studied by Stent and Brenner (1961). They made two important observations: (a) that derivatives of the original strain respond in the relaxed manner to the removal of any amino acid for which they have become auxotrophic; and (b) that the "relaxed character" can be transferred to other strains by conjugation. The site of the relaxed mutation, the RC locus, was mapped at a position close to the streptomycin locus (Alföldi *et al.*, 1962), and the two known states of this locus were designated RC^{str} (or RC^+) and RC^{rel} (or RC^-). This map position, however, is not quite correct. Transduction experiments with phage 363 have shown the RC locus to be cotransduced with the *arg*A locus, and therefore located about "15 min." from the streptomycin-locus (R. Lavallé, personal communication). A number of independent relaxed strains have been isolated in this Laboratory (N. Fiil, personal communication), and by R. Lavallé, and shown to be mutations in the RC locus.

When a relaxed strain is starved of a required amino acid, it behaves like a CM-treated wild type, and preribosomal particles are accumulated (Dagley *et al.*, 1962). Like CM-particles, they are very sensitive to RNase, and have a low protein content. Preliminary experiments by Turnock and Wild (1965) indicated that the particles contain 90% RNA and only 10% protein.

As with the chloramphenicol experiments the indiscriminate action of different amino acids led to the notion that t-RNA is an effector in the regulation of RNA synthesis (Stent and Brenner, 1961). The uncharged t-RNA was supposed to be a repressor of RNA synthesis, and the amino acids acted as inducers by combining with and neutralizing the repressor. To explain the RC^{rel} mutations, Stent and Brenner (1961) suggested, in strict analogy with the Jacob–Monod model, that the RC locus might be either a regulator locus or an operator locus. The relaxed strains therefore should be constitutive-type mutants.

2. *Rôle of t-RNA*

The central rôle of t-RNA in this regulation was further emphasized by the work of Fangman and Neidhardt (1964a, b). They isolated mutants which allow continued growth in the presence of *p*-fluorophenylalanine. One of these mutants was shown to have an increased specificity of the phenylalanine-activating enzyme. The enzyme could no longer catalyse the coupling of the analogue to t-RNA, and in phenylalanine-requiring derivatives of this mutant, *p*-fluorophenylalanine could no longer replace the natural amino acid in promoting RNA formation.

Also *ts* mutants have been isolated with altered amino acid-activating enzymes (Eidlic and Neidhardt, 1965; Yaniv *et al.*, 1965). Among mutants induced by ethylmethane sulphonate, Eidlic and Neidhardt (1965) isolated one having a temperature-sensitive phenylalanine-activating enzyme and one with a sensitive valine-activating enzyme. Yaniv *et al.* (1965) induced their mutations by N-methyl-N′-nitro-N-nitrosoguanidine and, among mutants unable to grow at 41°, they obtained five with a temperature-sensitive valine-activating enzyme. This preponderance of the "valyl-mutants" is remarkable but unexplained. All "valyl-mutants" which were derived from stringent strains were unable to form proteins as well as RNA at the high temperature. This indicates that the activation of amino acids is an essential reaction in the regulation of RNA synthesis. The "phenylalanine" mutant was derived from a relaxed strain and, in this mutant, the raising of the temperature to 37° did not affect RNA synthesis. However, protein synthesis was not completely blocked at 37° either, and it is therefore not possible yet to draw too strong conclusions from the experiments with this strain. It is of importance to note that Eidlic and Neidhardt (1965), with their "valyl" mutant, found that addition of CM at high temperatures did not reconstitute RNA synthesis, indicating that CM as such has no other function in the wild types than making amino acids available for the regulatory mechanism.

The mechanism behind the t-RNA effect was suggested by Kurland and Maaløe (1962) and Stent and Brenner (1961) to involve an inhibition of the polymerase by uncharged t-RNA. Such an inhibition, indeed, was observed by Tissières *et al.* (1963). In their system, the activity of the DNA-dependent RNA polymerase from *Escherichia coli* could be decreased by 80% when uncharged t-RNA was added, whereas amino acid-charged t-RNA gave an inhibition of only 25%. These results were recently confirmed by Bremer *et al.* (1966) showing furthermore that the binding of t-RNA to the polymerase is virtually irreversible. The small difference in the effect of charged and uncharged t-RNA, although significant, is difficult to incorporate into any physiological scheme, even more since the concentrations of t-RNA which can be used *in vitro* are far below the actual concentrations within the cells.

According to the model, the level of regulation is set by the relative concentrations of charged and uncharged t-RNA molecules. Qualitatively this mechanism could explain why cells grown in an amino acid-containing medium are richer in RNA than cells grown in a minimal medium. In view of the fact that starvation of a single amino acid is able completely to shut off RNA formation, although about 95% of the t-RNA is still in the charged state, it seems difficult to visualize how a large number of different RNA concentrations in the cells can be established when bacterial cultures are grown at different rates.

That the relative concentrations of charged and uncharged t-RNA in stringent bacteria have little effect upon RNA formation was demonstrated by Morris and DeMoss (1965). RNA was isolated from a leucine auxotroph of *Escherichia coli* K 12 and divided into two portions; one portion served as a control and the other was treated with periodate to inactivate the uncharged t-RNA. The amounts of three species of t-RNA were measured in both samples by recharging with radio-active amino acids. In exponentially growing cells, leucine-specific t-RNA was found to be 70% charged, valine-specific 80%, and arginine-specific t-RNA only 30% charged. Starvation of leucine decreased the concentration of charged leucine-specific t-RNA to about 40% of the total. For the other two species of t-RNA, the percentage of the charged form did not change.

To separate samples of a culture starved of leucine, CM and puromycin were added at a concentration of 100 μg./ml. and, in both cases, RNA formation was re-initiated. In the culture supplemented with CM, 100% of the leucine-specific t-RNA was found to be charged, whereas the concentration in the puromycin-treated culture remained at 35% as in the non-supplemented culture. Both antibiotics inhibit protein synthesis completely at the concentrations used, but, whereas the point of attack of CM is virtually unknown, puromycin is known to release peptides from the ribosomes by forming a peptide bond to the free carboxyl group of the polypeptide chain (Nathans, 1964; Traut and Monro, 1964). Ezekiel (1963) found, for the isoleucine-specific t-RNA, that starvation of this amino acid decreased the charged concentration in an auxotrophic strain to about 15%, from about 90% in exponentially growing cells.

Addition of CM in varying concentrations to a starved culture gave evidence for a correlation between RNA formation and the concentration of isoleucine-charged t-RNA. It should be emphasized, however, that, at the low concentrations of CM used in this experiment, protein synthesis was probably not completely inhibited. With methionine-specific t-RNA, Martin *et al.* (1963) found that starvation for methionine did not result in changes in the degree of charging in methionine-requiring *Escherichia coli* strains.

Although these data provide rather powerful arguments against an RNA regulation at the level of activity of the enzyme, we may approach the problem differently. If we are right in our credo that all species of RNA are synthesized by the same enzyme, one would expect that a regulation affecting the activity of the enzyme would be equally effective on r-RNA, t-RNA and m-RNA.

3. *Regulation of General m-RNA Formation*

Most information available about RNA regulation is based on measurements of the stable species of RNA, namely r-RNA and t-RNA. It is much more difficult to obtain sound information about the metabolically unstable m-RNA since measurements of the rate of formation of this fraction are complicated by the presence in cells of a nucleotide pool. Figure 10 illustrates this point by representing the situation when radio-active uracil is fed to a *ura*$^-$ strain of *Escherichia coli*. At this stage, the cells contain a non-radio-active pool of nucleotides from which triphosphates are drawn for the synthesis of all RNA species. The incorporation of nucleotides into r-RNA and t-RNA is a virtually irreversible reaction, whereas the nucleotides incorporated into m-RNA are fed back into the pool. At equilibrium, the overall rate of decay is equal to the rate of incorporation.

The radio-active uracil is therefore taken up into the pool at a rate determined by the irreversible drainage of the nucleotides. If we now consider the extreme situation where the synthesis of r-RNA and t-RNA is blocked, it is obvious that the speed of cycling of the nucleotides through the m-RNA fraction can have any value without being reflected in the rate of uptake of radio-activity. This situation obtains when a stringent cell is starved of a required amino acid, and the rate of synthesis of r-RNA and t-RNA is decreased to about 10% of the control (Fig. 8).

In the experiments of Edlin and Maaløe (1966) on the 5-fluorouracil (FU)-induced expression of an amber mutant in the alkaline phosphatase gene, the influence of the pool is clearly expressed. This mutant isolated by Rosen (1965) is genetically constitutive, but forms no alkaline phosphatase unless FU is added to the medium. The responses of this stringent strain (*arg*$^-$, *pho*$^-$, R_2^- RC^{str}), and of a relaxed derivative of it, were compared during amino acid starvation in the presence of FU. In the relaxed strain, it was found that a subsequent incubation without FU in a complete medium led to the formation of alkaline phosphatase as an indication of the incorporation of FU in the phosphatase m-RNA. No enzyme formation was found under similar conditions in the stringent strain. Taken at face value, this can be interpreted to mean that no m-RNA is formed in the stringent cells during amino acid starvation. However, a closer examination of the nucleotide pool showed that FU was not taken up by the stringent cells during the starvation. The results of the enzyme measurements are therefore inconclusive.

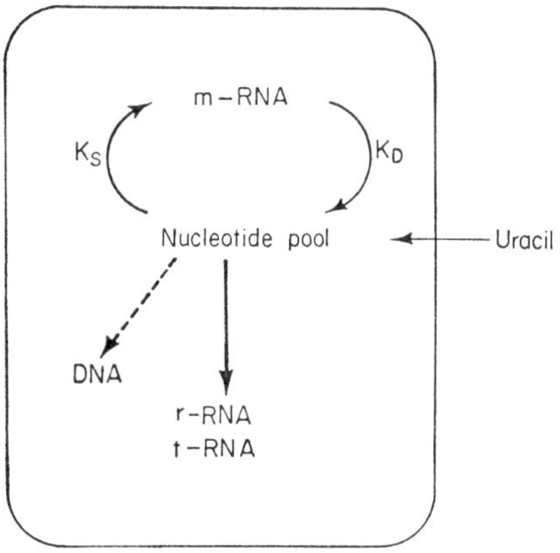

FIG. 10. Flow diagram for the incorporation of uracil into RNA.

Great care has to be taken in accepting evidence of a decreased turnover of m-RNA from experiments involving uptake of RNA precursors by pulsed or continuous labelling under conditions of restricted overall synthesis of RNA, unless the specific activity of the radio-active material in the pool is accounted for.

To obviate this problem we have tried to obtain a more direct measure of m-RNA synthesis making use of the property of this fraction to stimulate protein synthesis *in vitro* (Forchhammer and Kjeldgaard, 1967).

From experiments with *Escherichia coli* and *Bacillus subtilis* on the pulsed induction or derepression of the formation of β-galactosidase and alkaline phosphatase, it is known that the specific m-RNA fractions undergo an exponential decay when the inducer is removed from, or the corepressor added to, the culture (Kepes, 1963; Nakada and Magasanik, 1964; Hartwell and Magasanik, 1963, 1964; Fan, 1966). Similarly, it is shown that the radio-activity incorporated into RNA during a short pulse is lost from the cells in an exponential fashion when RNA synthesis is inhibited by actinomycin (Fan *et al.*, 1964; Leive, 1965), proflavin

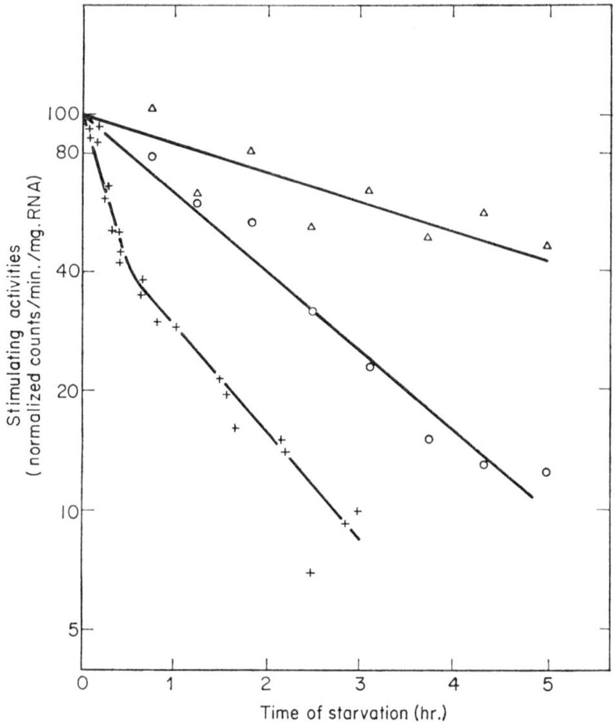

FIG. 11. Decay of the stimulating activity of RNA, isolated from uracil-starved *Escherichia coli*, on *in vitro* amino acid incorporation. The RNA was isolated from a culture of the stringent *E. coli* 15 *thy*⁻, *ura*⁻, *arg*⁻, *pro*⁻, *try*⁻, *met*⁻ during uracil starvation at 25°. The cells were starved (a) in the presence of the required amino acids, +——+; (b) after addition of chloramphenicol, ○——○; and (c) in the absence of arginine, proline and tryptophan, △——△. The incorporation of [6-^{14}C]leucine by a subcellular protein-forming system was followed as a function of the amount of RNA added. The normalized responses measured as counts/min./mg. RNA are plotted against time of starvation.

(Woese *et al.*, 1963) or 2,4-dinitrophenol (Soffer and Gros, 1964). In all these experiments, a half life of 1–2 min. at 37° was found for the messenger fraction.

In a uracil auxotroph, starvation of the required pyrimidine leads to a complete block of RNA synthesis. That this condition, as expected, leads to a breakdown of the m-RNA was established by isolating RNA from a culture of *Escherichia coli* 15 TAU bar (*thy*$^-$, *ura*$^-$, *arg*$^-$, *pro*$^-$, *try*$^-$, *met*$^-$, RC^{str}) after different times of uracil starvation at 25°. The stimulating activity, judged as amino acid-incorporation by a cell-free protein-forming system prepared according to Nirenberg and Matthaei (1961), was measured for these RNA preparations. The results of these experiments clearly demonstrated an exponential decay of the stimulating activity, and gave evidence for the existence of two RNA fractions with widely different rates of decay (Fig. 11). When properly corrected, the decay corresponded to a half-life at 37° of about 3 and 30 min. respectively. The implications of the presence of these two fractions will not be discussed further in this connection. It is assumed in these experiments that neither r-RNA nor t-RNA, which together account for the overwhelming part of the total RNA, shows any stimulating activity. This has been verified using purified RNA fractions.

When our TAU bar strain was starved simultaneously of both uracil and amino acids, a very slow decay of the stimulating activity was observed. This confirms the results of Fan *et al.* (1964) that m-RNA breakdown to the nucleotide level is linked to peptide bond formation by the ribosomes.

To study the rate of m-RNA synthesis, a TAU bar culture was starved of uracil in the presence of an amino acid to allow the decay of the major portion of the m-RNA. The culture was filtered on a membrane filter and the bacteria resuspended in uracil-containing media with or without a required amino acid. The re-appearance of the stimulating activity of the RNA isolated from such cultures was followed, and preliminary results show that the amino acid-starved culture, at all points, contains about 60% of the stimulating activity of the control culture. This is in contrast to the strongly repressed rate of total RNA formation in these stringent cells during amino acid starvation. The results suggest that the polymerase is still active in spite of the amino acid starvation. On the other hand, measurements of the stimulating activity of RNA isolated from a stringent culture during amino acid starvation in the presence of uracil showed slight increases only over several hours at 25°.

4. *Importance of the Ribosomes in Regulation of RNA Synthesis*

To reconcile these two apparent contradictory experiments we have to search for candidates other than the polymerase as effectors of the regulation of RNA synthesis, candidates which, at the same time, account for the important rôle of amino acids and their coupling to t-RNA. Ribosomes indeed satisfy all of these requirements. The implication of these particles in RNA formation was first suggested by Ames and Hartman (1963) and Stent (1964) mostly to explain the behaviour of polar mutations. Do we now have any evidence for such a function for the ribosomes?

Bremer and Konrad (1964) found that, when RNA polymerase was acting *in vitro*, the RNA formed stayed associated with the DNA, and they suggested that, in the absence of ribosomes, the RNA cannot be released. Working with a DNA-dependent protein-forming system, Byrne *et al.* (1964) and Bladen *et al.* (1965) found evidence for the existence of a complex of DNA, RNA, and ribosomes. Furthermore, it was mentioned above that, during balanced growth in different media, the amount of m-RNA formed is always proportional to the amount of r-RNA. Also, during Mg^{2+} starvation when the majority of ribosomes are degraded, the rate of m-RNA formation seems to be proportional to the number of ribosomes (Gros *et al.*, 1965). Shin and Moldave (1966) have recently demonstrated a stimulation of RNA synthesis *in vitro* by addition of ribosomes to the polymerase-DNA incubation mixture. The RNA formed was largely found in polysomal structures.

In general terms, a regulation by the ribosomes would mean that RNA is formed only if free ribosomes are available. When the polysomal structures are "frozen", as might be the case during amino acid starvation, and ribosomes are not released in an active form from the m-RNA molecules, RNA is formed only at a rate corresponding to the turnover of proteins and the residual rate of protein formation. The action of puromycin can easily be explained following the same scheme since it is known that this antibiotic permits a continued but abortive protein synthesis and movement of the ribosomes along the m-RNA (Villa-Trevino *et al.*, 1964). The mode of action of chloramphenicol (CM) is not known, but it has recently been found by Das *et al.* (1966) that the movement of ribosomes along m-RNA strands continues even when protein synthesis is inhibited. This is in agreement with the findings by Kepes (1963), Hartwell and Magasanik (1963), and Nakada and Magasanik (1964) that breakdown of specific m-RNA continues in the presence of CM, but is in contradiction to the findings that CM gives a protection of total m-RNA towards decay (Fig. 11; Fan *et al.*, 1964).

Very few positive results have been obtained from studies of relaxed mutants to illuminate the problem of RNA regulation. However, the data obtained might be explained by an abnormal behaviour of the ribosomes during amino acid starvation, allowing for the liberation of ribosomes from the m-RNAs. Such a behaviour could also explain why relaxed strains behave normally during upshifts, and only during downshifts manifest their abnormal behaviour.

Several suggestions have been put forward to explain the function of the ribosomes in the regulation of RNA synthesis. According to Stent (1964) the movement of the ribosomes during protein synthesis is essential to create a pulling force to release the m-RNA from the DNA. Other possibilities were suggested by Byrne *et al.* (1964) including the suggestion that the ribosomes function by stabilizing the RNA either against breakdown by nucleases or against structural changes. As mentioned above, we know that the synthesis of RNA *in vitro* (Bremer *et al.*, 1965), and *in vivo* (Goldstein *et al.*, 1966), occurs from the 5′ end of the polynucleotide strand. It is also known that the translation of the message into proteins also takes place from the 5′ end towards the 3′ end of the polynucleotide chain (Thach *et al.*, 1965; Salas *et al.*, 1965; Smith *et al.*, 1966). This relationship is a necessary condition for the release theory, but is equally important for all models.

The idea of a regulation by the ribosomes has mainly been argued on the basis of m-RNA. However, since it is observed that the r-RNA isolated from the preribosomal particles (e.g. those formed during CM treatment) show a stimulating activity on amino acid incorporation *in vitro* (Otaka *et al.*, 1964), it is not unlikely that r-RNA can act as a messenger for some ribosomal proteins. If this is so, the regulation of r-RNA formation might be completely analogous to that of m-RNA.

To obtain information about the time of insertion of a nucleotide into a nascent polynucleotide chain, i.e. the step time, Goldstein *et al.* (1965) studied the rate of incorporation of radio-active uridine in an *Escherichia coli* culture at 0°. They found a step time of 13 sec. corresponding to about 40 sec. per codon. This value has to be compared to the 5 sec. step time for amino acid addition to polypeptide chains at 0° (Goldstein *et al.*, 1964). From the number of uridine molecules incorporated per cell, it could be calculated that there were about 1000 nascent RNA chains per cell. Goldstein *et al.* (1964) found that the rate of protein synthesis at 0° was about 350 times slower than at 37°. If the same ratio holds true for RNA synthesis, one can calculate the time of synthesis for one 23S r-RNA molecule to be about 2 min. This rate seems rather slow, but it can account for the observed rates of synthesis of about one 23S r-RNA molecule per 2 sec. per r-RNA locus if the copying of the r-RNA locus takes place about 50 nucleotides apart. Such a distance between replicating units is close to the possible limit imposed by the large size of the polymerase (Fuchs *et al.*, 1964). Since the rate at which m-RNA is formed from a fully derepressed genetic locus seems to be of the same order of magnitude as that of r-RNA, the same arguments might hold true. It is anyhow very difficult from known data to draw any conclusions about a reasonable relationship between the rates of RNA and protein synthesis in support of the notion of the release function of the ribosomes.

The m-RNA fraction is extremely sensitive to the action of nucleases. As mentioned above, m-RNA is protected against this degradation when protein synthesis is blocked by amino acid starvation or by anaerobiosis (Fan *et al.*, 1964; Fan, 1966). It is conceivable that this protection is afforded by the ribosomes. During CM treatment, r-RNA is formed and is incorporated into CM-particles together with ribosomal proteins formed in the cells before the inhibition of protein synthesis by the drug. The amount of r-RNA formed decreases with time, and it was suggested by Kurland and Maaløe (1962) that the formation continues as long as a pool of ribosomal proteins is available to stabilize the r-RNA towards degradation. That the RNA in the CM-particles also might be unstable is suggested by the experiments of Dubin and Elkort (1965) who measured the rate of turnover of [^{14}C]-labelled methyl groups in RNA from CM-particles.

These findings seem to be in accordance with the notion of regulation of RNA formation based on a fixed rate of production of all RNA fractions, but with variations in the degree of protection. Although this model is not favourable for a strict energy economy of the cells, there are no data available as yet which permit a choice to be made between the alternatives for ribosomal regulation of RNA formation.

V. Regulation of Enzyme Synthesis

The general regulation of protein synthesis has been discussed above as a consequence mainly of the concentration of ribosomes in the cells. In this section, the regulation of the synthesis of specific proteins will be discussed as an extension and a special case of RNA regulation.

A. The Operator Model

The operator model proposed by Jacob and Monod (1961a) to explain the regulation of enzyme synthesis has ever since been in the focus of all discussions of regulation. Numerous experiments performed over the last five years have given circumstantial evidence in support of the model, but very little decisive information has been obtained to prove or disprove the model beyond doubts.

In this section, we shall use the Jacob–Monod model as a guide for our discussion. According to this model, the bacterial genome is functionally divided into a number of units of expression, the operons, containing the structural genes of one or more enzymes. The transcription of this information into m-RNA is under the control of two genetic loci, the regulator gene, R, and the operator gene, O. The R locus is the structural determinant of a repressor with two functional sites, one interacting with an inducer or corepressor, the other with steric specificity towards the

O locus. Attachment of the repressor to the O locus is believed to block the transcription of the entire operon.

With inducible enzymes, addition of inducer is thought to result in a steric modification of the repressor, leading to a decrease in its affinity for the operator locus. This gives a derepression of the operon and a production of m-RNA. With repressible enzymes, a corepressor is supposed to be essential for the maintenance of the repressor in a configuration with a steric fit for the O locus and therefore necessary for the blocking of transcription (Jacob and Monod, 1961a; Monod *et al.*, 1963).

1. *Specific m-RNA Formation*

It is an essential feature of the Jacob–Monod model that induction and derepression are accompanied by a transcription of the operon in question into m-RNA. The possibility of forming hybrids between m-RNA and heat-denatured DNA (Hall and Spiegelman, 1961; Bolton and McCarthy, 1962) offered an experimental approach for testing this prediction. The obvious requirement for the isolation of DNA molecules containing only the operon in question has been met for the inducible lactose and galactose operons and for the repressible tryptophan operon. With the *lac* operon, the specific DNA was isolated from Flac particles carried by a *Serratia* strain, and with the galactose and tryptophan operons, from the high-frequency transducing phages, λdg and $\phi 80\, dt_0$.

To follow the production of m-RNA from the *lac* region, pulse-labelled RNA, isolated from induced and non-induced cells, was annealed first with DNA from cells carrying a deletion of the *lac* locus, as a measure of total messenger, and then to the specific Flac DNA. Although the method is not very sensitive, it clearly shows that the relative amount of pulse-labelled material hybridizable with Flac DNA is higher in induced cells than in non-induced cells (Attardi *et al.*, 1962, 1963). Similarly for the *gal* operon, induced cells were found to contain more pulse-labelled RNA which can hybridize with λdg DNA than RNA from non-induced cells (Attardi *et al.*, 1962). For the repressible tryptophan locus, it was found by Imamoto *et al.* (1965a) that pulse-labelled RNA from cells grown in the absence of tryptophan could hybridize with $\phi 80\ dt_0$ DNA to a larger extent than RNA from cells grown in the presence of tryptophan. It is thus evident that derepression of an operon is accompanied by the production of specific m-RNAs.

2. *Multicistronic Messages*

Next, let us consider the size of the messenger produced from operons containing more than a single structural gene. The hypothesis of one operator–one messenger was first postulated by Martin (1963) on the basis of size analysis of pulse-labelled RNA from cells with a derepressed histidine operon. Since the histidine operon contains the structural genes of nine enzymes participating in the synthesis of histidine, a messenger

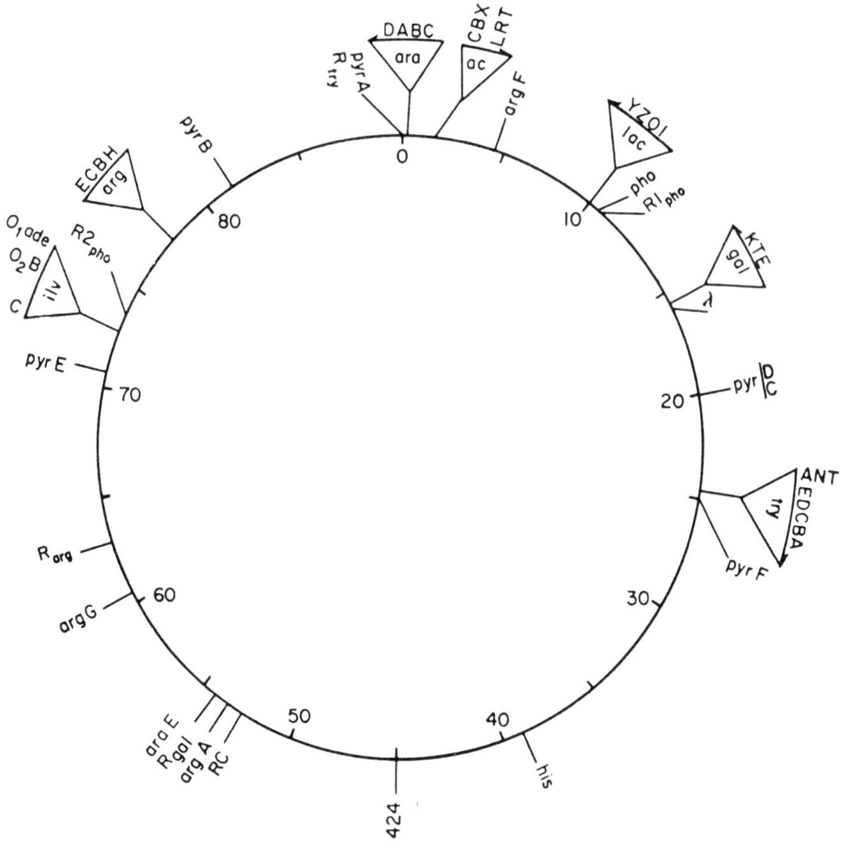

FIG. 12. Chromosomal map of *Escherichia coli*, modified from Taylor and Thoman (1964). The arrows indicate the most probable direction of transcription in a few operons. λ and 424 indicate points of attachment of these two prophages. The regulator loci are designated by R with a subscript to indicate the pathway involved. Other abbreviations are defined in the footnote on p. 52.

corresponding to all enzymes must be of an appreciable size. Although such calculations are beset with difficulties, it has been found that the size of the histidine m-RNA encountered in derepressed cells corresponds approximately to the sum of all cistrons in the operon. It is therefore a polycistronic message (Martin, 1963).

Three genes are known in the lactose operon, specifying the structure of β-galactosidase (z), β-galactoside permease (y), and transacetylase (a). RNA from non-induced *Escherichia coli*, pulse-labelled with [^{14}C]uracil, and from induced bacteria pulse-labelled with [^{3}H]uridine, were fractionated on a methylated albumin column (Attardi *et al.*, 1963). It was found that the highest $^{3}H/^{14}C$ ratio was in the fractions containing m-RNA chains, sedimenting between 20S and 30S. The m-RNA from the *lac* operon therefore seems to contain about 4000 nucleotides, corresponding to 1300 amino acids. It is not possible from this estimate to argue in favour of a polycistronic message since the monomers of β-galactosidase contain about 1200–1300 amino acids (Steers *et al.*, 1965a).

Measuring the size of specific polysomes in induced bacteria, by following the β-galactosidase activity associated with polysomes, Kiho and Rich (1964) obtained evidence for the participation of polysomes containing thirty to fifty ribosomes in the synthesis of β-galactosidase. Since haemoglobin monomers having about 140 amino acids in the polypeptide chain are formed by polysomes containing five ribosomes (Warner *et al.*, 1962), it seems permissible to conclude that the observed large polysomes contain a m-RNA corresponding to a protein of the size of 1000–1400 amino acids. This, however, is based on the assumption that the distance between ribosomes is the same all along the messenger. It is difficult to assess if this value is large enough to accommodate all enzymes in the *lac* operon. The galactoside permease is not characterized, whereas the transacetylase is found to be a dimeric protein with a molecular weight 65,000 (Zabin, 1963b). In a recent study, Kiho and Rich (1965) analysed the polysomes isolated from a large number of different *Escherichia coli* strains and obtained evidence for the existence of smaller "β-galactosidase polysomes" in cells carrying a deletion of, or a nonsense mutation in, the y gene. This is to be expected if transcription is starting from the operator, since the genetic order is found to be O–z–y (Jacob and Monod, 1965). On the other hand, in the wild-type strain, no transacetylase could be demonstrated in association with the large polysomes as might have been expected if the enzyme is produced by the same polysomal complex as the β-galactosidase (Kiho and Rich, 1965).

The structural genes for the six enzymes of the biosynthetic pathway from chorismic acid to tryptophan are located as an operon, with the operator locus close to the gene for the first enzyme in the pathway, the anthranilate synthetase (Matsushiro *et al.*, 1965). By hybridization of pulse-labelled RNA from derepressed cells with DNA from the phage $\phi 80\ dt_0$, the size of the specific m-RNA was determined by sucrose gradient centrifugation (Imamoto *et al.*, 1965b). Sedimentation coefficients of 20–35S were found for this fraction. Analysis of pulse-labelled RNA, isolated from an *Escherichia coli* strain carrying a deletion of the entire tryptophan region except for the gene for anthranilate synthetase, showed that the m-RNA for this enzyme is distinctly smaller in size than that for the complete operon. These data are clearly in support of a multicistronic message, but in no instance are they good enough to be taken as a proof.

3. *Co-ordinate Enzyme Production*

The production of m-RNAs from all structural genes within an operon during derepression obviously must result in the synthesis of all of the enzymes of this operon in a co-ordinate manner. Such co-ordinate expression was demonstrated for the histidine operon (Ames and Garry, 1959), and has been shown to exist for the *lac* operon (Zabin, 1963a), and for the tryptophan operon (Ito and Crawford, 1965).

A co-ordinate regulation of enzymes was originally believed to indicate that the enzymes involved belong to the same operon. This, however, is not so. In *Escherichia coli*, the synthesis of the six enzymes that catalyse reactions on the pyrimidine pathway is repressed by uracil (Yates and Pardee, 1956). Synthesis of four of the enzymes catalysing a sequence in the biosynthetic pathway is found to vary in a co-ordinate manner, whereas synthesis of the two remaining enzymes follows the same pattern qualitatively but not quantitatively (Beckwith *et al.*, 1962). The structural genes for the four co-ordinately expressed enzymes were thought to be located in a cluster (Beckwith *et al.*, 1962), but later findings have shown that only two of the genes are closely linked. Of the two misplaced genes, one maps relatively close to the cluster, while the other is about 180° removed on the circular chromosomal map (Signer *et al.*, 1965; Taylor *et al.*, 1964). The four enzymes induced in *E. coli* by arabinose also show a co-ordinate expression, although the locus for one of them, the arabinose permease, is far removed from the cluster of the structural genes for the other three enzymes (Englesberg *et al.*, 1965).

As regards the co-ordinate expression of the *lac* operon, Zabin (1963a) found that the molar ratio of β-galactosidase and transacetylase was about 8 to 1, assuming a molecular weight for the monomers of the two enzymes of 130,000 (Steers *et al.*, 1965a) and 32,500 (Zabin, 1963b) respectively. As such determinations require a detailed knowledge of the specific activities of the enzymes involved, it is only in a few cases that similar calculations have been attempted.

In the tryptophan operon, a comparison was made between the amount of the first (with respect to the operator locus) enzyme, anthranilate synthetase, and the last two, the A and B proteins of tryptophan synthetase. Molar ratios of 1:1 between the A and B proteins and about 4:1 between the A protein and the anthranilate synthetase (Ito and Crawford, 1965) were found. It therefore seems that a polarity with decreasing amounts of enzymes produced along the m-RNA is not an obligatory element of the expression of an operon.

4. *Polarity Mutants*

In many different operons, however, certain types of mutants, called polarity or dual-effect mutants, have been observed. In these strains, a mutation in one structural gene affects the quantities produced of other enzymes on the same operon (Jacob and Monod, 1961b; Lee and Englesberg, 1962; Ames and Hartman, 1963). An extensive analysis of the histidine operon has shown that genes situated between the operator and the gene carrying the polarity mutation are normally expressed, whereas the expression of all genes distal to the mutated site is greatly decreased. Of the histidine mutants tested, about half were of this type, and the polarity effect could be observed in the repressed as well as in the derepressed state.

The polarity effect could be accounted for by assuming the existence of a polycistronic message, and by assuming that the translation process by the ribosomes is hampered by certain codons, and that in most cases the ribosomes dissociate from the messenger at such points. Codons with this property might, if placed between neighbouring genes, account for the non-equimolar production of different enzymes belonging to the same operon. Polarity mutations are found in all genes within an operon. In the tryptophan operon, it is especially interesting that, among the mutants affecting the B protein of tryptophan synthetase, polarity effects on the A protein are only observed in mutants *not* producing an enzymatically inactive, antigenically cross-reacting B protein (Ito and Crawford, 1965).

It is therefore likely that the polarity effect is the result of a nonsense mutation in the structural genes that prevents the translation of the messenger beyond the mutated codon. This effect of nonsense mutations has been directly observed by demonstrating that *amber* mutants of the head protein gene of phage T4 make fragments of the head protein chain which terminate at a point corresponding to the mutant site (Sarabhai *et al.*, 1964; Stretton and Brenner, 1965).

A large number of nonsense mutations in the *z* gene of the *lac* operon have been analysed for their capacity to form β-galactoside permease and transacetylase (Newton *et al.*, 1965). All of the mutants studied were found to be more or less polar, and the amount of enzyme formed depended on the position of the nonsense mutation with respect to the *y* gene. The closer the mutation was to the *y* gene, the higher the activities of the permease and the transacetylase. Polar mutants have been found in the *y* gene as well. They show a normal production of β-galactosidase, but a decreased synthesis of the transacetylase. The position effect of the nonsense mutation might indicate that a point of attachment for the ribosomes could exist at the beginning of each gene and not only at the beginning of a polycistronic m-RNA.

Special types of mutants, designated O^0, are strongly polar. They are characterized by an extremely low basal production of enzyme, and following derepression, none of the enzymes coded for by the operon is produced. This type of mutation has been found in several operons including the *lac* operon (Jacob and Monod, 1961a), the *gal* operon (Adler and Kaiser, 1963), the histidine operon (Ames and Hartman, 1963), the tryptophan operon (Ito and Crawford, 1965), the arabinose operon (Helling and Weinberg, 1962), and in the pyruvate dehydrogenase locus (Henning *et al.*, 1965). In the penicillinase system of *Staphylococcus aureus*, mutants with the characteristics of O^0 mutants have recently been isolated (Richmond, 1966).

The O^0 mutants which were first obtained in the *lac* region were believed to have a mutation in the operator locus (Jacob and Monod, 1961a). However, it was shown by Beckwith (1964) that certain deletions of the O^0 mutation result in an inducible production of β-galactoside permease. The O^0 mutation, therefore, cannot be located in the O-locus, but is now recognized as a mutation in part of the *z* gene situated next to the operator region. It was further shown that some of the O^0 mutations were sensitive to a suppressor and that induction of the O^0 su^+ strains leads to the production of all enzymes of the *lac* operon (Beckwith, 1963; Brenner and Beckwith, 1965). Similar results were obtained by Orias and Gartner (1964) and Schwartz (1964).

It is most remarkable that, in the O^0 mutants, induction does not result in the production of *lac* messenger (Attardi *et al.*, 1963). The O^0 mutants in the *gal* region have similarly been shown not to produce any specific messenger even in the presence of inducer (Hill and Echols, 1965). These *gal* mutants map between the transferase and the epimerase genes and not at the extreme end of the gene cluster (Adler and Kaiser, 1963). This extraordinary property of the O^0 mutants obviously puts strong emphasis on the existence of a close correlation between messenger formation and translation.

In accordance with the above discussion, polar effects could be due to a failure of release of the m-RNA from the DNA because of the absence of translation of the nonsense codon (Stent, 1964). Alternatively, the polar effect might be due to a rapid degradation of the m-RNA. This requires that the reading of the message starts from the operator end and, because of the halt of ribosomal progression at the nonsense codon, the distal parts of the m-RNA are not protected against degradation or structural changes. No information is as yet available to permit a choice to be made between these two possibilities.

B. Regulatory Genes

1. *The* lac *Operon*

Thanks mainly to the work of the school at the Pasteur Institute, our information about the *lac* operon is by far the most complete and we shall base our discussion of the regulation of production of specific m-RNA on this system. As mentioned above, two genetic loci, R (in the *lac* operon called *i*) and O are implicated in this regulation. In wild-type strains, $i^+O^+z^+$, the basal level of β-galactosidase is found to 5–10 units per mg. dry weight. A large number of different galactosides are able to induce the formation of enzyme, the degree of induction showing large variations when different inducers are used. A maximum level of induction of about 5000–8000 units of β-galactosidase per mg. dry weight is reached by induction with isopropylthiogalactoside (IPTG) at a concentration above 5×10^{-4} *M*.

Two types of mutations in the *i* locus have been characterized, the constitutive i^- mutations and the superrepressed i^s mutations. With both types, the mutations seem to affect the properties of the repressor molecules.

In the i^- mutants, the affinity of the repressor towards the operator locus is believed to be changed either by alterations of the appropriate site in the repressor or by completely abolishing the production of repressor. In the i^- mutants, the rate of enzyme production is maximum and unaffected by the addition of inducers. Dominance tests in diploid cells, e.g. bacteria carrying an i^- mutation on the chromosome, and the wild-type allele, i^+, on a Flac episome, have shown that the i^+ allele is dominant over the i^- allele. This is to be expected if the i^+ gene is responsible for the production of a cytoplasmic product.

The i^s mutations are thought to result in an altered structure of the site on the repressor responsible for the reaction with the inducer. A number of different *i* mutants have been isolated, all of which are characterized by not producing β-galactosidase when grown in the presence of lactose (Willson *et al.*, 1964; S. Bourgeois, personal communication). In diploid cells, the i^s allele is dominant over both the i^+ and the i^- alleles. All i^s mutants isolated have a normal basal level of β-galactosidase, but vary in their induction pattern when tested with different inducers. Some mutants are only weakly induced by 0·1 M-IPTG, others are induced to near normal levels by 10^{-3} M-IPTG, but only weakly induced by $10^{-3}M$-methylthiogalactoside (S. Bourgeois, personal communication).

The properties of the repressor have been analysed by Sadler and Novick (1965) using mutants, derived from i^{s} strains, which as a result of a second mutation become constitutive when grown at 41°. Two types of mutants were found, one $i^{s,TL}$, in which the repressor itself is thermolabile, and another $i^{s,TSS}$ in which synthesis of the repressor is affected by the high temperature. Several important observations appeared during these studies. The rate of β-galactosidase formation seems to be inversely proportional to the repressor concentration. The repressor is unstable during growth, with a mean life of one-tenth to one-fifth of a generation. That the repressor, indeed, does react with the inducer is indicated by the observation that the addition of IPTG increases the temperature sensitivity of the thermolabile repressor and augments the amount of repressor formed by the TSS mutant at 41°. Furthermore, evidence was obtained which seems to indicate that the repressor is assembled from subunits.

All these observations, taken together with the variations in the effect of different inducers, can best be explained by assuming that the repressor is a protein. That this indeed is so was demonstrated by Bourgeois *et al.* (1965) in experiments involving the suppression of nonsense mutations in the *i* gene. From a diploid and su^{-} strain carrying an i^{s} mutation on a Flac particle and a *lac* deletion on the chromosome, a large number of constitutive mutants were isolated. The Flac particles were transferred into F^{-} strains carrying different suppressors and the reappearance of the i^{s} character was examined. Among 975 i^{s}, i^{-} mutants, twenty-four were found to be sensitive to suppression. Since the action of the suppressors was shown to be at the level of translation of the m-RNA into proteins (Brenner and Stretton, 1965), these experiments obviously furnish proof of the protein nature of the *i* gene product. Analogous results were obtained by Müller-Hill (1966).

It is important to note that the production of the repressor is not induced by the addition of IPTG (Novick *et al.*, 1965). Indeed, this was to be expected from earlier results. If the addition of inducer to a culture results in an increased concentration of repressor, the maintenance of a steady-state level of enzyme production would require a continuously increasing concentration of the inducer. Such a course of induction has never been observed. The question of what determines the rate of synthesis of the *i*-gene messenger remains unsettled.

The O locus is genetically located between the *i* and the *z* loci and its size is ill-defined. It is certainly a rather small region, since the map distance between the i_{3}^{-} mutation and one of the early markers in the *z* gene, namely z_2^-, is less than 15% of the map distance for the *z* gene (Willson *et al.*, 1964). Since the position of the i_3^- mutation within the *i* gene is not known, the O locus might be much smaller than the 15% calculated.

The O locus is defined by one class of mutations, called O^c. The strains carrying these mutations are constitutive and produce, without inducer, an amount of enzyme varying between 2% and 100% of the maximum (Jacob *et al.*, 1964). Mutants with a low constitutive level can be fully induced. These mutations are believed to result in an altered structure of the O locus which gives a decrease in affinity between the repressor protein and the locus, Since it is difficult to visualize that changes in a single base pair should suffice to give a substantial alteration in structure, it is in accordance with the general picture that all O^c mutants seem to be deletions (Jacob *et al.*, 1964). O^c mutants were obtained as spontaneous mutations, by treatment with X-rays, and with ethylmethane sulphonate, but none was isolated after u.v. irradiation or after treatment with 2-aminopurine. It is in agreement with this picture that revertants of O^c mutants to O^+ have never been isolated (Jacob *et al.*, 1964).

It is important in this discussion of the operator model to know if the O locus governs the synthesis of a protein. Two experiments are relevant in this respect. In the experiments of Bourgeois *et al.* (1965), it was found that, of the constitutive mutants isolated, 125 were i^sO^c, and none of these was sensitive to suppression. Compared to the 2–3% suppressor-sensitive i^- mutations, these results are not significant, but only suggestive. To consider the possibility that the operator region is coding for a terminal part of the β-galactosidase molecule (although a part without effect on the catalytic properties of the protein), Steers *et al.* (1965b) analysed and compared the β-galactosidases isolated from an i^- strain and an O^c strain. No differences in the properties of the two enzymes were found which indicates that the O locus is not a cistron within the *z* gene.

The unique properties of the *i* and the O genes in the regulation of the *lac* operon have been demonstrated by the isolation of large deletions covering both of these genes (Jacob *et al.*, 1964, 1965). To avoid the possibility that such large deletions might affect other vital functions of the cells, the deletions were induced in F′ particles of different sizes but all carrying an i^s mutation in the *lac* operon. From diploid cells (i^s/F′i^s), revertants were isolated that were able to grow on melibiose at 40°. At this temperature, the melibiose is penetrating into the cells solely by the action of the β-galactoside permease. A small percentage of the revertants were shown to have a deletion extending from inside the *z* gene beyond the *i* gene. They have a rather low level of permease and transacetylase activities, but are not inducible by β-galactosides. By using a very large

F′ episome containing both the *lac* and the purine regions, a few mutants were isolated in which the deletion extended from the *z* gene into the purine region. Production of β-galactoside permease and transacetylase were repressed by the addition of adenine to the medium (Jacob *et al.*, 1965). That a genetic region can thus be re-arranged to be under the control of another operator has also been shown in other organisms. Part of the *lac* operon has been incorporated into the tryptophan operon and the production of β-galactoside permease is now repressible by tryptophan (Beckwith and Signer, 1966). With the tryptophan operon (Matsushiro *et al.*, 1962) as well as the histidine operon (Ames *et al.*, 1963), deletions of the operator end of the regions have made the remaining enzymes of the operon insensitive to repression by addition of the appropriate amino acids to the medium.

The *lac* operon is unique also in the sense that both the R and the O loci are closely linked. No other system which has been studied offers this advantage which, of course, is especially convenient when dominance relationships are studied.

2. *Other Inducible Operons*

A few other inducible operons have been analysed in detail. The utilization of galactose by *Escherichia coli* depends upon the action of three enzymes, galactokinase, galactose transferase and galactose epimerase. They are all induced by galactose or fucose (Kalckar *et al.*, 1959; Buttin, 1961), and are under co-ordinate control (Yarmolinski *et al.*, 1961; Buttin, 1963a). Among the constitutive mutants, one class, which may be of the R type, is not linked to the *gal* region but is found close to the *arg* A locus. Other mutants are of the O^c type and seem to map at the end of the region close to the structural gene for the epimerase (Buttin, 1963b). By transduction experiments with λ *dg* phages, the dominance of the O^c mutations over the O^+ wild-type has been established (Buttin, 1961). These O^c mutants have been shown to produce a specific m-RNA also in the absence of inducer (Attardi *et al.*, 1963). It should again be mentioned that the O^o type mutants of the *gal* region seem to map between the transferase and epimerase genes (Adler and Kaiser, 1963).

As mentioned above, the four enzymes responsible for the utilization of arabinose by *Escherichia coli* are co-ordinately induced, but only three are genetically linked (Englesberg *et al.*, 1965). A fourth gene, the C gene, is found at the extremity of the cluster and has been characterized by two mutant alleles, C^- and C^c. The C^c mutants are constitutive, whereas the C^- mutants are deficient in all three enzymes and the unlinked permease. In transient diploid strains obtained by mating of Hfr and F^- cells, it has been shown that C^c is dominant over C^- but recessive to C^+. This behaviour is difficult to put into a simple scheme in analogy with other systems. It is certain that the C gene is neither an R nor an O locus, but it is remarkable that all the constitutive mutants isolated map in the C gene (Englesberg *et al.*, 1965).

The genetic analysis of the inducible penicillinase of *Staphylococcus aureus* has recently been greatly facilitated by the demonstration that the genetic determinants for the penicillinase are incorporated into an extra-chromosomal element (Novick, 1963; Novick and Richmond, 1965). This element can be transferred by transduction, and it is thus possible to form heterodiploids in which, furthermore, the two structural genes may code for antigenically different penicillinases. Constitutive mutants have been isolated in which the mutations are situated close to the structural gene. In diploid cells, it is found that the wild-type inducible allele is dominant over the constitutive (Richmond, 1965).

3. *Repressible Operons*

In *Salmonella typhimurium*, the structural genes for the ten enzymes that catalyse reactions on the histidine pathway are clustered in a small region of the bacterial chromosome (Hartman *et al.*, 1960; Ames *et al.*, 1960; Ames *et al.*, 1961). Mutants which can grow in the presence of triazolealanine are found to be permanently derepressed for synthesis of the histidine enzymes. Certain of these mutants were found to have a mutation near the end of the histidine region and are believed to be of the O^c type although a dominance test has been impossible (Ames and Hartman, 1963). Other constitutive mutations map outside the region and seem to affect several functions of importance for the repression, e.g. the efficient coupling of histidine to t-RNA (Hartman *et al.*, 1964; Roth 1965).

The alkaline phosphatase of *Escherichia coli* is formed only when the concentration of inorganic phosphate in the growth medium is very low. Non-repressible mutants fall into two groups, both genetically and with respect to the amount of enzyme they produce. Mutants with alterations in the R1 locus, which is very closely linked to the structural gene of the phosphatase, are "low level" constitutives and produce about 20% of the maximum amount of enzyme in the presence of inorganic phosphate. Mutations in the R2 locus, which maps far away from the structural gene, result in complete loss of repressibility (Echols *et al.*, 1961). Both the R1 and the R2 constitutive mutations are recessive to the wild-type allele in heterozygotes, indicating that a product of the two regions is required for the repression to be active. Since certain $R1^-$ and $R2^-$ mutants are sensitive to a suppressor, and such R^-su^+ strains behave phenotypically like the wild-type, it was concluded that the R1 and R2 gene products are proteins (Garen and Garen, 1963). This was partially verified by the isolation of a protein (Garen and Otsuji, 1964) which seems to be structurally specified by part of the R2 region. It is remarkable, however, that the R2 protein was found to be inducible together with the phosphatase, and was found in high concentrations in constitutive strains. This apparent co-ordinate production of the two proteins is not obligatory. Mutants have been isolated in the R1 region which are phenotypically phosphatase-negative, but have an intact *pho*-gene. These $R1^c$ mutants are recessive to the $R1^+$ allele, but although they produce no phosphatase, the R2 protein is formed in a low-phosphate medium (Garen and Otsuji, 1964).

The inducibility makes the R2 protein fundamentally distinct from the *i* protein of the *lac* region. However, as for the *lac* operon, it has been found that the repressor, which in this system is the R1 gene product, is metabolically unstable (Gallant and Spottswood, 1964). The same conclusion was reached in experiments in which protein synthesis was stopped by the addition of CM or of canavanine (Gallant and Stapleton, 1964).

In *Escherichia coli* the seven enzymes involved in the biosynthesis of arginine are repressed when arginine is added to a growing culture. Repression, however, seems not to be co-ordinate (Vogel, 1961; Maas, 1961). The structural genes have been mapped by recombination and transduction (Gorini *et al.*, 1961; Maas, 1961; Glansdorff, 1965). Four of the structural genes form a cluster, while the remaining ones map singly and at great distance from each other and from the cluster. Non-repressible mutants have been isolated by selecting for the ability to grow in the presence of canavanine. These mutations map apart from the structural genes in the R_{arg} locus and, since they affect all of the enzymes on the pathway collectively, they are classified as R mutations. In dominance studies, it has been found that the R^+ allele is dominant in zygotes and diploid cells (Maas *et al.*, 1964; Maas and Clark, 1964).

C. Models of Regulation

In the preceding section, our discussion of the regulation of enzyme synthesis has been based on the terminology and the model of operator control at the level of m-RNA transcription. It is evident from the diversity of the regulatory circuits discussed that this model in its simplest form can satisfy the disparity only with difficulty. On the other hand, it is very likely that many of the loci that are classified today as regulatory genes eventually will be recognized as structural genes for enzymes involved in the production or degradation of inducers or co-repressors. It is also obvious that the question of a specific regulation of repressor formation creates problems that severely augment the complexity of regulation models.

The existence of O^0 mutants stresses the close correlation between m-RNA transcription and the process of translation of the messenger into proteins. Accordingly we have to invoke alternate steps in the regulation of m-RNA production other than the mere opening and closing of a region of the DNA. The implication of the ribosomes in this function and the possible mechanisms have been discussed in this article.

In bacteria, the m-RNA fraction is metabolically unstable. The possible existence of messenger fractions with different half-lives is naturally of great importance for the expression of a regulatory mechanism on the level of messenger production. Nothing is at present known about the messenger decay process, nor about the direction in which it occurs and the enzymes involved.

The degradation of specific m-RNA, as measured by enzyme formation, seems not to be influenced by an inhibition of protein synthesis, contrary to the findings for total messenger. These findings could indicate that degradation of m-RNA takes place in the same direction as the translation process, and that the destruction of a region for the initiation of protein formation is sufficient to stop enzyme formation. Such a limited degradation might not influence the less specific measurements of decay, e.g. by a cell-free protein-forming system. However, it cannot be excluded that the degradation takes place in a random fashion and thus stops enzyme formation. In any event, the degradation of the total m-RNA seems to be dependent on protein synthesis, or at least upon the movement of ribosomes along the messenger strand. In the case of the O^0 mutants, no specific m-RNA is formed although the locus is fully derepressed. Some of the O^0 mutants contain a nonsense mutation which is not translated into an amino acid, and the polypeptide chain is accordingly broken at this position. It is not known if a nonsense codon has any effect which might lead to an obligatory disengagement of the ribosomes from the messenger or if the movement of the ribosomes continues beyond the nonsense codon. A special codon, corresponding to formylmethionine, has been suggested to be essential for polypeptide chain initiation (Clark and Marcker, 1965; Adams and Capecchi, 1966; Webster *et al.*, 1966). Under these conditions, a movement of the ribosomes beyond a nonsense codon would not give rise to polypeptide formation. It is not unlikely that such an abortive protein synthesis would lead to a frequent detachment of the ribosomes from the messenger.

A continued ribosome progression could explain the strong position effect of nonsense mutations in the *z* gene (Newton *et al.*, 1965). With O^0 mutants, an abortive translation combined with a low rate of transcription of the messenger could explain the observed lack of m-RNA.

Following speculations along these lines, it is possible to incorporate the data discussed above, pertinent to the operator and the regulator loci. This can be done by implying that the operator region is copied as part of the messenger immediately before the codon for chain initiation. The repressor might then show affinity towards the O region on the messenger and, by attachment, block the initiation of transcription. This blocking would be analogous to the situation with the O^0 mutants, and the messenger produced, would be degraded too rapidly to be observed in a hybridization test. Induction, and the reaction between inducer and repressor, would lead to an unmasking of the initiator-codon and to enzyme synthesis.

That a regulation at the level of the translation is not completely improbable is seen from the work of McAuslan (1963) on pox virus-infected HeLa cells. In these experiments, the messenger for virus-induced thymidine kinase was shown to be stable for 18 hr., although a repression of the formation of thymidine kinase normally takes place after 6 hr.

It is difficult from the information available at the end of April, 1966, when this review was being finished, to advocate any one regulation hypothesis without reserve. It is clear that, contrary to a regulation at the level of transcription, a regulation at the level of translation requires a continuous and often futile synthesis of RNA. We are not in a situation to judge if such a process of turnover is energetically feasible.

We have often been tempted to argue that the most logical and economical solution from our human point of view also is the best solution for the cell. Although this attitude in many cases has been a helpful guide in selecting an experimental approach, it might easily lead us astray, since we do not know if our logic, under all conditions, is the "logic" of the cells.

References

Adams, J. M. and Capecchi, M. R. (1966). *Proc. nat. Acad. Sci. Wash.* **55**, 147.

Adler, J. and Kaiser, A. D. (1963). *Virology* **19**, 117.

Alföldi, L., Stent, G. S. and Clowes, R. C. (1962). *J. molec. Biol.* **5**, 348.

Ames, B. N. and Garry, B. (1959). *Proc. nat. Acad. Sci. Wash.* **45**, 1453.

Ames, B. N., Garry, B. and Herzenberg, L. A. (1960). *J. gen. Microbiol.* **22**, 369.

Ames, B. N. and Hartman, P. E. (1963). *Cold Spr. Harb. Symp. quant. Biol.* **28**, 349.

Ames, B. N., Hartman, P. E. and Jacob, F. (1963). *J. molec. Biol.* **7**, 23.

Ames, B. N., Martin, R. G. and Garry, B. J. (1961). *J. biol. Chem.* **236**, 2019.

Aronson, A. I. and Spiegelman, S. (1961). *Biochim. biophys. Acta* **53**, 70.

Attardi, G., Naono, S., Gros, F., Brenner, S. and Jacob, F. (1962). *Compt. rend. Acad. Sci.* **255**, 2303.

Attardi, G., Naono, S., Rouvière, J., Jacob, F. and Gros, F. (1963). *Cold Spr. Harb. Symp. quant. Biol.* **28**, 363.

Beckwith, J. R. (1963). *Biochim. biophys. Acta* **76**, 162.

Beckwith, J. R. (1964). *J. molec. Biol.* **8**, 427.

Beckwith, J. R. and Signer, E. R. (1966). *J. molec. Biol.* **19**, 254,

Beckwith, J. R., Pardee, A. B., Austrian, B. and Jacob, F. (1962). *J. molec. Biol.* **5**, 618.

Berns, K. I. and Thomas, C. A. (1965). *J. molec. Biol.* **11**, 476.
Bladen, H. A., Byrne, R., Levin, J. G. and Nirenberg, M. W. (1965). *J. molec. Biol.* **11**, 78.
Blinkova, A. A., Bresler, S. E. and Lanzov, V. A. (1965). *Z. Vererbungsl.* **96**, 267.
Bolton, E. T. and McCarthy, B. (1962). *Proc. nat. Acad. Sci. Wash.* **48**, 1390.
Bonhoeffer, F. (1966). *Z. Vererbungsl.* **98**, 141.
Bonhoeffer, F. and Gierer, A. (1963). *J. molec. Biol.* **7**, 534.
Bonhoeffer, F. and Schaller, H. (1965). *Biochem. Biophys. res. Commun.* **20**, 93.
Borek, E. and Ryan, A. (1958). *J. Bact.* **75**, 72.
Bouck, N. and Adelberg, E. A. (1963). *Biochem. Biophys. res. Commun.* **11**, 24.
Bourgeois, S., Cohn, M. and Orgel, L. E. (1965). *J. molec. Biol.* **14**, 300.
Bremer, H. and Konrad, M. W. (1964). *Proc. nat. Acad. Sci. Wash.* **51**, 801.
Bremer, H., Konrad, M. W., Gaines, K. and Stent, G. S. (1965). *J. molec. Biol.* **13**, 540.
Bremer, H., Yegian, C. and Konrad, M. (1966). *J. molec. Biol.* **16**, 94.
Brenner, S. and Beckwith, J. R. (1965). *J. molec. Biol.* **13**, 629.
Brenner, S. and Stretton, A. O. W. (1965). *J. molec. Biol.* **13**, 944.
Buttin, G. (1961). *Cold Spr. Harb. Symp. quant. Biol.* **26**, 213.
Buttin, G. (1963a). *J. molec. Biol.* **7**, 164.
Buttin, G. (1963b). *J. molec. Biol.* **7**, 183.
Byrne, R., Levin, J. G., Bladen, H. A. and Nirenberg, M. W. (1964). *Proc. nat. Acad. Sci. Wash.* **52**, 140.
Cairns, J. (1963). *J. molec. Biol.* **6**, 208.
Campbell, A. (1957). *Bact. Rev.* **21**, 263.
Chamberlin, M. and Berg, P. (1962). *Proc. nat. Acad. Sci. Wash.* **48**, 81.
Clark, B. F. C. and Marcker, K. A. (1965). *Nature, Lond.* **207**, 1038.
Cuzin, F. (1962). *Compt. rend. Acad. Sci.* **254**, 4211.
Cuzin, F. and Jacob, F. (1965a). *Compt. rend. Acad. Sci.* **260**, 2087.
Cuzin, F. and Jacob, F. (1965b). *Compt. rend. Acad. Sci.* **260**, 5411.
Dagley, S., White, A. E., Wild, D. G. and Sykes, J. (1962). *Nature, Lond.* **194**, 25.
Das, H., Goldstein, A. and Kanner, L. (1966). *Mol. Pharmacol.* **2**, 158.
Doi, R. H. and Igarashi, R. T. (1964). *Nature, Lond.* **203**, 1092.
Dubin, D. T. and Elkort, A. T. (1965). *Biochim. biophys. Acta* **103**, 355.
Echols, H., Garen, A., Garen, S. and Torriani, A. (1961). *J. molec. Biol.* **3**, 425.
Ecker, R. E. and Schaechter, M. (1963). *Ann. N.Y. Acad. Sci.* **102**, 549.
Edlin, G. and Maaløe, O. (1966). *J. molec. Biol.* **15**, 428.
Eidlic, L. and Neidhardt, F. C. (1965). *J. Bact.* **89**, 706.
Englesberg, E., Irr, J., Power, J. and Lee, N. (1965). *J. Bact.* **90**, 946.
Ezekiel, D. H. (1963). *Biochem. Biophys. res. Commun.* **14**, 64.
Fan, D. P. (1966). *J. molec. Biol.* **16**, 174.
Fan, D. P., Higa, A. and Levinthal, C. (1964). *J. molec. Biol.* **8**, 210.
Fangman, W. L. and Neidhardt, F. C. (1964a). *J. biol. Chem.* **239**, 1839.
Fangman, W. L. and Neidhardt, F. C. (1964b). *J. biol. Chem.* **239**, 1844.
Forchhammer, J. and Kjeldgaard, N.O. (1967). *J. molec. Biol.* in press.
Fraenkel, D. G. and Neidhardt, F. C. (1961). *Biochim. biophys. Acta* **59**, 96.
Franzen, J. S. and Binkley, S. B. (1961). *J. biol. Chem.* **236**, 515.
Fuchs, E., Zillig, W., Hofschneider, P. H. and Preuss, A. (1964). *J. molec. Biol.* **10**, 546.
Fuhs, G. W. (1965). *Bact. Rev.* **29**, 277.
Fuller, W., Wilkins, M. H. F., Wilson, H. R. and Hamilton, L. D. (1965). *J. molec.* Biol. **12**, 60.
Fulton, C. (1965). *Genetics* **52**, 55.
Gallant, J. and Spottswood, T. (1964). *Proc. nat. Acad. Sci. Wash.* **52**, 1591.

Gallant, J. and Stapleton, R. (1964). *J. molec. Biol.* **8**, 442.
Garen, A. and Garen, S. (1963). *J. molec. Biol.* **6**, 433.
Garen, A. and Otsuji, N. (1964). *J. molec. Biol.* **8**, 841.
Geiduschek, E. P., Nakamoto, T. and Weiss, P. (1961). *Proc. nat. Acad. Sci. Wash.* **47**, 1405.
Geiduschek, P., Tocchini-Valentini, G. P. and Sarnat, M. T. (1964). *Proc. nat. Acad. Sci. Wash.* **52**, 486.
Gellert, M., Smith, C. E., Neville, D. and Felsenfeld, G. (1965). *J. molec. Biol.* **11**, 445.
Giacomoni, D. and Spiegelman, S. (1962). *Science* **138**, 1328.
Glansdorff, N. (1965). *Genetics* **51**, 167.
Goldberg, I. H. and Rabinowitz, M. (1962). *Science* **136**, 315.
Goldberg, I. H., Rabinowitz, M. and Reich, E. (1962). *Proc. nat. Acad. Sci. Wash.* **48**, 2094.
Goldstein, A., Goldstein, D. B. and Lowney, L. I. (1964). *J. molec. Biol.* **9**, 213.
Goldstein, A., Kirschbaum, J. B. and Roman, A. (1965). *Proc. nat. Acad. Sci. Wash.* **54**, 1669.
Goodman, H. M. and Rich, A. (1962). *Proc. nat. Acad. Sci. Wash.* **48**, 2101.
Gorini, L., Gundersen, V. and Burger, M. (1961). *Cold Spr. Harb. Symp. quant. Biol.* **26**, 173.
Gros, F. and Gros, F. (1958). *Exp. Cell Res.* **14**, 104.
Gros, F., Naono, S., Rouvière, J., Hayes, D. and Cukier, R. (1965). *Proc. FEBS 2nd. Meet., Vienna.*
Gross, J. D. and Caro, L. (1965). *Science* **150**, 1679.
Gross, J. D. and Caro, L. G. (1966). *J. molec. Biol.* **16**, 269.
Hall, B. D., Green, M., Nygaard, A. P. and Boezi, J. (1963). *Cold Spr. Harb. Symp. quant. Biol.* **28**, 201.
Hall, B. D. and Spiegelman, S. (1961). *Proc. nat. Acad. Sci. Wash.* **47, 137.**
Hanawalt, P. C. and Maaløe, O. (1961). *J. molec. Biol.* **3**, 144.
Hartman, P. E., Loper, J. C. and Šerman, D. (1960). *J. gen. Microbiol.* **22**, 323.
Hartman, P. E., Roth, J. R. and Ames, B. N. (1964). *Bact. Proc.* p. 18.
Hartwell, L. H. and Magasanik, B. (1963). *J. molec. Biol.* **7**, 401.
Hartwell, L. H. and Magasanik, B. (1964). *J. molec. Biol.* **10**, 105.
Hayashi, M., Hayashi, M. N. and Spiegelman, S. (1964). *Proc. nat. Acad. Sci. Wash.* **51**, 351.
Helling, R. B. and Weinberg, R. (1962). *Genetics* **48**, 1397.
Helmstetter, C. E. and Cummings, D. J. (1963). *Proc. nat. Acad. Sci. Wash.* **50**, 767.
Henning, U., Dennert, G., Szolyvay, K. and Deppe, K. (1965). *Z. Vererbungsl.* **97**, 1.
Hill, C. W. and Echols, H. (1965). *Biophys. J.* **5**, 91.
Hurwitz, J., Evans, A., Babinet, C. and Skalka, A. (1963). *Cold Spr. Harb. Symp. quant. Biol.* **28**, 59.
Hurwitz, J., Furth, J. J., Malamy, M. and Alexander, M. (1962). *Proc. nat. Acad. Sci. Wash.* **48**, 1222.
Imamoto, F., Morikawa, N. and Sato, K. (1965a). *J. molec. Biol.* **13**, 169.
Imamoto, F., Morikawa, N., Sato, K., Mishima, S., Nishimura, T. and Matsushiro, A. (1965b). *J. molec. Biol.* **13**, 157.
Ito, J. and Crawford, I. P. (1965). *Genetics* **52**, 1303.
Jacob, F., Brenner, S. and Cuzin, F. (1963). *Cold Spr. Harb. Symp. quant. Biol.* **28**, 329.
Jacob, F. and Monod, J. (1961a). *J. molec. Biol.* **3**, 318.
Jacob, F. and Monod, J. (1961b). *Cold Spr. Harb. Symp. quant. Biol.* **26**, 193.

Jacob, F. and Monod, J. (1963). *In* "Cytodifferentiation and Macromolecular Synthesis", (M. Locke, ed.), p. 30, Academic Press, New York.
Jacob, F. and Monod, J. (1965). *Biochem. Biophys. res. Commun.* **18**, 693.
Jacob, F., Ullman, A. and Monod, J. (1964). *Compt. rend. Acad. Sci.* **258**, 3125.
Jacob, F., Ullman, A. and Monod, J. (1965). *J. molec. Biol.* **13**, 704.
Jacob, F. and Wollman, E. (1957). *Compt. rend. Acad. Sci.* **245**, 1840.
Kahan, E., Kahan, F. M. and Hurwitz, J. (1963). *J. biol. Chem.* **238**, 2491.
Kalckar, H. M., Kurahashi, K. and Jordan, E. (1959). *Proc. nat. Acad. Sci. Wash.* **45**, 1776.
Kennell, D. and Magasanik, B. (1962). *Biochim. biophys. Acta* **55**, 139.
Kepes, A. (1963). *Biochim. biophys. Acta* **76**, 293.
Kiho, Y. and Rich, A. (1964). *Proc. nat. Acad. Sci. Wash.* **51**, 111.
Kiho, Y. and Rich, A. (1965). *Proc. nat. Acad. Sci. Wash.* **54**, 1751.
Kjeldgaard, N. O. (1961). *Biochim. biophys. Acta* **49**, 64.
Kjeldgaard, N. O. (1963). "Dynamics of Bacterial Growth", Nyt Nordisk Forlag, Arnold Busck, Copenhagen.
Kjeldgaard, N. O. and Kurland, C. G. (1963). *J. molec. Biol.* **6**, 341.
Kjeldgaard, N. O., Maaløe, O. and Schaechter, M. (1958). *J. gen. Microbiol.* **19**, 607.
Koch, A. L. (1965). *Nature, Lond.* **205**, 801.
Kohiyama, M., Cousin, D., Ryter, A. and Jacob, F. (1966). *Ann. Inst. Pasteur* **110**, 465.
Kohiyama, M., Lamfrom, H., Brenner, S. and Jacob, F. (1963). *Compt. rend. Acad. Sci.* **257**, 1979.
Kurland, C. G. and Maaløe, O. (1962). *J. molec. Biol.* **4**, 193.
Kurland, C. G., Nomura, M. and Watson, J. D. (1962). *J. molec. Biol.* **4**, 388.
Lark, C. and Lark, K. G. (1964). *J. molec. Biol.* **10**, 120.
Lark, K. G. (1966a). *In* "Cell Synchrony", (I. L. Cameron, ed.), p. 54, Academic Press, New York.
Lark, K. G. (1966b). *Bact. Rev.* **30**, 3.
Lark, C. (1966c). *Biochim. biophys. Acta* **119**, 517.
Lark, K. G. and Bird, R. (1965a). *J. molec. Biol.* **13**, 607.
Lark, K. G. and Bird, R. E. (1965b). *Proc. nat. Acad. Sci. Wash.* **54**, 1444.
Lark, K. G. and Lark, C. (1965). *J. molec. Biol.* **13**, 105.
Lark, K. G., Repko, T. and Hoffman, F. J. (1963). *Biochim. biophys. Acta* **76**, 9.
Lee, N. and Englesberg, E. (1962). *Proc. nat. Acad. Sci. Wash.* **48**, 335.
Leive, L. (1965). *J. molec. Biol.* **13**, 862.
Levinthal, C. and Davison, P. F. (1961). *J. molec. Biol.* **3**, 674.
Maaløe, O. and Kjeldgaard, N. O. (1966). "Control of Macromolecular Synthesis", W. A. Benjamin, New York.
Maaløe, O. and Rasmussen, K. W. (1963). *Colloques Intern. Recherche Scientifique (Paris)*, No. 124, p. 165.
Maas, W. K. (1961). *Cold Spr. Harb. Symp. quant. Biol.* **26**, 183.
Maas, W. K. and Clark, A. J. (1964). *J. molec. Biol.* **8**, 355.
Maas, W. K., Maas, R., Wiame, J. M. and Glansdorff, N. (1964). *J. molec. Biol.* **8**, 359.
McAuslan, B. R. (1963). *Virology* **21**, 383.
McCarthy, B. (1962). *Biochim. biophys. Acta* **55**, 880.
Martin, F. M., Yegian, C. and Stent, G. S. (1963). *Biochem. J.* **88**, 46 P.
Martin, R. G. (1963). *Cold Spr. Harb. Symp. quant. Biol.* **28**, 357.
Maruyama, Y. and Yanagita, T. (1956). *J. Bact.* **71**, 542.
Massie, H. R. and Zimm, B. H. (1965). *Proc. nat. Acad. Sci. Wash.* **54**, 1636.
Matney, T. S. and Achenbach, N. E. (1962). *J. Bact.* **84**, 874.

Matsushiro, A., Kida, S., Ito, J., Sato, K. and Imamoto, F. (1962). *Biochem. Biophys. res. Commun.* **9**, 204.
Matsushiro, A., Sato, K., Ito, J., Kida, S. and Imamoto, F. (1965). *J. molec. Biol.* **11**, 54.
Meselson, M. and Stahl, F. W. (1958). *Proc. nat. Acad. Sci. Wash.* **44**, 671.
Mitsui, H., Ishihama, A. and Osawa, S. (1963). *Biochim. biophys. Acta* **76**, 401.
Monod, J., Changeux, J-P. and Jacob, F. (1963). *J. molec. Biol.* **6**, 306.
Morris, D. W. and DeMoss, J. A. (1965). *J. Bact.* **90**, 1624.
Müller-Hill, B. (1966). *J. molec. Biol.* **15**, 374.
Nagata, T. (1963). *Proc. nat. Acad. Sci. Wash.* **49**, 551.
Nakada, D. and Magasanik, B. (1964). *J. molec. Biol.* **8**, 105.
Nathans, D. (1964). *Proc. nat. Acad. Sci. Wash.* **51**, 585.
Neidhardt, F. C. (1963). *Biochim. biophys. Acta* **68**, 365.
Neidhardt, F. C. (1964). *Progr. nucleic acid Res.* **3**, 145.
Neidhardt, F. C. and Fraenkel, D. G. (1961). *Cold Spr. Harb. Symp. quant. Biol.* **26**, 63.
Neidhardt, F. C. and Magasanik, B. (1960). *Biochim. biophys. Acta* **42**, 99.
Neuhard, J. and Munch-Petersen, A. (1966). *Biochim. biophys. Acta* **114**, 61.
Newton, W. A., Beckwith, J. R., Zipser, D. and Brenner, S. (1965). *J. molec. Biol.* **14**, 290.
Nirenberg, M. W. and Matthaei, J. H. (1961). *Proc. nat. Acad. Sci. Wash.* **47**, 1588.
Nomura, M. and Watson, J. D. (1959). *J. molec. Biol.* **1**, 204.
Novick, A., McCoy, J. M. and Sadler, J. R. (1965). *J. molec. Biol.* **12**, 328.
Novick, R. P. (1963). *J. gen. Microbiol.* **33**, 121.
Novick, R. P. and Richmond, M. H. (1965). *J. Bact.* **90**, 467.
Orias, E. and Gartner, T. K. (1964). *Proc. nat. Acad. Sci. Wash.* **52**, 859.
Otaka, E., Osawa, S. and Sibatani, A. (1964). *Biochem. Biophys. res. Commun.* **15**, 568.
Pardee, A. B. and Prestidge, L. S. (1956). *J. Bact.* **71**, 667.
Pritchard, R. H. and Lark, K. G. (1964). *J. molec. Biol.* **9**, 288.
Reich, E., Franklin, R. M., Shatkin, A. J. and Tatum, E. L. (1961). *Science* **134**, 556.
Richardson, C. C., Schildkraut, C. L., Aposhian, H. V. and Kornberg, A. (1964). *J. biol. Chem.* **239**, 222.
Richmond, M. H. (1965). *J. Bact.* **90**, 370.
Richmond, M. H. (1966). *Biochem. Biophys. res. Commun.* **22**, 38.
Rosen, B. (1965). *J. molec. Biol.* **11**, 845.
Rosset, R., Monier, R. and Julien, J. (1964). *Biochem. Biophys. res. Commun.* **15**, 329.
Rosset, R., Julien, J. and Monier, R. (1966). *J. molec. Biol.* **18**, 308.
Roth, J. R. (1965). *Fed. Proc.* **24**, 416.
Ryter, A. and Jacob, F. (1963). *Compt. rend. Acad. Sci.* **257**, 3060.
Sadler, J. R. and Novick, A. (1965). *J. molec. Biol.* **12**, 305.
Salas, M., Smith, M. A., Stanley, W. M., Wahba, A. J. and Ochoa, S. (1965). *J. biol. Chem.* **240**, 3988.
Sarabhai, A. S., Stretton, A. O. W., Brenner, S. and Bolle, A. (1964). *Nature, Lond.* **201**, 13.
Schaechter, M., Bentzon, M. W. and Maaløe, O. (1959). *Nature, Lond.* **183**, 1207.
Schaechter, M., Maaløe, O. and Kjeldgaard, N. O. (1958). *J. gen. Microbiol.* **19**, 592.
Schaechter, M., Previc, E. P. and Gillespie, M. E. (1965). *J. molec. Biol.* **12**, 119.
Schwartz, N. M. (1964). *J. Bact.* **88**, 996.
Shin, D. H. and Moldave, K. (1966). *Biochem. Biophys. res. Commun.* **22**, 232.

Signer, E. R., Beckwith, J. R. and Brenner, S. (1965). *J. molec. Biol.* **14**, 153.
Silver, S. D. (1963). *J. molec. Biol.* **6**, 349.
Smith, M. A., Salas, M., Stanley, W. M., Wahba, A. J. and Ochoa, S. (1966). *Proc. nat. Acad. Sci. Wash.* **55**, 141.
Smith, R. C. and Maaløe, O. (1964). *Biochim. biophys. Acta* **86**, 229.
Soffer, R. L. and Gros, F. (1964). *Biochim. biophys. Acta* **87**, 423.
Steers, E., Craven, G. R., Anfinsen, C. B. and Bethune, J. L. (1965a). *J. biol. Chem.* **240**, 2478.
Steers, E., Craven, G. R. and Anfinsen, C. B. (1965b). *Proc. nat. Acad. Sci. Wash.* **54**, 1174.
Stent, G. S. (1964). *Science* **144**, 816.
Stent, G. S. and Brenner, S. (1961). *Proc. nat. Acad. Sci. Wash.* **47**, 2005.
Stretton, A. O. W. and Brenner, S. (1965). *J. molec. Biol.* **12**, 456.
Sud, I. J. and Schaechter, M. (1964). *J. Bact.* **88**, 1612.
Taylor, A. L. and Adelberg, E. A. (1961). *Biochem. Biophys. res. Commun.* **5**, 400.
Taylor, A. L., Beckwith, J. R., Pardee, A. B., Austrian, R. and Jacob, F. (1964). *J. molec. Biol.* **8**, 771.
Taylor, A. L. and Thoman, M. S. (1964). *Genetics* **50**, 659.
Thach, R. E., Cecera, M. A., Sundarajan, T. A. and Doty, P. (1965). *Proc. nat. Acad. Sci. Wash.* **54**, 1167.
Tissières, A., Bourgeois, S. and Gros, F. (1963). *J. molec. Biol.* **7**, 100.
Tocchini-Valentini, G. P., Stodolsky, M., Aurisicchio, A., Sarnat, M., Graziosi, F., Weiss, S. B. and Geiduschek, E. P. (1963). *Proc. nat. Acad. Sci. Wash.* **50**, 935.
Traut, R. R. and Monro, R. E. (1964). *J. molec. Biol.* **10**, 63.
Turnock, G. and Wild, D. G. (1965). *Biochem. J.* **95**, 597.
Villa Trevino, S., Farber, E., Staehlin, T., Wettstein, F. O. and Noll, H. (1964). *J. biol. Chem.* **239**, 3826.
Vogel, H. J. (1961). *Cold Spr. Harb. Symp. quant. Biol.* **26**, 163.
Warner, J. R., Rich, A. and Hall, C. (1962). *Science* **138**, 1399.
Webster, R. E., Engelhardt, D. L. and Zinder, N. D. (1966). *Proc. nat. Acad. Sci. Wash.* **55**, 155.
Willson, C. and Gros, F. (1964). *Biochim. biophys. Acta* **80**, 478.
Willson, C., Perrin, D., Cohn, M., Jacob, F. and Monod, J. (1964). *J. molec. Biol.* **8**, 582.
Woese, C., Naono, S., Soffer, R. L. and Gros, F. (1963). *Biochem. Biophys. res. Commun.* **11**, 423.
Yaniv, M., Jacob, F. and Gros, F. (1965). *Bull. Soc. Chim. Biol.* **47**, 1609.
Yankofsky, S. A. and Spiegelman, S. (1962a). *Proc. nat. Acad. Sci. Wash.* **48**, 1069.
Yankofsky, S. A. and Spiegelman, S. (1962b). *Proc. nat. Acad. Sci. Wash.* **48**, 1466.
Yankofsky, S. A. and Spiegelman, S. (1963). *Proc. nat. Acad. Sci. Wash.* **49**, 538.
Yarmolinsky, M. B., Jordan, E. and Wiesmayer, H. (1961). *Cold Spr. Harb. Symp. quant. Biol.* **26**, 217.
Yates, R. A. and Pardee, A. B. (1956). *J. biol. Chem.* **221**, 757.
Yoshikawa, H. (1965). *Proc. nat. Acad. Sci. Wash.* **53**, 1476.
Yoshikawa, H., O'Sullivan, A. and Sueoka, N. (1964). *Proc. nat. Acad. Sci. Wash.* **52**, 973.
Yoshikawa, H. and Sueoka, N. (1963a). *Proc. nat. Acad. Sci. Wash.* **49**, 559.
Yoshikawa, H. and Sueoka, N. (1963b). *Proc. nat. Acad. Sci. Wash.* **49**, 806.
Zabin, I. (1963a). *Cold Spr. Harb. Symp. quant. Biol.* **28**, 431.
Zabin, I. (1963b). *J. biol. Chem.* **238**, 3300.

Biochemical Aspects of Extreme Halophilism

HELGE LARSEN

Department of Biochemistry, The Technical University of Norway, Trondheim, Norway

I. Introduction 97
II. Extreme Halophiles and their Growth Relations 98
III. Metabolic Apparatus of the Extreme Halophiles 101
A. Intracellular Salt Content 101
B. Salt Relations of Individual Enzymes 103
C. Salt Relations of the Ribosomes 107
D. Salt Relations of the Uptake Mechanism 108
E. Nucleic Acids 108
F. Metabolic Pathways 110
IV. Cell Envelope of the Halobacteria 111
A. Requirement for Salt to Maintain the Cell Envelope . . 112
B. Structure and Chemical Composition 114
C. Function of Salts in Maintaining the Envelope . . . 123
V. On the Rise of Extreme Halophilism 126
References 130

I. Introduction

Around the turn of the century several bacteriologists recognized the reddening of heavily salted protein products (fish, hides, viscera) as bacterial growth. Years passed, however, before it was clearly realized, mainly through the work of Klebahn (1919) and Harrison and Kennedy (1922), that the red-coloured bacteria living in this very salty environment are highly specialized organisms. Not only can these bacteria manage to live and proliferate in a concentrated sodium chloride brine; they absolutely require a very strong—almost saturated—salt brine for best maintenance and growth. This property of "extreme halophilism" is found only among the bacteria, and is to our knowledge confined to very few bacterial types. Through the years bacteriologists have given fairly detailed descriptions of these organisms, and have also studied their salt relations. Studies on the biochemical basis for the extreme salt requirement were, however, largely neglected until about 10 years ago. At that time N. E. Gibbons in Canada started to explore the biochemical peculiarities of the extreme halophiles, and a few other investigators have since joined in this research. The present writer reviewed the biochemical work on extreme halophilism five years ago, and stated that, although valuable information had been obtained, the work was still in its initial phase (Larsen, 1962). During the period that has elapsed since the appearance of that review, a good deal of new and interesting information has been gained on the biochemistry of the extreme halophiles. This pertains in particular to the ways in which salt influences the structural and functional components of these fascinating organisms which Baas Becking (1928) characterized as representing "the borderland of physiological possibilities".

II. Extreme Halophiles and their Growth Relations

All organisms which have been shown with certainty to grow best in the presence of sodium chloride concentrations above 20% can be assigned to one of two types of bacteria. In the following review, these organisms are referred to as "the halobacteria" and "the halococci", respectively. They are all obligate aerobes, and are usually red-to-orange in colour.

The halobacteria are rod-formed or irregular in shape, Gram-negative, non-spore-formers, lophotrichously flagellated when motile, and lyse when exposed to hypotonic solutions. They grow optimally in media containing about 25% (w/v) NaCl. A minimum of 12–20% NaCl is required for growth to occur, a figure differing considerably from strain to strain (Kurochkin, 1962). In the 7th. edition of "Bergey's Manual" (1957), organisms belonging to this type have been collected in the genus *Halobacterium* and assigned to the family Pseudomonadaceae.

The halococci are coccoid in shape and occur as single cells, pairs, irregular cell clusters or sarcina-like packets. The morphology varies from strain to strain as well as with the conditions of culture. The cells are Gram-variable, non-motile, and do not form spores. They do not display the lysis phenomenon in hypotonic solutions which is so characteristic for the halobacteria. They grow optimally in media containing 20–25% NaCl, and they require at least 5–15% NaCl for growth. In the 7th. edition of "Bergey's Manual" (1957), the halococci are assigned to the family Micrococcaceae, in part to the genus *Micrococcus* and in part to the genus *Sarcina*.

The extreme halophiles are widely distributed in nature in brines containing very high salt concentrations. They are especially conspicuous in the solar evaporation ponds of salt works, where they frequently occur in such large numbers that they give a red colour to the concentrated brine.

The characteristic red colouration of the extreme halophiles is due to carotenoids. However, the pigmentation bears no direct relation to the halophilic property of the cells. Its physiological significance rather lies in the ability of the carotenoids to protect the cells against deleterious effects of strong sunlight, as shown in comparative experiments with an artificially produced colourless mutant and its carotenoid-containing progenitor (Dundas and Larsen, 1962, 1963). The strongly saline waters in which the extreme halophiles occur are usually found under a bright sun. Under such conditions, carotenoid-containing cells will grow and multiply more freely than non-pigmented ones, and will outgrow the non-pigmented cells in the course of time. The rather consistent reports in the literature that the extreme halophiles are pigmented can reasonably be explained on this ground. One might, however, expect the occurrence in nature also of colourless extreme halophiles, namely in saline soils or in deeper waters and bottom sediments of salt lakes where sunlight is less intense or absent. To the author's knowledge, the microflora of such localities has not been systematically investigated. Henis and Eren (1963) reported lack of success in finding halophilic bacteria in highly saline soil from near the Dead Sea. However, these investigators tested their soil samples for halophiles growing in soil extract-agar containing 10% NaCl, and in this medium one can hardly expect extreme halophiles to develop.

The extremely halophilic bacteria are generally cultured on a protein digest (tryptone, peptone, or yeast autolysate) medium containing a suitable amount of sodium chloride. Eimhjellen (1965) has recently reviewed the methods of culturing, including the methods of obtaining enrichment cultures and isolating pure cultures.

The extreme halophiles have a quite complex nutritional requirement. They do not generally metabolize carbohydrates; proteins and amino acids seem to be the carbon sources most readily utilized. Dundas *et al.* (1963) composed a chemically defined medium for a strain of *Halobacterium salinarium*. Besides inorganic salts, this medium contained ten amino acids of which four (valine, methionine, leucine, isoleucine) were essential for growth while the others stimulated growth. In addition cytidylic acid was stimulatory to growth. This medium supported good growth, not only of the strain for which it was designed, but also of some other strains of halobacteria. Multiplication was, however, not quite as rapid as on a yeast extract medium and proceeded with a considerably longer lag phase. Strains of halococci and also some strains of halobacteria would not grow on this synthetic medium.

Onishi *et al.* (1965) designed a chemically defined medium which supported good growth of all the extreme halophiles tested by them, both halobacteria and halococci. Besides inorganic salts this medium contained fifteen amino acids, two nucleotides (adenylic and uridylic acids), and glycerol. In special tests on a strain of *Halobacterium cutirubrum*, which grew just as well on the synthetic as on a complex medium, four of the amino acids (arginine, leucine, lysine, valine) proved essential; each of the others was stimulatory when in mixture. Especially interesting is the stimulatory effect of glycerol which the authors suggest acts as a readily utilizable precursor of cell lipids.

Concerning the inorganic constituents of the culture media it should be emphasized that the extreme halophiles have a specific requirement for sodium chloride for growth. Numerous experiments have been reported in which no growth was obtained when sodium chloride was replaced by other salts or compounds in the culture medium; for a review see Larsen (1962). Sodium ions can, however, be partially replaced by potassium ions (Brown and Gibbons, 1955; Christian, 1956), and to a very small extent by magnesium ions (Weber, 1949).

The report by Brown and Gibbons (1955) that the extreme halophiles require unusually high concentrations of magnesium ions (0·1–0·5 *M*) in the culture medium for good growth is particularly interesting. Growing in a medium low in Mg^{2+} the normally rod-shaped halobacteria take on a coccoid form which may be maintained even after transfer back to a medium containing a high concentration of Mg^{2+}. The same authors reported that the requirement of the extreme halophiles for potassium ions is only slightly higher than that reported for many other organisms (about 2 m*M*). This is particularly noteworthy in view of the fact that the intracellular K^+ concentration of extreme halophiles is extremely high (Section III A, p. 102).

Onishi *et al.* (1965) reported that ammonium chloride at an optimum concentration of 0·2–0·5% led to a considerable decrease in the duration of the lag period for growth of *H. cutirubrum*. In a later paper, Onishi and Gibbons (1965) reported experiments showing that the requirement for ammonium ions is not specific, and that the stimulatory effect is related to the utilization of amino acids.

Sehgal and Gibbons (1960) reported a minimum requirement of 10 p.p.m. ferrous ions for good growth of various halobacteria, and a stimulatory effect of manganous ions on *H. cutirubrum* at a concentration of 0·05 p.p.m. Systematic investigations to establish the optimum concentrations of other inorganic components normally present in the culture media have not been reported.

The extreme halophiles multiply relatively slowly even under the most favourable conditions designed. The shortest generation time obtained for the halobacteria is not much less than 7 hr., and for the halococci about 15 hr. For much biochemical work, a high cell yield in cultures is desirable. Holmes *et al.* (1965) have reported that a strain of *H. salinarium* yielded 8–10 g. (wet weight) of cells per litre in a yeast extracttrypticase culture medium, provided the use of about a 25% inoculum. The large inoculum also decreased the length of the lag phase.

III. Metabolic Apparatus of the Extreme Halophiles

The requirement of the extreme halophiles for a very high concentration of sodium chloride in their growth medium naturally arouses an interest in the intracellular salt content of these organisms. If the intracellular salt concentration were comparable to that of other bacteria, this would imply that, unless some very special mechanisms are operating, the cell content of the extreme halophiles must be at an exceptionally low level of hydration due to the dehydrating effect of the external salt. In addition, the cells would possibly be exposed to a great osmotic strain. If, on the other hand, the salt concentration inside the cells were about equal to that in the environment, this would counterbalance the dehydrating effect of the external salt and there would be no unusual osmotic problem. Under any circumstances, the water activity (a_w) inside the cells must be expected to be exceptionally low, and the metabolic apparatus thus exposed to conditions quite different from those existing in other bacteria. How can the metabolic apparatus of the extreme halophiles operate under the extreme conditions which these organism require? And how is the metabolic apparatus constructed? Research on these problems over the past 10 years has provided a good deal of information, but many questions are still left unanswered.

A. Intracellular Salt Content

Experimental evidence has accumulated showing that the intracellular salt concentration of the extreme halophiles is indeed very high and reaches values comparable to that in the medium in which the cells are grown. This evidence has been furnished in part on the basis of chloride determinations in packed cells (Gibbons and Baxter, 1953; Holmes, 1964), and in part on the basis of cryoscopic measurements on heated cell pastes (Christian 1956; Christian and Ingram, 1959). Cells grown in the presence of different concentrations of salt have a salt content that depends upon the concentration in the medium.

It appears that besides Na^+ and Cl^-, K^+ is a dominating component of the internal salt. This was originally indicated by Brown and Gibbons (1955) and Christian (1956), and was rather unexpected since K^+ is only a minor component of the media in which the cells are normally cultured. The very high intracellular K^+ content of the extreme halophiles has, however, been thoroughly confirmed by Christian and Waltho (1962). They tested both a halobacterium, *H. salinarium*, and a halococcus,

Sarcina morrhuae, and determined the apparent intracellular concentration of a number of constituents by analysing packed cells and correcting for interstitial space using phosphate as an indicator. Their data for Na^+, K^+ and Cl^- are given in Table 1 and show that, in *H. salinarium*, the apparent concentration of K^+ is considerably higher than that of Na^+, whereas in *S. morrhuae* it is somewhat lower. It is noteworthy that K^+ is concentrated in the cells against a steep gradient, and indeed in *H. salinarium* to such an extent that the intracellular concentration of KCl reaches a value close to the solubility limit of this salt. Taken together,

TABLE 1. Concentration of Intracellular Constituents of Extreme Halophiles (from Christian and Waltho, 1962)

	Bacterium	
	Halobacterium salinarium	*Sarcina morrhuae*
Constituent (molal equivalents)		
Na^+	$1.37 \pm 0·21$	$3·17 \pm 0·28$
K^+	$4·57 \pm 0·12$	$2·03 \pm 0·36$
Cl^-	$3·61 \pm 0·07$	$3·66 \pm 0·25$
Ratio (molal concentrations)		
cell Na^+/medium Na^+	0·30	0·70
cell K^+/medium K^+	143	63
cell $(Na^+ + K^+)$/ medium Na^+	1·32	1·15
cell Cl^-/medium Cl^-	0·80	0·81

Cells were grown in a medium containing 4·0 *M*- (i.e. about 4·5 molal) NaCl + 0·32 *M*-K^+.

Na^+ and K^+ exceeded somewhat the Na^+ concentration in the medium. The internal chloride concentration, however, appeared to be somewhat lower than the external. Other low-molecular weight compounds (amino acids, phosphates) were found within the cells in modest amounts only.

The very high K^+ content of the extreme halophiles bears an interesting relationship to another finding of Christian and Waltho (1961). Testing a number of different non-halophilic bacteria, they established a relation between the K^+ content of the cells and their tolerance towards sodium chloride. The higher the salt tolerance the higher was the K^+ content when the cells were grown in ordinary media. A high K^+ content thus seems to be a general characteristic of bacteria possessing the ability to live in strong NaCl brines, and seemingly regardless of whether the organisms normally live in the brine or just possess a potential ability to do so. Christian and Waltho suggested that the high K^+ content of the non-halophiles might impart a resistance to plasmolysis and dehydration, thus overcoming the detrimental effects of the external salts. One can envisage the K^+ playing a similar rôle in the extreme halophiles, but it should be noted that in these bacteria Na^+ also penetrates readily and might fill the same functions. Other possible functions of the internal K^+ of the extreme halophiles are discussed in later sections.

B. Salt Relations of Individual Enzymes

To Baxter and Gibbons (1954, 1956, 1957) we owe the important discovery that the enzymes of the extreme halophiles are adapted to function at the very high salt concentrations found within the cells. Using crude extracts of *H. salinarium*, these investigators tested the response of a number of enzymes towards sodium chloride at different concentrations. In all cases they found good activity at salt concentrations corresponding to those presumably existing within the cells under normal culture conditions; in most cases the enzymes displayed a relatively high, and even maximum, activity at these extreme salt concentrations. At low salt concentrations, most of the enzymes showed little or no activity. These enzymes can rightly be designated halophilic.

The investigations on the salt response of the enzymes of extreme halophiles were extended by O. Aasmundrud and H. Larsen to comprise also a number of enzymes of the halococcus, *S. morrhuae*. The curves obtained were in accordance with those found for *H. salinarium*, and it was especially striking that, although the various types of enzymes could differ quantitatively from each other in their salt response, enzymes of the same type from the two different bacteria displayed similar salt response (Larsen, 1962). The list of enzymes tested has recently been extended by Holmes *et al.* (1965). Altogether about twenty different enzymes have been reported on, and all data fit the general contention that the enzymes of the extreme halophiles are extremely halotolerant, and in most cases even strikingly halophilic.

In Fig. 1 are given some typical examples of salt response curves of enzymes from *H. salinarium*. Cysteine desulphydrase (curve a) is about the least salt-tolerant enzyme of those tested in this organism, but still it shows a fairly good activity at 4·3 *M*-NaCl. Cytochrome oxidase (curve b) has a salt response typical of several of the enzymes tested. $NADH_2$ dehydrogenase (curve c) is an example of an enzyme with an extreme salt requirement. Interestingly enough succinate dehydrogenase, which is believed to be functionally closely related to $NADH_2$ dehydrogenase, displays the same extreme salt response as the latter.

The enzymes of the extreme halophiles clearly differ in their salt response from the corresponding enzymes of biological systems which are not exposed to high concentrations of salt. Many reports have appeared in the literature on the effect of salt on the activity of enzymes from such sources, and it appears that, although some enzymes are rather salt-tolerant, others are quite salt-sensitive. Fig. 2 gives as example the

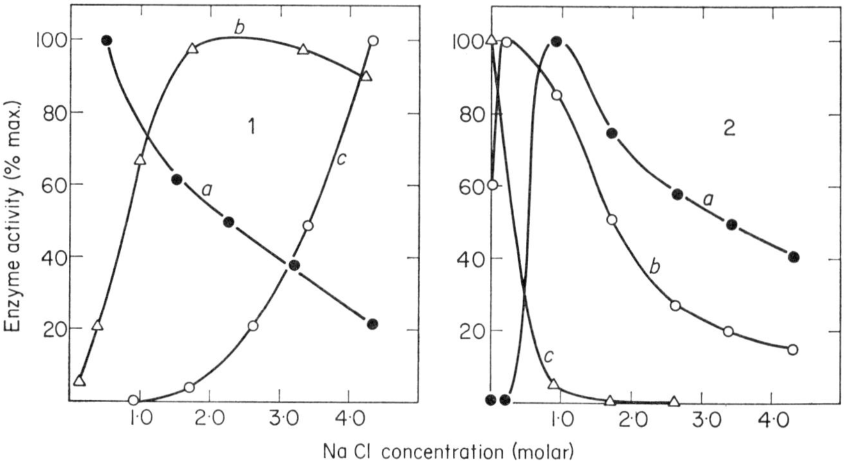

FIG. 1. Effect of sodium chloride on the activities of enzymes extracted from *Halobacterium salinarium*. a: cysteine desulphydrase (Baxter and Gibbons, 1957), b: cytochrome oxidase (Baxter and Gibbons, 1956), c: $NADH_2$, dehydrogenase (H. Larsen and T. Langsrud, unpublished observations).

FIG. 2. Effect of sodium chloride on the activities of malate dehydrogenases extracted from different sources. a: *Halobacterium salinarium*, b: halotolerant rod, c: pig liver. From Holmes and Halvorson (1965a).

effect of salt on malate dehydrogenases from different biological sources showing different responses to salt (Holmes and Halvorson, 1965a). The enzyme from *H. salinarium* (curve a) displays a considerably higher activity at high salt concentrations than malate dehydrogenase from a halotolerant, rod-shaped bacterium (curve b). Malate dehydrogenase from liver (curve c) is relatively very salt-sensitive.

In their early experiments, Baxter and Gibbons (1954, 1956, 1957) tested the effect of a number of salts other than sodium chloride on the enzymes of *H. salinarium*. It appeared that several of the alkaline cations and the anions bromide and nitrate could replace Na^+ and Cl^-, respectively, in affecting the activity of the halophilic enzymes. Potassium, as the chloride was especially potent in this respect. Mole for mole, K^+ was in all cases at least as effective as Na^+ in activating the halophilic enzymes; in most cases, potassium chloride gave about double the activity as compared to that obtained with sodium chloride at the same molar concentrations. Similar results have been found for the enzymes of *S. morrhuae* (O. Aasmundrud and H. Larsen, unpublished observations).

The strongly stimulatory effect of both sodium chloride and potassium chloride upon the enzymes of the extreme halophiles brings to mind the finding that, besides Na^+ and Cl^-, K^+ is also a major intracellular constituent in these bacteria (Section III A, p. 101). It seems obvious that these three intracellular constituents play an important rôle in the cells by keeping the enzymes in an activated state.

The function of the sodium and potassium chlorides is not limited to an activation of the enzymes. These salts also act as stabilizers of the enzymes. It has been shown in numerous experiments that, when the salt is removed from the enzymes, for example by dialysis, they tend to become irreversibly denatured quite rapidly (Baxter and Gibbons, 1954, 1957, 1959). Holmes and Halvorson (1963, 1965b) have studied this phenomenon in some detail using enzymes from *H. salinarium* and have arrived at some quite interesting results. After removal of salt from enzyme preparations by dialysis against water, the enzyme activity could not be restored or only erratically so, by addition of solid NaCl. If, however, the salt-free preparations were dialysed against 25% NaCl so that they were gently exposed to an increase in the NaCl concentration, a number of enzymes regained activity even if they had been kept in a salt-free environment for a relatively long period of time. It is particularly interesting that under these conditions some enzymes reproducibly regained an activity which was a constant fraction of the original activity before the removal of the salt. Malate dehydrogenase, for example, could be restored to about 60% of the original activity; and alanine dehydrogenase to about 70% of the original activity. The restorable fraction was stable under prolonged exposures to salt-free environments. Some enzymes (ethanol dehydrogenase, aspartate aminotransferase) did not show a restoration of activity upon dialysis of salt-free preparations against 25% NaCl; these enzymes became completely and irreversibly denatured in the absence of salt. Alkaline phosphatase regained full activity even after exposure to a salt-free environment for 48 hr.

The discovery that enzyme activity can be restored from salt-free preparations by dialysis against 25% NaCl has provided a very important means of effecting the purification of enzymes from extreme halophiles. Such purifications were earlier an ungrateful task. The enzymes had to be kept in the presence of high salt concentrations in order to prevent permanent loss of activity, and under these conditions conventional methods of protein purification can be rather difficult. Thus ammonium sulphate may not act as a salting-out agent, and conventional materials for column chromatography may not retain proteins. Working with salt-free preparations and conventional techniques (acetone and ammonium sulphate precipitations, chromatography on DEAE cellulose, electrophoresis on polystyrene particles), Holmes and Halvorson (1965a) could effect an almost 1000-fold purification of malate dehydrogenase from *H. salinarium*.

Holmes and Halvorson (1965b) carried out a detailed study of the purified malate dehydrogenase thus obtained and confirmed its halophilic character. Studies of the enzyme activity at lowered salt concentrations suggested that most of the enzyme molecules were equally sensitive to salt deprivation. The inactivation by removal of the salt was markedly inhibited by the presence of the reduced co-enzyme, $NADH_2$. The re-activation of enzyme activity by dialysis of the salt-free preparation against 25% NaCl was a slow process. Activity was regained more slowly than the rate at which salt entered the dialysis bag, suggesting that the re-activation is a complex process, probably comprising conformational changes in the protein molecules. Kinetic studies on the return of activity suggested either that re-activation did not follow a first-order reaction, or that more than one species of malate dehydrogenase, each with its characteristic re-activation rate, was present. KCl was equally effective as NaCl in re-activating the enzyme.

On the basis of his studies on the halophilic lactate dehydrogenase of *H. salinarium*, Baxter (1959) proposed that a function of the salt is to decrease the electrostatic repulsion between ionized groups within the enzyme molecule. Only at high salt concentrations will the molecule attain the conformational structure in which it is active as a catalyst. By lowering the salt concentrations, the molecule will expand to an inactive conformation. The studies of Holmes and Halvorson (1965b) suggest a similar explanation for a function of salt with malate dehydrogenase. They observed a relation between salt concentration and pH value suggesting that ionized groups within the enzyme molecule play a rôle both for the activity of the enzyme and for its re-activation from the salt-free state. They also observed that the sedimentation rate, determined by centrifugation in sucrose gradients, decreased upon lowering the salt concentration. This was difficult to reconcile with a disaggregation of the molecule since virtually all of the enzyme activity could be regained by increasing the salt concentration. It seemed more probable that the lowering of the sedimentation rate reflected an increase in the partial specific volume of the molecule.

C. Salt Relations of the Ribosomes

Bayley and Kushner (1964) have isolated the ribosomes of the halobacterium, *H. cutirubrum*, and studied extensively their relations to salts. In their native state, the ribosomes of *H. cutirubrum* appear to be similar to the ribosomes of *Escherichia coli* with respect to the size of the particles and the gross chemical composition, i.e. 70S particles consisting of 60% RNA and 40% protein. In striking contrast to the *E. coli* (and other) ribosomes, however, the *H. cutirubrum* ribosomes are stable only in solutions containing KCl close to saturation (4*M*) and Mg^{2+} in the concentration range 0·1–0·4 *M*. Ribosomes from non-halophilic organisms tend to break down at concentrations of alkali chlorides above 0·15 *M*. When exposed to lower concentrations of K^+ and Mg^{2+}, the ribosomes of *H. cutirubrum* dissociate into 52S and 31S subunits, and these may be further degraded to give 42S, 33S and 22S particles and soluble protein. The latter three types of particles contain 80% RNA which represents the bulk of the RNA in the native 70S ribosomes. The protein dissolved constitutes about 75% of the protein of the ribosomes and displays acidic properties as shown by electrophoresis.

The requirement for KCl to maintain the 70S particles is quite specific. When KCl was replaced by iso-osmolar NaCl, dissociation to 52S particles and also complex aggregation phenomena occurred. Replacing KCl by other chlorides of the alkaline series resulted in an extensive degradation of the 70S particles. A requirement for magnesium ions to maintain integrity is a well-known property of ribosomes from other organisms. However, the ribosomes of *H. cutirubrum* require 10–100 times the Mg^{2+} concentration required by *E. coli* for stability.

Bayley (1966) has recently investigated further the *H. cutirubrum* ribosomes and their dissociation products. Amino acid analysis and gel electrophoresis showed the acidic nature of the protein released by lowering the salt concentration. The protein remaining in the particle fraction is relatively basic. Analysis of K^+ and Mg^{2+} showed that potassium ions were easily removed from the particles by washing with dilute salt solutions; a certain amount of Mg^{2+} was more firmly bound to the particles. Bayley suggested that a function of the magnesium ions is to stabilize the configuration of the ribosomal RNA rather than to take part in the binding of protein to RNA. This suggestion is in accordance with that for the function of Mg^{2+} in ribosomes of other organisms. The unusually high concentration of magnesium ions required by the ribosomes of *H. salinarium* is possibly to overcome the disturbing effect of excess potassium ions.

The main chemical feature that distinguishes the ribosomes of *H. salinarium* from those of non-halophilic organisms seems to be the high content of acidic protein. The dominating ribosomal protein in *E. coli*, for example, is clearly basic in nature. Bayley pointed out that the potassium ions, which are normally present in the halophilic cells in extremely high concentrations and indeed in concentrations needed to stabilize the ribosomes, may neutralize the negative charges on the acidic protein and RNA, thereby enabling these components to aggregate and form stable and functionally active ribosome structures. No function has yet been found for Na^+ in the ribosomal apparatus.

D. Salt Relations of the Uptake Mechanism

Stevenson (1966) studied the uptake of glutamate by *H. salinarium* and found it to be specifically dependent upon a high concentration of NaCl in the medium in which the cells were suspended. Replacing the Na^+ by equimolar concentrations of K^+ (as the chloride or the acetate), or the Cl^- by equimolar concentrations of acetate (as the sodium salt), gave a negligible uptake of the amino acid by the cells.

In the other functions thus far assigned to NaCl in the extreme halophiles, this salt has been found to be non-specific in the sense that other salts may fill the same functions when replacing NaCl. Thus, alkali halides, and notably KCl which is present in high concentrations within the cells, activate all the individual enzymes so far tested (Section IIIB, p. 104); many salts which interact weakly with proteins, notably KCl and sodium acetate, protect the halobacteria from structural transformations and lysis when replacing NaCl in iso-osmolar concentrations (Section IVA, p. 113). KCl, or any other salt, cannot, however, replace NaCl for growth of these organisms, which suggests some specific function for NaCl in the cells. Stevenson's finding that only NaCl stimulates the mechanism for the uptake of glutamate by *H. salinarium* points to such a specific function. Others may well be found.

Uptake of glutamate by *H. salinarium* is by "active transport"; in other words, the mechanism requires a supply of energy to mediate the uptake. Stevenson found that the uptake was severely retarded under anaerobic conditions and also in the presence of 2,4-dinitrophenol. This suggests that NaCl might be required for the energy-supplying mechanism and not necessarily for the process of migration across the cell membrane. Since KCl can replace NaCl in maintaining respiration in *H. salinarium*, the possibility exists that the specific requirement for NaCl in the uptake mechanism is connected with the process of oxidative phosphorylation.

E. Nucleic Acids

Very few data are available on the nucleic acid content of the extreme halophiles. Smithies *et al.* (1955) analysed a strain of *Pseudomonas salinaria* (*H. salinarium*) and found, on a salt-free dry weight basis, 15·0% total nucleic acids of which 4·3% was DNA. These values are comparable to those commonly found for other bacteria.

The cells of many halophilic bacteria may be sticky when handled. Smithies and Gibbons (1955) found that the stickiness of *P. salinaria* (*H. salinarium*) was probably due to DNA adhering to the cells, and suggested that DNA may leak out of the cells under certain nutritionally unfavourable conditions. It is also possible that the reason for the apparent stickiness of the halobacteria should be sought in the extreme fragility of these cells; they are extraordinarily sensitive to mechanical stress. Experiments carried out by I. D. Dundas in the author's laboratory showed that the most careful packing in the centrifuge, followed by an equally careful resuspension, resulted in a killing of about half the population. If the cells are not handled with the utmost care, suspensions may appear abnormally viscous, presumably due to DNA released from the damaged cells.

The base composition of bulk RNA and DNA of a strain of *H. salinarium* has been determined chemically by K. Ormerod in the author's laboratory. The results are given in Table 2. These show that the composition of the RNA is as commonly found in bacteria; the composition of the DNA is as commonly found in members of the genus *Pseudomonas* (Belozersky and Spirin, 1960) to which the halobacteria bear much resemblance in many other respects. In a table compiled by Marmur *et al.* (1963) giving the DNA base composition of bacteria are listed data for six organisms assigned to the genus *Halobacterium*. There is good reason to believe that four of these strains are not halobacteria or, for that matter, extreme halophiles, since Joshi *et al.* (1963), in work on organisms with exactly the same strain designations, specifically state that "none of these organisms requires the high salt media". The other two organisms, *H. salinarium* and *H. cutirubrum*, are both listed with a DNA–GC content of 66–68% which fit well with the data in Table 2.

Baxter (1961, 1964) investigated further the nucleic acids of the same strain (No. 1) of *H. salinarium* as reported in Table 2. He chromatographed cell-free extracts on two different cellulose ion exchange materials (DEAE- and ECTEOLA-celluloses) using an eluent with increasing NaCl concentration. The elution patterns obtained for DNA and RNA were identical to those for a non-halophilic pseudomonad. Melting curves for DNA preparations gave T_m values which, according to the relationship found by Marmur and Doty (1959), fitted well with the base data listed in Table 2. Baxter concluded that, on the basis of the criteria tested, there is no reason to believe that the nucleic acids of the halobacteria differ in their fine structure from those of non-halophilic organisms.

Joshi *et al.* (1963) have investigated the DNA of *H. salinarium* strain 1. Centrifugation in a CsCl density gradient gave one major and one minor band at positions corresponding to GC contents of 67% and 58%, respectively. The major, higher density band did not represent a denatured form of the lower density band. The minor, lower density band was estimated to contain about 20% of the total DNA. Thus, the two components pooled should display a GC content of about 65% if the position of each band truly reflected their GC content according to the general relationship established by Sueoka *et al.* (1959) and Rolfe and Meselson (1959). A value of about 65% GC has actually been found for the bulk DNA (Table 2). Further, the melting curves of the bulk DNA determined by Joshi *et al.* suggested the presence of two components. A strain of *H. cutirubrum* gave similar results, but the satellite DNA represented in this case less of the total, namely about 10%.

TABLE 2. Base Composition of Nucleic Acids of *Halobacterium salinarium* strain 1 (K. Ormerod, unpublished observations)

Nucleic acid	Base composition (moles %)					$\frac{Pu}{Py}$	$\frac{A+T}{G+C}$
	Guanine	Adenine	Cytosine	Thymine	Uracil		
DNA	31·0 ± 2·7	17·5 ± 0·8	34·5 ± 1·5	17·0 ± 1·7	—	0·94	0·53
RNA	33·4 ± 0·4	24·0 ± 0·5	22·0 ± 1·3	—	20·0 ± 0·9	1.37	0·16

Cultures were harvested in the late logarithmic phase of growth. The values represent an average of nine determinations for DNA, and six for RNA. ± Values represent standard deviations.

DNA satellite bands are rare in bacteria and may be due to foreign episomal elements (Sueoka, 1964). Joshi *et al.* emphasize that this possibility cannot be ruled out for the halobacteria tested by them. The amount of satellite (10–20% of the total DNA) was, however, surprisingly large. At present the significance of the findings of Joshi *et al.* awaits further elucidation.

Detailed work has not yet been carried out on the RNAs of the extreme halophiles. In their work on the ribosomes of *H. cutirubrum*, Bayley and Kushner (1964) reported that about 50% of the total RNA was ribosomal, about 30% soluble, and about 15% present in a fraction that contained mainly cell envelope fragments.

F. METABOLIC PATHWAYS

There is at present no reason to believe that the general metabolic patterns of the extreme halophiles differ basically from those of the non-halophiles. Very little has been done, however, to elucidate the metabolic pathways of the extreme halophiles; the feeling that their metabolic patterns are essentially the same as those of the non-halophiles is mainly based on our knowledge of their nutritional characteristics (Section II, p. 99) and chemical composition (Section IV, p. 117).

In the laboratory, the extreme halophiles are cultured best on a protein digest. They seem to have a well-developed apparatus for metabolizing amino acids whereas carbohydrates are generally not attacked. These organisms probably also feed mainly on proteins and protein digests in nature. Gibbons (1957) found that, out of forty-nine strains tested, forty-five had gelatinolytic and twenty-two caseinolytic activities. Most of the strains had been kept on laboratory media (protein digests) for several years and, during this period, Gibbons observed that there was a gradual decrease in the proteolytic ability, particularly the ability to decompose casein. Such laboratory cultures might thus deserve the designation "physiological artefacts" as coined by Kluyver and Baars (1932), but whatever are their nutritional characteristics in the laboratory it seems clear that metabolizing amino acids is a most common habit of these organisms.

The only detailed study on amino acid metabolism of the extreme halophiles is that on arginine metabolism of *H. salinarium* by Dundas (1965) and Dundas and Halvorson (1966). Arginine was shown to be essential for growth, and was degraded by an arginine deiminase to citrulline, and citrulline further degraded by an ornithine carbamoyltransferase to carbamoyl phosphate and ornithine. Citrulline could replace arginine for growth, and seemed to be converted to arginine by the way of argininosuccinate. These metabolic conversions are quite similar to those found in most non-halophilic bacteria capable of degrading arginine; the uniqueness of these conversions in *H. salinarium* seems to lie solely in the halophilic nature of the enzymes.

IV. Cell Envelope of the Halobacteria

Exposure of halobacteria to sufficiently dilute salt solutions results in a lysis of the cells. If the exposure is gradually effected by a step-wise lowering of the NaCl concentration from the normal 25%, the rod-shaped cells change their shape concomitantly, from rods through irregular transition forms to spheres, and the spheres undergo sudden lysis when the NaCl concentration reaches 5–10%. Changes in structure also take place upon heating a suspension of the cells. When suspended in 25% NaCl and heated to 60–65°, the rods are transformed to spheres in a rather sudden conversion; upon prolonged incubation at this temperature, the spheres lyse.

Lysis in hypotonic solutions is well-known also for many of the less halophilic organisms, as for example bacteria indigenous to the sea. The marine bacteria, however, are much less susceptible to dilute salt environment than the halobacteria. Their normal environment contains about 3% NaCl; at this salt concentration, the halobacteria cannot even exist. Only at NaCl concentrations well below 1% will the marine bacteria lyse.

The organisms belonging to the second group of the extreme halophiles, the halococci, do not lyse in hypotonic solutions. These organisms require a high concentration of NaCl for growth. Their structure, however, is not much affected even by exposure to pure water, as observed in the light microscope.

The fact that the halobacteria have a lower salt limit of 5–10% NaCl for their very existence is an outstanding phenomenon in the living world and has aroused the curiosity of several microbial physiologists. The lysis phenomenon was first believed to be solely osmotic in nature; in hypotonic solutions the osmotic pressure inside the cells would cause the cell envelope to rupture. Recent investigations have shown, however, that osmosis alone certainly does not explain the lysis of the halobacteria in hypotonic solution. An equally, or possibly more important, factor lies in the unique chemical properties of their cell envelope which requires a high salt concentration for its maintenance; at low salt concentrations, it looses its rigidity and also disintegrates spontaneously into very small sub-units. The evidence for these phenomena, and the chemical basis for their occurrence, will now be discussed in some detail. The term "cell envelope" is used in accordance with Salton's (1964) implication of the term, namely to designate the structure (or structures) enveloping the cytoplasm and comprising both the principle conferring shape and rigidity upon the cell ("wall") and the principle of the cytoplasmic membrane. As will be discussed, it is not possible at present to distinguish clearly between the "wall" principle and the "membrane" principle in the halobacteria, as is commonly done for most other bacteria.

A. Requirement for Salt to Maintain the Cell Envelope

The first indications that the halobacteria require a high concentration of NaCl in their environment to maintain their cell envelope were provided by Abram and Gibbons (1960, 1961). These investigators observed that the halobacteria lysed at a fixed salt concentration, independent of that at which the cells were grown. The structural transformations from rods to spheres obtained by diluting the saline environment were not accompanied by an increase in cell volume, indicating a rapid adjustment to osmotic equilibrium. A number of cations of the alkaline series did not protect the rod-form of the cells as well as Na^+; the cells took on irregular and spherical forms and lysed at molar concentrations higher than that found for NaCl. Urea caused a transformation of the rods to spheres, and even lysis, at NaCl concentrations which protected the rod-form. On the basis of their observations, Abram and Gibbons (1961) put forward the hypothesis that the cell envelopes of the halobacteria "are held together rather loosely, as by hydrogen bonds, Coulomb forces, or 'salt' linkages, and in the presence of high concentrations of NaCl the electrostatic forces are screened so that the bonds hold the organism in a rod-shape".

Testing *H. salinarium* and the ability of a large number of different chemicals to protect its normal rod-form when replacing NaCl, Mohr and Larsen (1963) established the following relationships. Chemicals which do not possess a strong net charge in aqueous solutions do not protect the structure of the organisms when replacing NaCl in iso-osmolar concentrations. This is contrary to what might be expected if osmotic phenomena were solely responsible for the structural transformations and lysis of the cells in hypotonic solutions. The rod-shape of the organisms is only maintained by very high concentrations of ions known to interact weakly with common proteins. Sodium, chloride and acetate ions have these properties. Ions known to interact strongly with common proteins, such as cadmium, trichloroacetate and thiocynate ions, and chemicals believed to break secondary bonds between protein molecules, such as urea and guanidine, fail completely to protect the rod-shape when replacing Na^+ and/or Cl^-. These chemicals actually effect a transformation of rods to spheres, and frequently cause a lysis of the spheres, even when brought in contact with cells suspended in strong NaCl solutions,

Boring *et al.* (1963) reported that, in *H. cutirubrum*, $MgCl_2$, $CaCl_2$, $Na_2S_2O_3$ and sodium acetate, as well as NaCl, would protect the structural integrity when present in high concentrations. Upon transformation of the rods to spheres by diluting the NaCl-containing environment with water, the cells lose their viability. However, this transformation does not involve a rupture of the cells since no leakage of material from the spherical cells to the medium takes place. Kushner and Bayley (1963) found that *H. cutirubrum* is more resistant to lysis under acid than under neutral conditions. In 0·1–1% acetic acid, the cells will not lyse even in the complete absence of salt, but they become spherical and permeable.

The studies reported above on the behaviour of whole cells support the contention of Abram and Gibbons (1961) that the cell envelope of the halobacteria is very sensitive, and that a function of sodium chloride is to maintain chemical bonds in the envelope structure. Further and final evidence for this contention has been provided from studies of isolated cell envelopes and envelope fragments. It was shown independently by the three groups most actively engaged in this research that isolated cell envelopes would spontaneously "dissolve", i.e. disintegrate to very small subunits, upon removal of NaCl from their environment (Brown and Shorey, 1962; Brown, 1963; Larsen, 1962; Mohr and Larsen, 1963; Kushner *et al.*, 1964; Kushner, 1964). In these studies evidence has been presented that the disintegration of the cell envelopes in hypotonic solution is not enzymically mediated (Brown, 1963; Mohr and Larsen, 1963; Kushner, 1964; Onishi and Kushner, 1966), and that it is pH-dependent in a way similar to that of the lysis of whole cells (Brown 1963; Kushner *et al.*, 1964). The disintegration process is also strongly temperature-dependent (Brown, 1963; Kushner *et al.*, 1964), and seems to involve at least two separate processes, one being more rapid than the other in the complete absence of salt (Kushner *et al.*, 1964). Brown (1963) found that $CaCl_2$, in relatively very low concentrations, would effectively prevent the disintegration when replacing NaCl. In experiments comparing intact cells, cells made permeable, and isolated cell envelopes, Kushner (1964) found that NaCl is much more effective than KCl or NH_4Cl in preserving the integrity of intact cells, but only slightly more effective in preserving the integrity of the cell envelopes and the permeable cells. $MgCl_2$ is much more effective in preserving the integrity of the envelopes and of the permeable cells than of the intact cells.

The above-mentioned studies on the behaviour of isolated cell envelopes have contributed significantly to interesting hypotheses concerning their maintenance by salt in the intact cell. These hypotheses are, however, also rooted in studies on the structure and chemical composition of the envelope. It is therefore expedient to give an outline of the latter studies before discussing the function of salt in preserving the integrity of the envelope.

B. Structure and Chemical Composition

1. *Electron Microscope Studies*

a. *Surface Pattern.* A number of investigations have shown that a characteristic of the halobacteria is a textured surface with a regular hexagonal pattern (Houwink, 1956; Mohr and Larsen, 1963; Kushner and Bayley, 1963; Kushner *et al.*, 1964). This pattern is present also in the isolated envelopes when prepared in NaCl solutions of sufficient strength, and appears on electron micrographs of replicated envelopes as a texture of regular, pebble-like structures of about 130 Å in diameter (Fig. 3). A similar texture is also found on the surface of other bacteria (Salton, 1964). Studying whole cells of *H. cutirubrum*, Kushner and Bayley (1963) found that upon acidification to pH 2–3 the regular pattern would disappear; under these conditions the surface of the cell became wrinkled and irregular. Upon slight acidification (pH 4), the regular, hexagonal pattern was maintained, although the cells took on a spherical shape. Isolated cell envelopes lost their regular pattern and became wrinkled upon slight as well as stronger acidification (Kushner *et al.*, 1964). It was suggested that acidification tended to contract the cell envelope, and that upon slight acidification such contraction would occur readily in the isolated envelope

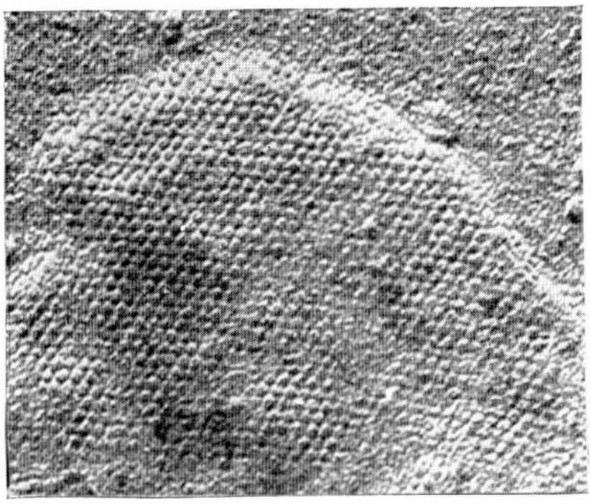

FIG. 3. A carbon replica of an envelope fragment of *Halobacterium salinarium*. × 100,000. Made by V. Mohr.

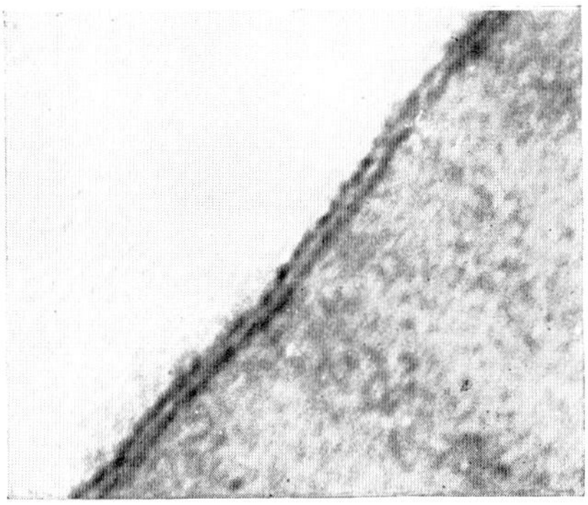

FIG. 4. A thin section of *Halobacterium salinarium* fixed with osmium tetroxide. × 140,000. Made by R. Svardal.

but not in the whole cell. The studies also revealed that the hexagonal pattern of the envelope was maintained upon lowering the NaCl concentration from the normal 4·5 *M* (26%) to 2·5 *M* (14%); i.e. a salt concentration at which most of the cells take on a spherical or bulgy form.

b. *Cross Section.* Brown and Shorey (1962, 1963) sectioned whole cells and isolated envelopes of *H. halobium* and *H. salinarium*, using permanganate as the fixing agent. Sections of whole cells of *H. halobium* revealed one three-layered envelope resembling the "unit membrane" as described by Robertson and others. A five-layered structure was seen in some sections of *H. halobium* and in all sections of whole cells of *H. salinarium*. The isolated cell envelopes of both species appeared characteristically as a three-layered structure of the "unit membrane" type. The total thickness of the structure was, however, somewhat larger than that of the common "unit membrane", namely about 110 Å as compared with 75–80 Å reported for the "unit membrane" of different types of organisms including many Gram-negative bacteria. In a more recent publication, Brown *et al.* (1965) report on the occasional observation of a five-layered structure in isolated envelopes of *H. halobium*, giving the envelope a "compound" appearance. In no case, however, could two separable three-layered structures be demonstrated, such as is commonly found in other Gram-negative bacteria.

Another strain of *H. salinarium* (No. 1) has been sectioned by R. Svardal of the Electron Microscope Laboratory, The Technical University of Norway, using both permanganate and osmium tetroxide as fixing agents (unpublished observations). In this case, the characteristic picture of the envelope in whole cells was one three-layered structure (Fig. 4). No evidence was found for a second "membrane" structure or a "compound membrane". In Svardal's pictures, the thickness of the envelope structure, measured as the distance between the outer sides of the two electron-dense layers, corresponds to 130–140 Å; the unstained middle layer is about 50 Å and the stained layers each about 40 Å thick.

The discrepancies between the findings of Brown *et al.* and of Svardal might be ascribed to strain differences or conditions of culturing; this applies also to the difference in the thickness of the three-layered structure. What stands out as noteworthy in all investigations carried out until now, however, is the lack of success in demonstrating the presence in the halobacteria of two clearly separable membrane structures, the outer three-layered structure commonly referred to in other Gram-negative bacteria as the "wall", and the inner three-layered structure referred to as the "cytoplasmic membrane". Now, negative results should be used cautiously; it might be that the native envelope structure of the halobacteria is exceptionally sensitive to the preparation methods for thin sectioning so that the treatments lead to dissolutions and distortions not met with in other bacteria. Or possibly the envelope layers of the halobacteria are of a chemical nature different from those in other Gram-negative bacteria, so that they are not equally well revealed by the common staining techniques. Systematic investigations of this problem ought to be carried out. At this stage in our knowledge, however, the results suggest that the halobacteria might have a simpler envelope structure than commonly found in other Gram-negative bacteria. As first pointed out by Brown and Shorey (1963), the simple envelope structure may be a feature having a bearing upon the life of these organisms in the strong brines.

2. *Chemical Composition*

Smithies *et al.* (1955) were the first to analyse the gross chemical composition of the cell envelope of a halobacterium and reported it to consist of lipoprotein. In recent years more detailed investigations have been made and have led to a number of interesting findings.

a. *Lipids.* Sehgal *et al.* (1962) found that almost all of the lipids of *H. cutirubrum* cells are phosphatides, and that these are unusual in that they contain little, if any, fatty acid ester groups, but appear to have long-chain alkyl groups joined by ether linkages to glycerol. One component, an ether analogue of diphosphatidyl glycerol, was dominant. Kushner *et al.* (1964) found a total lipid content of 22% in the envelope of the same

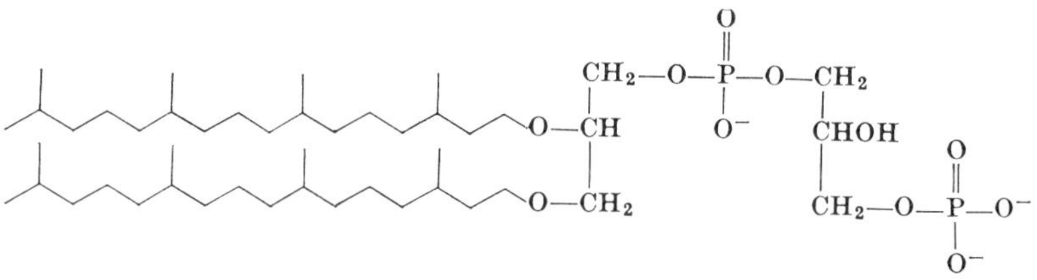

FIG. 5. The chemical structure of the dominating lipid component of the cell envelope of *Halobacterium cutirubrum* (Kates *et al.*, 1963, 1965).

organism; the ether component dominating the lipids of the whole cell was concentrated in the envelope. Kates *et al.* (1963, 1965) have elicited in detail the chemical constitution of this component (Fig. 5); they estimate it to account for about 13% of the salt-free dry weight of the envelope. The isoprenoid nature of the alkyl chains is noteworthy. As a matter of fact, no evidence was found for the presence in the whole cells of fatty acid groups or, indeed, normal straight-chain hydrocarbon groups. Kates *et al.* (1965) point out that *H. cutirubrum* appears to lack the malonyl–CoA system for the synthesis of the common, unbranched fatty acid chains; this organism seems to employ instead the mevalonate system for the synthesis of isoprenoid chains which may fill the same functions as the unbranched hydrocarbon chains in other organisms.

Brown (1965) reported that the cell envelope of *H. halobium* contained 17–23% lipid; Cho and Salton (1966) found only traces of, if any, fatty acids in hydrolysates of this lipid. The same has been found for

H. salinarium in the author's laboratory. A very low, or possibly zero, content of higher fatty acid residues therefore seems to be a common feature of the cell envelope of the halobacteria, and in this way they differ pronouncedly from other bacteria.

The carotenoids of the halobacteria are situated mainly, or wholly, in the cell envelope, and a function of these compounds seems to be to protect the envelope against photochemical damage. The mechanism of protection is not known, but it seems reasonable to suggest that the carotenoid molecules act by quenching electronic excitation impressed by light upon other pigment systems of the envelope. This implies that the carotenoid molecules are spatially located very close to the photo-activated system (Dundas and Larsen, 1963). The carotenoid content of the envelope is of the order of 0·1%, and Baxter (1960) found that it varied little with the culture conditions but increased with the age of the culture. Nandy and Sen (1963) have studied the influence of culture conditions on the carotenogenesis of a halococcus, *Sarcina litoralis*.

b. *Protein*. Kushner *et al.* (1964) estimated the protein content of the envelope of *H. cutirubrum* to be roughly 45–57%. Brown and Shorey (1963) reported protein values of 65% and 75% for the envelopes of *H. halobium* and *H. salinarium*, respectively. In none of the cases could α,ϵ-diaminopimelic acid (DAP) be detected in the envelope hydrolysates. Mohr and Larsen (1963), referring to unpublished work in the author's laboratory, mentioned that DAP has been found in hydrolysates of envelopes of a *H. salinarium* strain. I take this opportunity to state that the latter report was wrong; in a later investigation, H. Steensland in the author's laboratory was unable to find DAP. If present the content of this compound in the envelope is less than 0·001%.

The proteins of the cell envelope of the halobacteria appear to be quite acidic in nature. This was first pointed out by Brown (1963) who found an excess of acidic over basic amino acids of about 15 mole % for the envelope protein of *H. halobium*. The corresponding value for the envelope protein of a marine pseudomonad was only about 5 mole %. No correction was made for amide. Kushner and Onishi (1966) presented a complete analysis of the amino acid composition of the envelope of *H. cutirubrum*. Their data, which should be considered more accurate than those presented somewhat earlier by Kushner *et al.* (1964) on the same organism, show an excess of acidic over basic amino acids of about 14 mole %. Assuming all the ammonia to stem from the amide groups of glutamine and asparagine, the excess is about 7 mole %. There are not many data in the literature on non-halophilic Gram-negative bacteria for comparison, but the few reported support the contention that the envelope protein of the halobacteria might be more acidic than that occurring in the common Gram-negative bacteria. Thus, an excess of acidic over basic amino acids of only about 9 mole % was found for *E. coli* and three *Salmonella* spp.; these data were not corrected for amide (Howe *et al.*, 1965).

Interesting observations on the acidic nature of the envelope protein have recently been contributed by Brown (1965). He titrated *H. halobium* envelopes electrometrically with acid and base, and compared the results to similar titrations on the envelope material after dissolving the envelopes by removing sodium chloride from their environment. He found that all detectable carboxyl groups were titrated on envelopes suspended in NaCl solutions strong enough to keep the envelope intact, i.e. 4*M*-NaCl. However, at this NaCl concentration, many basic groups, which could be

TABLE 3. Sugars found in Hydrolysates of Cell Envelopes of Various Halobacteria

	Content (% salt-free dry wt.)		
	Halobacterium halobium	*Halobacterium salinarium*	*Halobacterium cutirubrum*
Reducing substances (as glucose)	6·2	5·2	7·0
Hexoses (as glucose)	1·7–4·3	4·9	5·9
Glucose	0·65	0·60	—
Amino sugars (as glucosamine)	0·5–1·2	1·3	0·45
Pentoses (as ribose)	—	—	0·98
Heptoses	traces	0	—
References	Brown and Shorey (1963); Brown *et al.* (1965)	Brown and Shorey (1963)	Kushner *et al.* (1964)

titrated in envelope material dissolved in water, would escape titration. He concluded that these basic groups are "buried" in the intact envelope, possibly by being bound to acidic groups of the phospholipid component. These experiments support and strengthen the idea that, at physiological pH values, the intact envelope is strongly negatively charged at or near its surface by the dominance at this location of ionized carboxyl groups.

c. *Carbohydrates.* Sugars are detected upon hydrolysis of isolated cell envelopes of the halobacteria (Table 3); however, the amounts are small compared to those commonly found in hydrolysates of envelopes of other Gram-negative bacteria (Salton, 1964). Muramic acid has not been found, either by Brown and Shorey (1963) in *H. salinarium* and *H. halobium* or by Kushner *et al.* (1964) in *H. cutirubrum*. The failure to demonstrate muramic acid in the halobacteria tested might be highly significant because it indicates the lack in these organisms of a mucopeptide "rigid layer" of the type found in other bacteria (Salton, 1964). Whether the other carbohydrates play a structural rôle is at present not known. A strain of *H. salinarium* investigated in the author's laboratory contained about 1·5% amino sugar (as glucosamine) in the cell envelope. About half of this was glucosamine; the rest was mainly a 2-amino sugar, substituted also in the 3-position, and giving the same Rondle and Morgan colour reaction (1955) as muramic acid. However, the latter compound did not behave chromatographically as muramic acid.

Mohr and Larsen (1963) reported that proliferating cells of *H. salinarium* were about equally sensitive to penicillin as those of other Gram-negative bacteria, and also displayed a similar pattern of structural transformation (sphere formation) and lysis following penicillin treatment. This was taken as an indication of a mucopeptide structure in the *H. salinarium* envelope. Since that time, investigations have been reported which cast some doubt upon the original belief that penicillin inhibits mucopeptide synthesis in Gram-negative bacteria in the same way as rather firmly established for the Gram-positive bacteria (Plapp and Kandler, 1965a). The penicillin sensitivity of *H. salinarium* might therefore reflect an interference with the formation of some other envelope component. In addition, it should be mentioned that a more recent investigation by S. Omang in the author's laboratory has shown that proliferating cells of *H. salinarium* are completely unaffected by D-cycloserine, a compound known to interfere similarly with mucopeptide synthesis both in Gram-positive and Gram-negative bacteria (Plapp and Kandler, 1965b).

Summing up the present situation, it can be stated that the evidence is against the view that the halobacteria contain in their envelope a mucopeptide structure of the type found in other bacteria; indeed, this may be an outstanding characteristic of these organisms. The possibility exists, however, that other carbohydrates may have structural functions in the envelope of the halobacteria, but no evidence has yet been presented to show that this is the case.

d. *Nucleic Acids.* Cell envelopes of the halobacteria, prepared by mechanical breakage of the cells and repeated fractional centrifugations, tend to contain nucleic acid, as shown by their content of the nucleic acid pentoses and by the ultraviolet absorption pattern (Kushner *et al.*, 1964). It is not known whether the nucleic acid is a natural component of the envelope or whether it is a contaminant derived from the cytoplasm. In the isolated envelope, the nucleic acid fraction appears to be quite tightly bound.

3. *Characteristics when Dissolved at Low Ionic Strength*

Upon dissolution of cell envelopes of *H. halobium* in water, Brown (1963) obtained a sedimentation pattern in the ultracentrifuge showing the presence of a major component with a sedimentation coefficient ($S_{20,w}$) of about 4·0. The material dissolved in water did not precipitate upon raising the sodium chloride concentration to a value at which the intact envelopes are stable. Ammonium sulphate, however, precipitated most of the dissolved envelope material at half saturation. In similar ultracentrifugation studies on envelopes of *H. cutirubrum* dissolved in 0·1 *M*-NaCl or KCl, Onishi and Kushner (1966) observed two components, one moving rapidly away from another slower moving, main component. The latter gave an uncorrected sedimentation coefficient of 5·2. By raising the salt concentration, the amount of the more rapidly moving material increased whereas the amount of the more slowly moving material decreased. Addition of salt thus caused an aggregation of the dissolved envelope material, but the aggregation appeared to be rather unspecific since the more rapidly moving material spread out during migration.

Some progress has been made in purifying and characterizing a lipoprotein particle fraction from *H. salinarium* envelopes dissolved in water (Urdahl *et al.*, 1966). Upon removal of NaCl from the envelopes by dialysis against water, the resulting solution was treated with deoxyribonuclease (DNase) and ribonuclease (RNase) to remove nucleic acids which are bound to the envelope material. After gel filtration on a Sephadex column, which retained material of molecular weight less than about 200,000, the front fraction containing the lipoprotein was chromatographed on a carboxymethyl cellulose column. The bulk of the protein migrated as one band when eluted with a weak phosphate buffer of pH 7·0. The fraction thus obtained contained practically all the protein of the intact cell envelope, all the pigment, but practically no nucleic acid. The material migrated as one sharp band by free boundary electrophoresis, displaying a negative charge in the pH region 5·5–10·0 and the rather high electrophoretic mobility of 9×10^{-5} cm.^{2} sec.$^{-1}$ V^{-1} at pH 7·5. In the electron microscope, the material appeared as particles of somewhat irregular size, measuring between 100 and 200 Å.

The particle fraction of the *H. salinarium* envelope has also been chemically characterized (Steensland *et al.*, 1966). The total lipid content was near 20%; the protein content, as estimated from an amino acid composition analysis, was 70–75%. The 3-substituted amino sugar present in the intact envelope (Section IV B 2 c, p. 120), was quantitatively retained in the particle fraction; the glucosamine of the intact envelope was lost during the purification. The material is obviously lipoprotein containing the bulk of the material of the intact envelope; besides nucleic acid only small amounts of protein and carbohydrate were removed by the purification procedure. The amino acid analysis showed an excess of acidic over basic components of about 20 mole %, not corrected for amide. The ammonia in the hydrolysate can reasonably be assumed to stem almost entirely from protein amide. Correcting for this constituent gives an excess acidic over basic amino acids of at least 5 mole%. In other words the bulk of the protein of *H. salinarium* envelopes is dominatingly acidic.

The homogeneous behaviour of the cell envelope particles of *H. salinarium*, as shown by ion exchange chromatography and electrophoresis, is striking and reflects chemical properties common to all particles. It would be rash, however, to conclude that all particles have the same chemical composition. Already their somewhat different size, as revealed by the electron microscope, suggests chemical differences. In addition it might be argued that it is hard to envisage each of these very small particles containing all the biochemical functions assigned to the bacterial envelope (including, for example, the respiratory mechanism and permeases). It seems more reasonable to assume that these functions are, at least in part, distributed between the particles. Possibly the particles are cut roughly to the same pattern as far as their chemical composition is concerned, but "ornamented" differently with regard to the biochemical functions possessed in the intact cell envelope. This seems an interesting problem for further investigation.

4. *Molecular Architecture*

Thin sections of halobacteria envelopes have revealed a three-layered structure similar to that characteristic of the "unit membrane" (Fig. 4). There has been some discussion about the chemical interpretation of the triple layering generally met with in cell membranes; model experiments now seem to have made it reasonably certain that it depicts a bimolecular layer of phospholipid sandwiched between two layers of protein, i.e. the Danielli–Davson model of the cell membrane (Stoeckenius, 1962). In the case of the halobacteria investigated until now, the envelope lipid does not contain fatty acid residues; in the one case thoroughly studied, the bulk of the lipid is an analogue of phosphatidyl glycerophosphate containing two 3,7,11,15-tetramethylhexadecyl groups joined by ether linkages to the glycerol (Section IV B 2a, p. 117). One can envisage the long-chained groups from two sets of phospholipid molecules being more or less integrated with each other so as to form a bimolecular lipid leaflet of the type proposed for the "unit membrane". The phosphate groups would be situated at the surface of the leaflet. Kates *et al.* (1965) suggest that the highly branched chains of this phospholipid may fill the same function as that proposed for the unsaturated fatty acids of plant and animal membranes, namely to prevent undesirable close packing of the phospholipid molecules. A highly branched chain structure of this type may help to account for the unusual thickness of the unstained lipid layer found in *H. salinarium* (Fig. 4).

On the basis of his electrometric titration studies (Section IV B 2b, p. 119), Brown (1965) has presented an interesting discussion on how the protein might be bound to a phospholipid leaflet in the *H. halobium* envelope. He finds it probable that, in the intact envelope, α- and ϵ-amino groups of the protein are bound to acidic groups of the phospholipid by hydrogen bonds, and that these bonds influence protein conformation so that carboxyl groups will dominate over amino groups at or near the envelope surface. Brown argues further that the hydrogen bonding between protein amino groups and lipid phosphoric acid groups does not account entirely for the attachment of protein to lipid. He envisages additional hydrogen bonding between unspecified groups or hydrophobic association between the non-polar side chains of the protein and the lipid.

Not much can be said at present about how the particles, which are liberated when the envelopes dissolve in water, are constructed into the intact envelope. Further research may make it possible to relate these particles to the characteristic hexagonal surface pattern of the envelope, but there is at present no serious evidence for such a relationship. Accepting the contention that the lipid forms a continuous layer in the envelope, and that this layer is perforated by a large number of tiny pores containing functional protein structures, it is possible that the disintegration of the envelope in water involves ruptures from pore to pore. In whichever way the rupture takes place, one might envisage that it leads to drastic changes in the fragments liberated. The lipid may tend to round up to form a globule; irreversible re-arrangements may occur in the protein components. This effect may add significantly to the denaturation of the protein which occurs upon removal of the salt independently of the ruptures. The liberated particles might thus represent a highly denatured form of the corresponding fragment in the intact envelope. Such denaturations may account for the lack of success in precipitating the material of the envelopes dissolved in water, when raising the NaCl or KCl concentration to a value at which the intact envelope is stable.

C. Function of Salts in Maintaining the Envelope

Of the properties of the halobacteria envelopes outlined in the preceding sections, the acidic nature of the envelope protein has attracted most attention in the search for an explanation of the salt requirement for their maintenance. It was first pointed out by Brown (1963) and Kushner and Bayley (1963) that the dependence of the envelopes for salt might be due to the presence of negatively charged groups in or on the envelope; an excess of negative charges would cause a dispersion of the envelope subunits if not effectively neutralized by cations. This view is supported by the findings of the excess acidic over basic amino acids in the envelope protein, the "masked" basic groups accentuating the effect of the excess acidic groups, and the ability of various cations, including H^+, to prevent envelope disintegration, as outlined above. The effect of the salt seems to reside primarily in the cation rather than the anion.

Further experiments to strengthen the contention that cations prevent a disintegration of the envelope by neutralizing negatively charged groups were provided by Brown (1964a). He introduced extra carboxyl groups into the envelopes of a marine bacterium by succinylating the ϵ-amino groups of the lysyl residues of the envelope protein. In this way he increased the excess acidic over basic groups from about 5 to about 15 mole %, i.e. to about the level found in halobacteria envelopes (not corrected for amide). In their natural state, the envelopes of the marine bacterium were "slightly" halophilic, disintegrating at NaCl concentrations below 1–2%. When succinylated, these envelopes became remarkably more halophilic in the way that they required more than 5% NaCl in the environment to prevent their disintegration.

The question as to whether negatively charged groups on the phospholipid may contribute to the halophilic character of the halobacteria envelopes has been studied in some detail by Kushner and Onishi (1966). They found that, upon removal by proteolytic enzymes of most of the protein from envelopes of *H. cutirubrum*, structures remained which apparently contained only phospholipid and which maintained their integrity at much lower salt concentrations than were required by undigested envelopes. On the other hand, upon removal of most of the phospholipid from the envelopes by ethanol extraction under acid conditions, structures remained which required at least as much NaCl or KCl as the native envelopes to prevent their disintegration. These observations give further strength to the view that the carboxyl groups on the envelope protein are primarily responsible for the halophilic character.

As mentioned in Section IV A (p. 114), Kushner (1964) found that NaCl, on a molar basis, is much more effective than KCl in preserving the integrity of the cells, but is only slightly more effective in preserving the integrity of isolated envelopes. On the strength of this observation, and of the knowledge that the halobacteria have a strikingly higher intracellular content of potassium than of sodium ions, he suggested that the external surface of the cell envelope may specifically require sodium ions to maintain its integrity, whereas the internal surface is supported at least as well by potassium as by sodium ions. This hypothesis has later been reconsidered (Onishi and Kushner, 1966); the authors now feel that sodium ions may act specifically on certain sites on the cell envelope to prevent leakage of material from the cells, in addition to the effect in preventing envelope disintegration. The last hypothesis is based on the observation that NaCl is much more effective than KCl in preventing protein leakage from the cells, when these are suspended in hypotonic solutions of a salt concentration (2 *M*) at which not much disintegration of the envelope occurs with either salt. Drapeau and MacLeod (1965) have demonstrated a similar effect in a marine bacterium. This organism is also a halophile, but with a very modest halophilic character compared to the halobacteria. A specific function of sodium ions in this organism appears to be the prevention of leakage of intracellular material.

The general ability of cations to "maintain the cell envelope" of the halobacteria applies to the effect of preventing total disintegration of the envelope into very small particles. However, only a few cations can replace Na^+ in maintaining the natural rod-form of these organisms. When Na^+ at 4·3 *M* is replaced by isomolar concentrations of other cations, the cells of *H. salinarium* are transformed to spheres or transition forms between rods and spheres, but the spheres do not lyse. The potassium ion is an exception to this rule, being almost as effective as Na^+ in protecting the rod-form; Mg^{2+} is somewhat less effective than K^+ (Mohr and Larsen, 1963). It thus appears that sodium and potassium ions, in addition to preventing disintegration of the envelope into small particles, have a rather specific effect in conferring rigidity upon the envelope so that the cells appear in the form of a rod. The latter effect is not clearly understood; Brown (1964b) suggests that sodium ions may stiffen the envelope by forming an oriented "atmosphere" along the envelope surface.

In other Gram-negative bacteria, the principle conferring rigidity and shape upon the cells is commonly ascribed to the mucopeptide component (Weidel and Pelzer, 1964). In the halobacteria, this component seems to be absent, its function being taken over by sodium and possibly potassium ions. However, other polysaccharide components may contribute to the shaping of the cell in the form of a rod; the penicillin sensitivity of *H. salinarium* points to such a possibility which ought to be looked into further. Also magnesium ions are in some way involved in this picture; when growing in media low in Mg^{2+} but normally high in Na^+, the halobacteria become spherical in shape (Brown and Gibbons, 1955). How the magnesium ions contribute to the maintenance of the rod-shaped structure is not clear; Kushner and Onishi (1966) point out that divalent cations such as Mg^{2+} may be primarily bound to the phosphate groups on the lipid and in this manner support the envelope structure.

Another component which may conceivably contribute to the rigidity of the cell envelope is the hydrocarbon chain structure of a bimolecular phospholipid leaflet. The rather sudden transformation of rods to spheres undergone by halobacteria when the temperature is raised to 60–65° may reflect a melting of a crystalline-like hydrocarbon structure. Altogether, a number of factors may be brought into the discussion of what confers shape and rigidity upon the halobacterium cell, but their relative importance remains to be elicited. What stands out as particularly interesting for the present discussion is the specific effect of sodium and potassium ions in maintaining the rod-form of the cells. How these ions contribute to the maintenance of this structure needs further attention in order to reach a thorough understanding of extreme halophilism.

V. On the Rise of Extreme Halophilism

Over the years a number of reports have appeared on the ability of non-halophiles to adapt to life in strong salt brines, and also on the ability of extreme halophiles to adapt to life in dilute salt brines or an environment devoid of NaCl. For a review of this work the reader is referred to Larsen (1962). It suffices here to summarize the findings, which are rather inconsistent. Some investigators have reported success in training extreme halophiles to proliferate in ordinary culture media devoid of NaCl; others have reported that they have completely failed in such attempts and have found very little change in the salt response of these organisms upon repeated subculture on media low in salt. Success has been claimed in the training of an ordinary, non-halophilic micrococcus to grow well at very high NaCl concentrations and even to become dependent upon NaCl like a true halococcus. Others have reported that they have succeeded in isolating or enriching for extreme halophiles when using inocula from salt-free or marine sources; however, it is difficult to believe that extreme halophiles normally live in salt-free or marine environments since numerous laboratory experiments have shown that these organisms are killed when exposed to such conditions. The reports rather nourish a feeling that non-halophilic and marine bacteria may acquire an extremely halophilic character. On the other hand, there are the many reports in the literature concerning the lack of success in adapting bacteria to higher salt concentrations; for the many different types of bacteria tested there seems to be a characteristic upper limit for the NaCl concentration at which they can grow.

Workers in the author's laboratory have attempted to adapt extreme halophiles to growth at lower salt concentrations. These experiments have all met with utter lack of success. K. Middleton, in the author's laboratory, has carried out extensive experiments in an attempt to adapt different types of non-halophilic and marine bacteria to growth at high salt concentrations (unpublished observations). These experiments included the use of a marine Gram-negative rod and a marine micrococcus which had many of the characteristics of the halobacteria and the halococci, respectively, but differed greatly from these in their salt response. Middleton worked with populations up to 10^{12} individuals in each experiment; in some experiments the organisms were exposed to mutagenic agents, in others not. All these experiments gave completely negative results in the sense that no evidence was obtained for the rise in these populations of organisms possessing the ability to grow at higher salt concentrations, or, for that matter, possessing an extremely halophilic character.

The discrepancies between the results of the various workers may be ascribed to the use of different bacteria. An appraisal of the available information on this subject makes one hesitant, however, to believe that bacteria can undergo spontaneous drastic changes in their salt response. The claims in the literature to this effect are based on experiments which are not altogether convincing. The most carefully conducted work supports a contention that extreme halophilism is a permanent character which is not easily acquired.

It remains to discuss the problem of how extreme halophilism has arisen in nature. Our knowledge on the extreme halophiles is at present far too meagre to give such a discussion much weight, but the information gained on the biochemistry of these organisms over the past few years has furnished a certain basis for attractive ideas. Further experimentation can support these ideas, but may also show that they are improbable.

As pointed out in Section III F (p. 110) there is at present no reason to believe that the metabolic pattern of the extreme halophiles differs basically from that of other bacteria. As to specific metabolic pathways, the absence of the malonyl-CoA system and the compensating use of the mevalonate system may possibly be a feature of the halobacteria, but there is no evidence to relate such a feature to the phenomenon of extreme halophilism. The absence of muramic acid is another feature of the halobacteria, but possibly not an absolutely unique one since muramic acid has been reported to be absent also in the envelope of pleuropneumonia-like organisms (Salton, 1964). As pointed out by Brown (1964b) the halobacteria have a reduced need for a structural polysaccharide in their envelope because of the rigidity conferred upon this structure by the monovalent cations. One might therefore assume that the halobacteria simply lost the ability to synthesize their mucopeptide component in the course of time, and that this loss did not seriously affect them since they had an alternative rigidity mechanism already developed.

The evidence available so far points to the proteins as being the unique components of the extreme halophiles and as being the determinants of their extremely halophilic character. The enzymes need salt for their activation; the protein components of the cell envelope and of the ribosomes are of such a character that they are held together in these structures only in the presence of salt. Concerning the proteins of the envelope and the ribosomes, their halophilic nature is ascribed to the excess of acidic residues as outlined above, and Bayley (1966) has offered an interesting explanation of how the acidic character of these proteins might have come about. He points out that the ribosomal proteins of *H. cutirubrum* differ from those of *E. coli* and other organisms in their relative content of acidic and basic amino acids, but the total content of these amino acids is about the same. The relatively high content of glutamic and aspartic acids in the *H. cutirubrum* ribosomal protein therefore means that there is a corresponding low content of basic amino acids. Comparing the recently elicited nucleotide sequences of RNA codons for acidic and basic amino acids (Nierenberg *et al.*, 1965), it appears that some of these codons differ only by the base in the 5′-terminal position. Thus, misreading guanine and adenine might result in the incorporation in a polypeptide chain of glutamate instead of lysine, or, for that matter, aspartate instead of asparagine. A series of such misreadings could develop a highly acidic protein and as a consequence cations would be required to neutralize the acidic groups in order to keep the protein in the proper conformational state. The protein would have become halophilic.

The acidic character has thus far only been shown for the proteins of the cell envelope and the ribosomes of the halobacteria. Whether this is a general phenomenon comprising also the other proteins (enzymes) in the cell, and the halococci as well as the halobacteria, is not yet known, but it could serve as a hypothesis for further experimentation. In this connection, it is obvious that the development of an acidic character in a protein is not necessarily dependent upon the substitution of an acidic for a basic amino acid. Substitution of an acidic for any non-acidic, or a non-basic for a basic, might serve the same purpose.

In order to develop a pronounced acidic character in a protein, a number of misreadings must occur within the same cistron. Possibly this will happen only through a series of mutational events. In order to give all or a sufficient number of the proteins an acidic character so that the whole cell becomes an extreme halophile, so many mutational events must occur that one can only assume this to have taken place in nature over a very long period of time. In addition, these events must have occurred under conditions particularly favourable for the development of the mutant cells, namely in a strongly saline environment.

If the hypothesis be accepted that a prerequisite for the development of an extremely halophilic character is the occurrence of many mutational events resulting in the transformation of a large number of proteins, then the lack of success in demonstrating in the laboratory the rise of an extreme halophile from an organism not possessing this property becomes more understandable. Possibly such events could be speeded up by the use of agents that will specifically cause misreadings of the RNA code, such as is believed to be a property of streptomycin (Anderson *et al.*, 1965). Under any circumstances the ability to synthesize acidic proteins may not alone explain the phenomenon of extreme halophilism. An ability to accumulate K^+ within the cell in unusual high concentrations seems to be part of the picture; the particulars of this mechanism are not known. The possibility that the extreme halophiles synthesize relatively more of the normal acidic proteins, and thus are affected in their regulation mechanism, should also be taken into account as it could be part of the complete story, even though this does not seem too likely at present.

Another problem which awaits clarification is the finding in nature of only two distinct types of extreme halophiles, namely the pseudomonad type and the micrococcaceae type. Viewed with the eyes of the traditional bacteriologist, these two types of bacteria are rather different from each other, and one is brought to wonder why just these two were selected by Nature for extreme halophilism. Possibly further work will reveal that there are other types of extremely halophilic bacteria, but the fact that only the two types have been found until now is rather striking and suggests that they may have some property in common selecting them before others for transformations to extreme halophiles. Possibly this could simply be a common preference for the ecological conditions otherwise prevailing at the sites where a high salt concentration would cause a selection of individuals that acquired a strongly halophilic character. If this were the case, pseudomonads and micrococcaceae would be present in greater numbers than other organisms at these sites, and thus stand the greatest chance of becoming transformed to extreme halophiles. A characteristic that we know these two types of bacteria have in common is the extremely high GC content of their DNA. They occupy in fact the extreme end of the GC scale. Whether this characteristic has a bearing on a tendency of the organisms to become extremely halophilic cannot be said at present.

Extreme halophilism stands out as a unique and curious phenomenon in the living world, and it is as such attractive to investigate. The organisms possessing this phenomenon should, however, be brought to the attention of physiologists and biochemists also because they may serve as very valuable objects in the study of basic problems in molecular biology. Their very extremeness offers an attractive possibility to apply the concept of comparative biochemistry, i.e. the search among different but related processes of a common denominator which may throw light upon problems otherwise inaccessible to experimentation. It suffices here to remind the reader of how at one time the study of the photosynthetic bacteria gave a better understanding of the mechanism of oxygen evolution in green plants. In the case of the extreme halophiles the study of protein synthesis appears particularly attractive because of the unusual properties of the protein components in these organisms. The cell envelope of the halobacteria is another attractive object for study. Its apparent simplicity may be of great value in the elucidation of cell envelope structure, synthesis and functions. In this article an attempt has been made to point out the peculiarities of the extreme halophiles and to describe what we at present know about them. If other physiologists and biochemists hence come to realize that these organisms could be useful in their work, then the aim of this article is achieved.

References

Abram, D. and Gibbons, N. E. (1960). *Canad. J. Microbiol.* **6**, 535.

Abram, D. and Gibbons, N. E. (1961). *Canad. J. Microbiol.* **7**, 741.

Anderson, W. F., Gorini, L. and Breckenridge, L. (1965). *Proc. nat. Acad. Sci. Wash.* **54**, 1076.

Baas Becking, L. G. M. (1928). *Tijdschr. Ned. Dierk. Ver.* (Ser. III D) **1**, 6.

Baxter, R. M. (1959). *Canad. J. Microbiol.* **5**, 47.

Baxter, R. M. (1960). *Canad. J. Microbiol.* **6**, 417.

Baxter, R. M. (1961). *Abstracts of Papers, Annual Meeting Canad. Soc. Microbiologists*, Kingston, Ontario, June, 1961.

Baxter, R. M. (1964). *Contr. Faculty Sci., Haile Sellassie I University.* Series C. (Zoology) No. 5. Addis Ababa.

Baxter, R. M. and Gibbons, N. E. (1954). *Canad. J. Biochem. Physiol.* **32**, 206.

Baxter, R. M. and Gibbons, N. E. (1956). *Canad. J. Microbiol.* **2**, 599.

Baxter, R. M. and Gibbons, N. E. (1957). *Canad. J. Microbiol.* **3**, 461.

Bayley, S. T. (1966). *J. molec. Biol.* **15**, 420.

Bayley, S. T. and Kushner, D. J. (1964). *J. molec. Biol.* **9**, 654.

Belozersky, A. N. and Spirin, A. S. (1960). *In* "The Nucleic Acids", (E. Chargaff and J. N. Davidson, eds.). Vol. 3, p. 147, Academic Press, New York.

"Bergey's Manual of Determinative Bacteriology", 7th. ed. (1957). (R. S. Breed, E. G. D. Murray and N. R. Smith, eds.), Williams & Wilkins Co., Baltimore.

Boring, J., Kushner, D. J. and Gibbons, N. E. (1963). *Canad. J. Microbiol.* **9**, 143.

Brown, A. D. (1963). *Biochim. biophys. Acta* **75**, 425.

Brown, A. D. (1964a). *Biochim. biophys. Acta* **93**, 136.

Brown, A. D. (1964b). *Bact. Rev.* **28**, 296.

Brown, A. D. (1965). *J. molec. Biol.* **12**, 491.

Brown, A. D. and Shorey, C. D. (1962). *Biochim. biophys. Acta* **59**, 258.

Brown, A. D. and Shorey, C. D. (1963). *J. cell Biol.* **18**, 681.

Brown, A. D., Shorey, C. D. and Turner, H. P. (1965). *J. gen. Microbiol.* **41**, 225.

Brown, H. J. and Gibbons, N. E. (1955). *Canad. J. Microbiol.* **1**, 486.

Cho, K. Y. and Salton, M. R. J. (1966). *Biochim. biophys. Acta* **116**, 73.

Christian, J. H. B. (1956). Ph.D. Thesis: University of Cambridge.

Christian, J. H. B. and Ingram, M. (1959). *J. gen. Microbiol.* **20**, 27.

Christian, J. H. B. and Waltho, J. A. (1961). *J. gen. Microbiol.* **25**, 97.
Christian, J. H. B. and Waltho, J. A. (1962). *Biochim. biophys. Acta* **65**, 506.
Drapeau, G. R. and MacLeod, R. A. (1965). *Nature, Lond.* **206**, 531.
Dundas, I. D. and Larsen, H. (1962). *Arch. Mikrobiol.* **44**, 233.
Dundas, I. D. and Larsen, H. (1963). *Arch. Mikrobiol.* **46**, 19.
Dundas, I. D., Srinivasan, V. R. and Halvorson, H. Orin (1963). *Canad. J. Microbiol.* **9**, 619.
Dundas, I. D. (1965). Ph.D. Thesis: University of Illinois.
Dundas, I. D. and Halvorson, H. Orin (1966). *J. Bact.* **91**, 113.
Eimhjellen, K. (1965). *In* "Anreicherungskultur und Mutantenauslese", (H. G. Schlegel, ed.), p. 126, Fischer Verlag, Stuttgart.
Gibbons, N. E. (1957). *Canad. J. Microbiol.* **3**, 249.
Gibbons, N. E. and Baxter, R. M. (1953). *Proc. Intern. Congr. Microbiol., 6th. Congr., Rome* **1**, 210.
Harrison, F. C. and Kennedy, M. E. (1922). *Trans. Roy. Soc. Canada* V, **16**, 101.
Henis, Y. and Eren, J. (1963). *Canad. J. Microbiol.* **9**, 902.
Holmes, P. K. (1964). Ph.D. Thesis: University of Illinois.
Holmes, P. K. and Halvorson, H. O. (1963). *Canad. J. Microbiol.* **9**, 904.
Holmes, P. K. and Halvorson, H. O. (1965a). *J. Bact.* **90**, 312.
Holmes, P. K. and Halvorson, H. O. (1965b). *J. Bact.* **90**, 316.
Holmes, P. K., Dundas, I. D. and Halvorson, H. O. (1965). *J. Bact.* **90**, 1159.
Houwink, A. L. (1956). *J. gen. Microbiol.* **15**, 146.
Howe, J. M., Featherston, W. R., Stadelman, W. J. and Banwart, G. J. (1965). *Appl. Microbiol.* **13**, 650.
Joshi, J. G., Guild, W. R. and Handler, P. (1963). *J. molec. Biol.* **6**, 34.
Kates, M., Sastry, P. S. and Yengoyan, L. S. (1963). *Biochim. biophys. Acta* **70**, 705.
Kates, M., Yengoyan, L. S. and Sastry, P. S. (1965). *Biochim. biophys. Acta* **98**, 252.
Klebahn, H. (1919). *Mitt. Inst. allgem. Botan. Hamburg* **4**, 11.
Kluyver, A. J. and Baars, J. K. (1932). *Proc. Koninkl. Akad. Wetenschap. Amsterdam* **35**, 370.
Kurochkin, B. I. (1962). *Mikrobiologiya* **31**, 72.
Kushner, D. J. (1964). *J. Bact.* **87**, 1147.
Kushner, D. J. and Bayley, S. T. (1963). *Canad. J. Microbiol.* **9**, 53.
Kushner, D. J. and Onishi, H. (1966). *J. Bact.* **91**, 653.
Kushner, D. J., Bayley, S. T., Boring, J., Kates, M. and Gibbons, N. E. (1964). *Canad. J. Microbiol.* **10**, 483.
Larsen, H. (1962). *In* "The Bacteria" (I. C. Gunsalus and R. Y. Stanier, eds.), Vol. 4, p. 297, Academic Press, New York.
Marmur, J. and Doty, P. (1959). *Nature, Lond.* **183**, 1427.
Marmur, J., Falkow, S. and Mandel, M. (1963). *Annu. Rev. Microbiol.* **17**, 329.
Mohr, V. and Larsen, H. (1963). *J. gen. Microbiol.* **31**, 267.
Nandy, S. C. and Sen, S. N. (1963). *Canad. J. Microbiol.* **9**, 601.
Nierenberg, M., Leder, P., Bernfield, M., Brimacombe, R., Trupin, J., Rottman, F. and O'Neal, C. (1965). *Proc. nat. Acad. Sci. Wash.* **53**, 1161.
Onishi, H. and Gibbons, N. E. (1965). *Canad. J. Microbiol.* **11**, 1032.
Onishi, H. and Kushner, D. J. (1966). *J. Bact.* **91**, 646.
Onishi, H., McCance, M. E. and Gibbons, N. E. (1965). *Canad. J. Microbiol.* **11**, 365.
Plapp, R. and Kandler, O. (1965a). *Arch. Mikrobiol.* **50**, 171.
Plapp, R. and Kandler, O. (1965b). *Arch. Mikrobiol.* **50**, 282.
Rolfe, R. and Meselson, M. (1959). *Proc. nat. Acad. Sci. Wash.* **45**, 1039.
Rondle, C. J. and Morgan, W. T. J. (1955). *Biochem. J.* **61**, 586.
Salton, M. R. J. (1964). "The Bacterial Cell Wall", Elsevier Publ. Co., Amsterdam.

Sehgal, S. N. and Gibbons, N. E. (1960). *Canad. J. Microbiol.* **6**, 165.
Sehgal, S. N., Kates, M. and Gibbons, N. E. (1962). *Canad. J. Biochem. Physiol.* **40**, 69.
Smithies, W. R. and Gibbons, N. E. (1955). *Canad. J. Microbiol.* **1**, 614.
Smithies, W. R., Gibbons, N. E. and Bayley, S. T. (1955). *Canad. J. Microbiol.* **1**, 605.
Steensland, H., Urdahl, N. and Larsen, H. (1966). Manuscript in preparation.
Stevenson, J. (1966). *Biochem. J.* **99**, 257.
Stoeckenius, W. (1962). *In* "The Interpretation of Ultrastructure", (R. J. C. Harris, ed.), Vol. 1, p. 349, Academic Press, New York.
Sueoka, N. (1964). *In* "The Bacteria", (I. C. Gunsalus and R. Y. Stanier, eds.), Vol. 5, p. 419, Academic Press, New York.
Sueoka, N., Marmur, J. and Doty, P. (1959). *Nature, Lond.* **183**, 1429.
Urdahl, N., Svardal, R. and Larsen, H. (1966). Manuscript in preparation.
Weber, M. M. (1949). *Biol. Rev. City College New York* **11**, 9.
Weidel, W. and Pelzer, H. (1964). *Advanc. Enzymol.* **26**, 193.

The Biochemistry of the Bacterial Endospore

W. G. MURRELL

Commonwealth Scientific and Industrial Research Organization, Division of Food Preservation, Ryde, New South Wales, Australia

I. Introduction

The spore is one of the most complex structures formed by single-celled bacteria. In its mature, resting state it is very resistant to adverse treatments. Very little was known, 10–15 years ago, about its method of formation or the biochemical steps involved. Considerable progress has been made in this time, but there are still many intriguing questions unanswered. Sporogenesis involves a special type of cell division and its further development involves an integrated sequence of new biochemical reactions under genetic control. These result in the production of enzymes, substances and structures new or different from vegetative

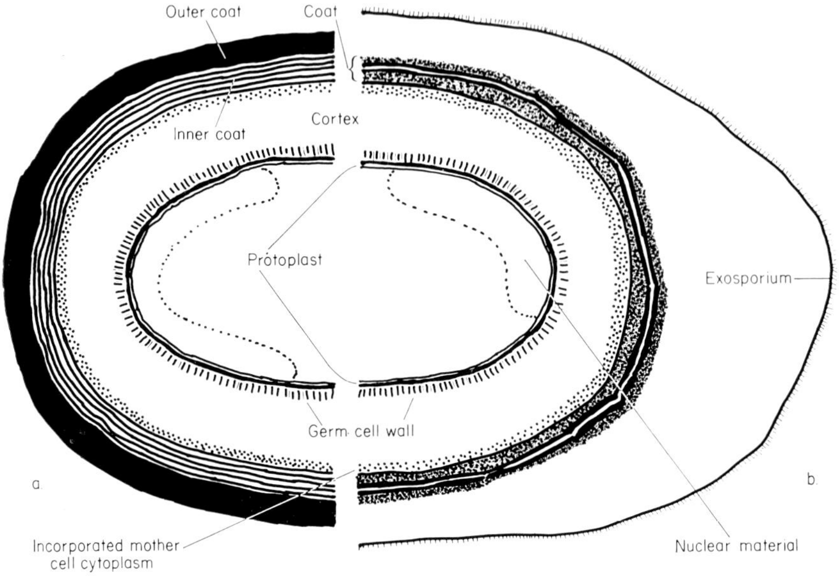

FIG. 1. Diagrammatic cross-section of two simple basic structural types of spores, showing various parts. Type a shows spores without an exosporium and usually with a distinctly laminated inner coat and a very electron-dense outer coat, e.g. *Bacillus coagulans* (Ohye and Murrell, 1962) and *Bacillus megaterium* (Robinow, 1960). Type b shows spores with an exosporium and with a coat usually consisting of one or two distinct laminae embedded in a less electron-dense layer; the coats often show incomplete fusion of the sections of laminae, e.g. *Bacillus cereus* (Figs. 5c, d and 7b). Various intermediate types (e.g. *Bacillus subtilis*, Fig. 5, Warth *et al.*, 1963b) and other more complex types (*Bacillus polymyxa*; Holbert, 1960) also occur.

material. During spore formation, the enzymes involved in the synthesis of such substances as the coat proteins and dipicolinic acid must be formed. The cortex, a loosely cross-linked layered matrix, is formed between the two plasma membranes of the forespore. About this differentiating cell, two to three coat layers, often with complex ridged patterns, are laid down, and in some species an outer loose envelope, the exosporium, is formed (Fig. 1). All these enveloping layers are synthesized around the spore protoplast within the mother cell.

The completed spore, which takes several generation times (6–8 hr.) to form and results from this relatively complex differentiation process, is remarkably resistant to heat, chemicals and adverse environmental changes. It can, however, revert rapidly (within minutes) to a heat-labile state in response to a variety of chemical and physical treatments. Complete regeneration to an active vegetative cell requires, however, several generation times (1–2 hr.).

Present chemical knowledge of the mature spore and the biochemistry of its formation is the subject of this review. First, however, to assist in understanding the biochemistry of the spore, a brief comment on definitions and on the cytological changes during sporogenesis and development is required.

A. Definitions

The terminology used in Figs. 1, 2 and 5 has been used throughout. Terms such as "spore wall" (see Robinow, 1960) and "spore core", which are in common use, have been abandoned. Spore wall (germ cell wall) implies that the germ cell is the spore. The term "coat" is preferred to "spore walls" to avoid confusion with vegetative cells. Since better resolution of spore components has been achieved, the term "spore core" has become confusing and unnecessary.

The sporulation process is now understood much better and so it has become necessary to define more accurately the various stages involved. This is essential for relating biochemical events to the various stages of development. The sporulation process has been divided into seven stages (Fig. 2) by Fitz-James (1963) and Schaeffer *et al.* (1963). It is recommended as highly desirable that all future biochemical studies should use synchronized cultures in which the biochemical events are related to the stage of morphogenesis as defined by the electron microscopy of thin sections. The terms post-logarithmic, pre-sporulation, and transitional are inadequate. Each stage has been given a common name to assist visualization (Fig. 2). The term "sporogenesis" ("the beginning of the spore"), where used, has been restricted to stages I and II (Fig. 2). For describing the part of the sporangium outside the developing spore or "non-spore part of the sporangium", the term "mother cell' has been found simplest and most convenient, although the difficulty of deciding whether the exosporium or coats at some stage are part of the mother cell has not been overlooked.

II. Cytological Development of the Spore

Sporogenesis is a logical process involving invagination of the protoplasmic membrane of the vegetative cell and enclosure of the genetic material (nucleus, nucleoid) by a double plasma membrane. This is diagrammatically outlined in Fig. 2. Before the elucidation of this process (Young and Fitz-James, 1959a, b; Fitz-James, 1960), various theories depicted spore formation as the aggregation or condensation of one or more densely staining granules within the cell (see Knaysi, 1948). Thin sectioning and improvements in electron microscope techniques

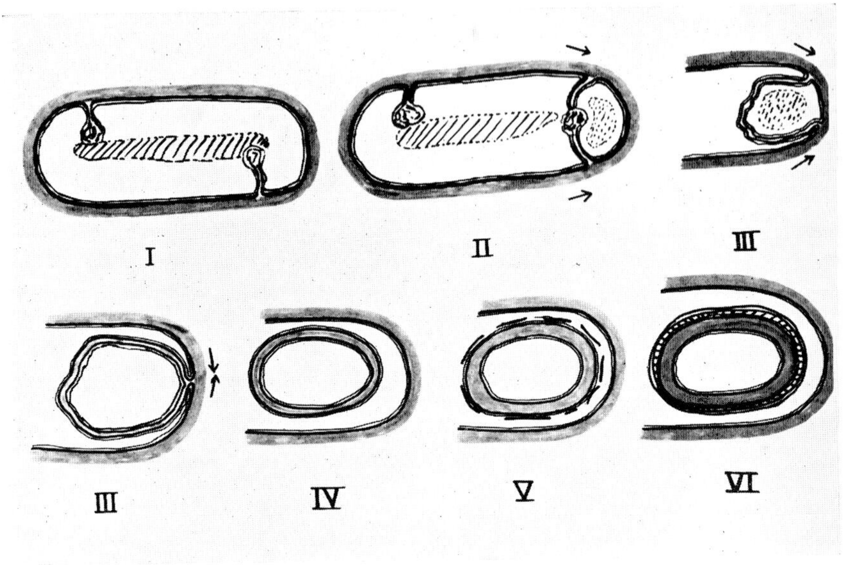

Fig. 2. Diagrammatic representation of six stages (I–VI) in sporulation, illustrating the invagination process, spore protoplast envelopment and further development. Stage I—preseptation or axial chromatin; II—septation; III—protoplast envelopment or protoplast development; IV—cortex formation; V—coat formation; VI—maturation; stage VII, the free spore, is not illustrated.

enabled the resolution of the membrane re-arrangements involved. Before this, certain nuclear events are considered essential (Young and Fitz-James, 1959a, b; Robinow, 1960; Ryter, 1965); these define the starting point of sporogenesis.

A. Nuclear Movements Associated with Sporogenesis

After the last nuclear division before sporogenesis, the two nuclei apparently condense, coalesce and form into an axial thread or cord of chromatin extending almost to the full length of the vegetative cell (Young and Fitz-James, 1959a; Fitz-James, 1960; Ryter and Jacob, 1964; Ryter, 1965; Ellar and Lundgren, 1966). Whether the axial thread of nuclear material arises this way or by normal DNA replication has genetic implications, and it would be desirable to confirm this with fully synchronized cultures and electron microscopy. The axial thread defines stage I (preseptation, Fig. 2). Only a small terminal part of the axial thread becomes enclosed in the forespore by the invaginating plasma membrane (Figs. 10, 14, in Fitz-James, 1960; plate 4 in Ryter and Jacob, 1964; Ryter, 1965). Chemical studies indicate that the DNA content of the spore is equivalent to the DNA content of a vegetative cell nucleoid (Fitz-James and Young, 1959; Young and Fitz-James, 1959a, b) and that the mother cell contains at the time of forespore septation an equal

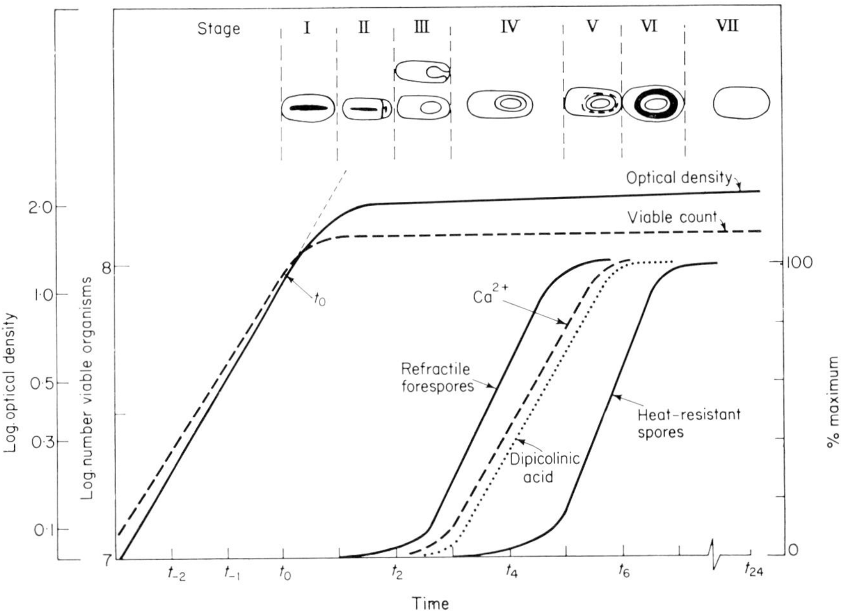

FIG. 3. Diagrammatic representation of the seven sporulation stages in relation to the cultural cycle, Ca^{2+} uptake and dipicolinic acid (DPA) formation. t_0, the end of exponential growth, arbitrarily defines the start of the sporulation process. t_1, t_2–t_{24} indicate the times in hours from t_0. Stage I to stage VI commonly takes 6–8 hr. in *Bacillus cereus*, *Bacillus megaterium* and *Bacillus subtilis*. (Based on Young and Fitz-James, 1959a; Fitz-James, 1963; and Schaeffer *et al.*, 1963).

amount of DNA. The division of nuclear material does not appear to be equal. Ryter (1965) estimates the sporal amount as only one-quarter by volume. The electron density of the spore nucleoid does not suggest that it is in a more condensed form. This inequality of nuclear division suggests that only a single strand of DNA from the mother cell may enter the spore. Replication during entry into the spore protoplast could explain some of the DNA-phosphorus content trends observed by Young and Fitz-James (1959a, b). On the other hand, some species may receive two strands or a more equable division of nuclear material (see p. 209).

If there is a different type of nuclear re-arrangement before axial thread formation, and an unequal distribution of nuclear material during sporogenesis, then the initial point in sporogenesis must originate with a genetic event before condensation of the nuclei and formation of the axial thread. The initial point is defined arbitrarily as occurring at t_0 (Schaeffer, 1961; Fig. 3). Until it can be detected simply, t_0 provides a useful starting point for the sporulation time-scale.

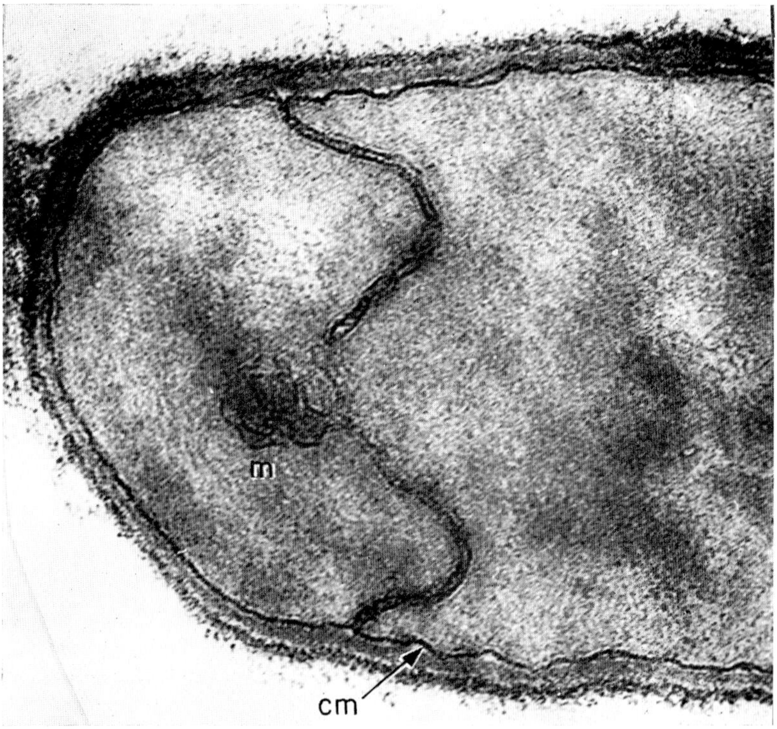

FIG. 4. Electron micrograph of a thin section of a cell of *Bacillus coagulans* showing septation nearly complete. m, mesosome; cm, cytoplasmic membrane. (From Ohye and Murrell, 1962).

B. SPORE SEPTUM FORMATION

The invagination of the plasma membrane to form the spore septum differs from that of vegetative cell division in that (i) it occurs usually very close to the end of the rod (Figs. 2, 4) and (ii) no cell wall material appears to be deposited between the invaginating layers of the membrane. Associated with the invaginating plasma membrane are mesosomes, connected both to the plasma membrane and the nuclear material (Fitz-James, 1963; Ryter and Jacob, 1964). Membrane growth thus draws the nucleoids apart as the mesosomal connections at each end become more distant to each other (Ryter and Jacob, 1964). Such a mechanism would require some growth of the sporangial membrane just before septation is complete. With completion of the septum by fusion of the membrane, the nucleoid of the spore, together with some cytoplasm, becomes surrounded by a single plasma membrane. This completes stage II (septation).

C. Envelopment of the Spore Protoplast by the Mother Cell

This occurs by an engulfing process (Fig. 2) due apparently to the directional growth of the plasma membrane of the sporangium at the points indicated by arrows. If growth of the plasma membrane of the spore protoplast but not that of the membrane of the mother cell is inhibited, and the growing area of the outer membrane is localized in a peripheral ring (arrows), then this growing membrane must push between the sporangial cell wall and the spore protoplast until the spore protoplast in engulfed. Fusion and separation of the enveloping membrane add a second membrane around the protoplast and free the spore protoplast into the cytoplasm of the mother cell. This process has been confirmed in several *Bacillus* and *Clostridium* species; in *Sporosarcina*, typical invagination occurs (Mazanec *et al.*, 1965) but evidence of the envelopment process is lacking. The mechanism suggested requires a repressor of membrane synthesis in the spore protoplast.

Examination of the diagrams in Fig. 2 and the micrographs in Fig. 4 shows that the outer sides of the plasma membrane come to face inward towards each other in the double membrane of the forespore. This means (i) that any membranal mechanism by which vegetative cell wall material is synthesized and built or excreted on the outside of plasma membranes is doubly provided for in the forespore, (ii) that any material laid down between the two forespore membranes is likely to resemble the vegetative cell walls in composition (Warth *et al.*, 1963b; Warth, 1965), and (iii) that this material is likely to be synthesized by both the forespore itself and by the mother cell or the outer membrane using intermediates formed in the mother cell. It may be that the germ cell wall is formed by the forespore and the main cortex by the mother cell. The relative positions of the two cortical structures to each cytoplasm, the two peak periods of synthesis and difference in degradation during germination (see p. 221), and the relative sizes of the synthesizing cells, would support this suggestion.

D. Cortex Formation

The cortex results from the deposition of material between the inner and outer membranes of the forespore probably by involvement of the membranes as already described (Young and Fitz-James, 1962; Ohye and

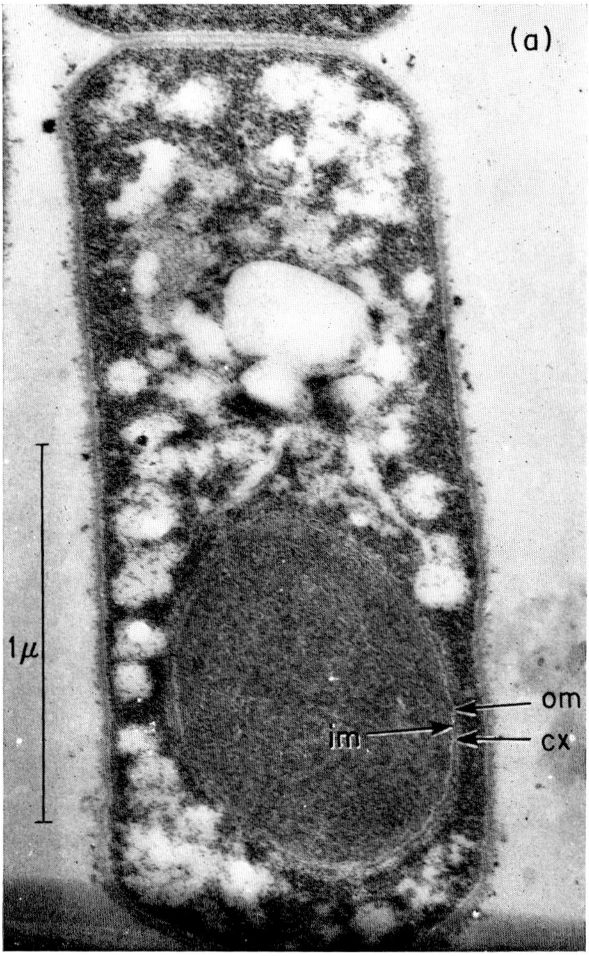

Fig. 5(a)

Fig. 5. Four stages in spore formation in *Bacillus cereus* T. (a) shows early stage IV and the beginning of cortex formation, and the many granules in the mother cell; (b) early stage V, showing the cortex developing and coat formation in progress. Note the cytoplasm of the mother cell enclosed by the coats; (c) stage VI showing the cortex well developed and the protoplast becoming characterless; (d) stage VII showing the free spore: cx, cortex; sc, spore coat; ex, exosporium; im and om, inner and outer membranes. Note depletion of the granular material.

Murrell, 1962). First, vesicular-like deposits may occur (Ohye and Murrell, 1962), then gradually a thin line of cortical material forms, similar in electron density to that of the vegetative cell wall (Fig. 5a). This gradually increases to about 0·1 μ thickness, about one-tenth of the diameter of the spore. The insoluble structural material of the cortex

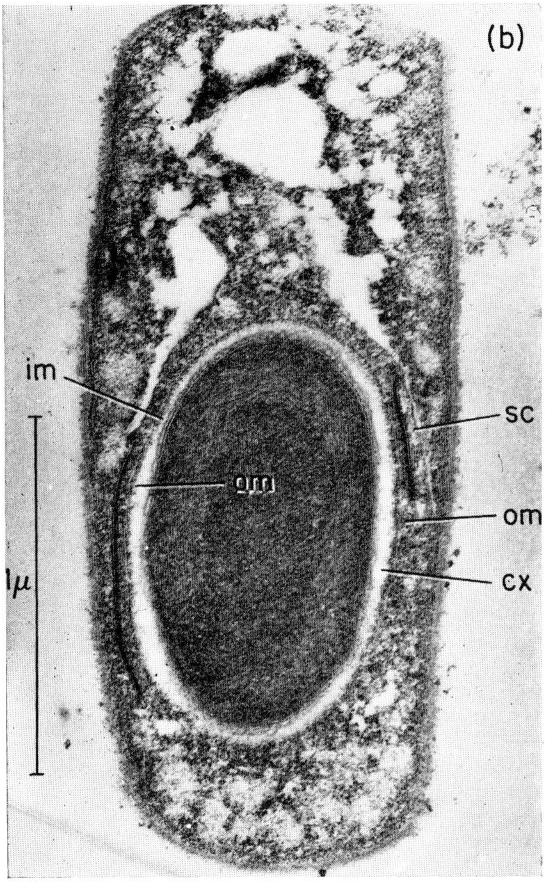

FIG. 5(b)

consists of a glycopeptide polymer (Warth *et al.*, 1962, 1963b; Warth, 1965). In later stages of spore formation and in the mature spore, the cortex appears electron-transparent. However, on acid or mechanical disruption of the spore, the cortex appears as a loosely cross-linked layered matrix (Fig. 7b; Mayall and Robinow, 1957; Warth *et al.*, 1963b; Murrell and Warth, 1965).

E. Coat Formation

The spore coat originates as a discontinuous deposit of coat material (very electron-dense) away from and around the outer membrane in the mother-cell cytoplasm (Fitz-James, 1962; Takagi *et al.*, 1960, 1965; Fig. 5b). These deposits increase in area, and coalesce to form a two- or

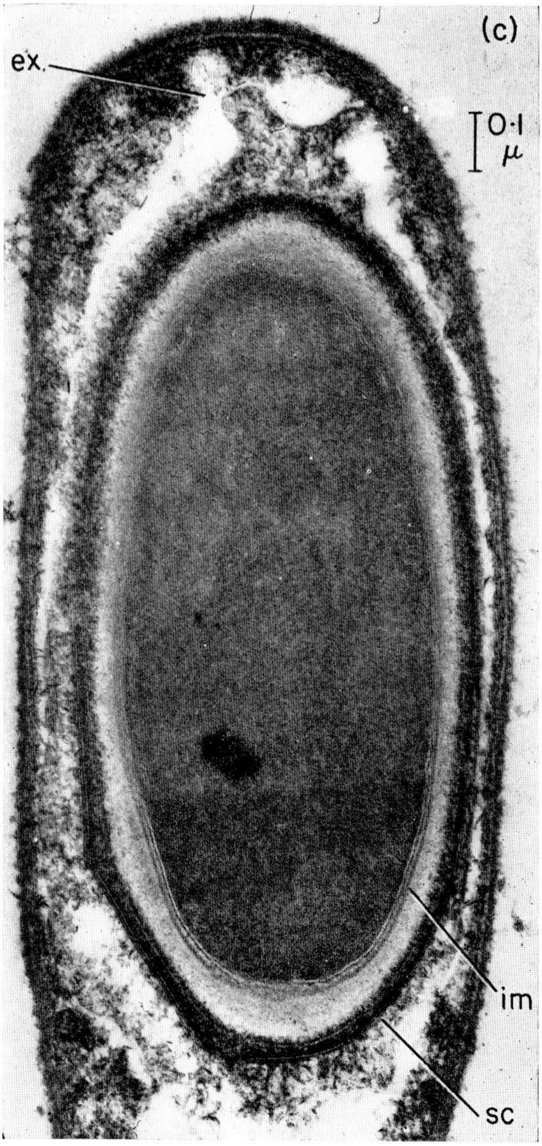

Fig. 5(c)

three-layered coat; the layers vary in electron density, degree of lamination (Fig. 1) and in the occurrence of ridge-like formations with definite patterns (Bradley and Franklin, 1958; Holbert, 1960). During coat

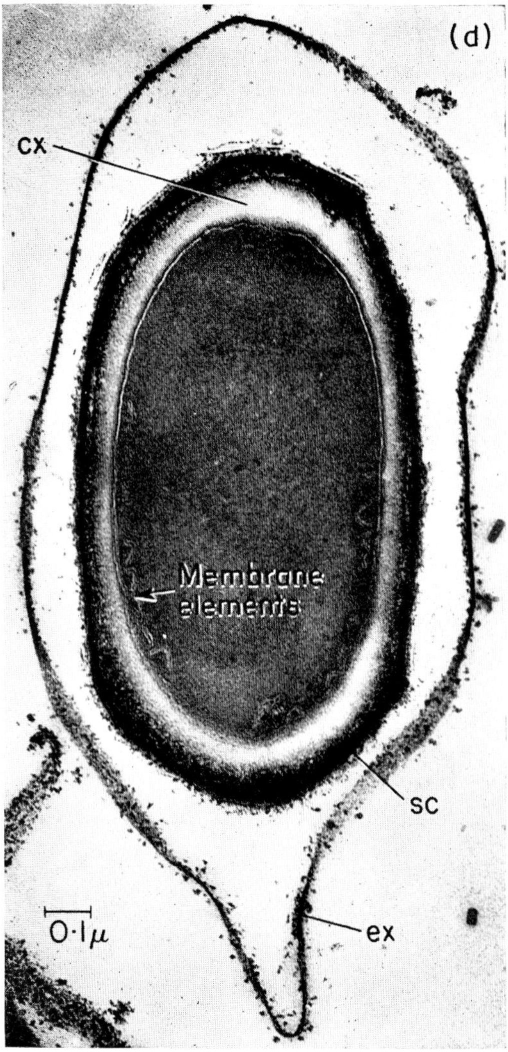

Fig. 5(d)

formation, some of the mother-cell cytoplasm becomes trapped between the outer membrane and the coat proper (Fig. 5b, c). This is probably the innermost layer of the coat described by Thomas (1964; see Fig. 9; p. 173).

F. Exosporium

In several *Bacillus* and *Clostridium* species, a membrane appears to originate apparently *de novo* in the mother-cell cytoplasm at an early stage of cortex formation (Hannay, 1961; Young and Fitz-James, 1962). In *Bacillus cereus* and *B. cereus* var. *anthracis*, this membrane has a complex ultrastructure (Gerhardt and Ribi, 1964). In *B. cereus*, it originates at the non-polar end of the forespore, and gradually encompasses the spore without any association with the spore coats or the mother-cell plasma membrane (D. F. Ohye and W. G. Murrell, unpublished observations).

When the spore is mature (fully refractile and heat resistant; stage VI), the mother cell lyses, and so does the cytoplasm enclosed by the exosporium. Lysis within the exosporium is not always complete, and certain granules may remain (Fig. 5d; Warth *et al.*, 1963b; Gerhardt and Ribi, 1964). When the spore becomes free (stage VII), the exosporium remains as a loose balloon-like sac about the spore.

G. Maturation or Ripening

Towards the end of stage V and during stage VI, the spore undergoes a number of changes which have been described as ripening. Its refractility increases, and it becomes heat resistant and impermeable to dilute basic stains. Ribosomes stained with Pb^{2+} become very faint in appearance, and their permeability to or affinity for osmium tetroxide is also greatly decreased. Also the mesosomes become more compact and practically invisible. During stages I–III, the nuclear material has a fibrous nature, and the nucleoid appears randomly arranged in an open often-branched network. During cortex formation, the nucleoid becomes more orderly and, in the resting spore, has a low density with no definite fibrous structure despite heavy metal "staining" (Fitz-James, 1962; Young and Fitz-James, 1962).

Certain anomalous types of spore development may occur (Robinow, 1960; Takagi *et al.*, 1960; Fitz-James, 1962) and certain mutants are blocked at recognizable stages in development (Fitz-James, 1963; Ryter *et al.*, 1961; Schaeffer *et al.*, 1963).

III. Chemical Composition of the Mature Spore

Early studies indicated little difference in the gross composition of spores and vegetative cells (e.g. Virtanen and Pulkki, 1933). The amino-acid composition of acid hydrolysates of cells and spores was also found to be similar (Pfennig, 1957). Tinelli (1955a) showed, however, that spores of *Bacillus megaterium* were rich in protein, and relatively low in carbohydrate and β-hydroxybutyrate compared with vegetative cells (Table 1). Analysis of exudates from spores and from the various layers of the spore has revealed considerable difference in the compositions of the spore and the vegetative cell. An important initial step came with the discovery of dipicolinic acid (pyridine-2,6-dicarboxylic acid, DPA; Powell, 1953) and muramic acid in spore germination exudates (Strange and Powell, 1954; Strange, 1956). Further progress has resulted from

TABLE 1. Gross Composition of Spores and Vegetative Cells of *Bacillus megaterium* (Data from Tinelli, 1955a)

	% dry weight	
Component	Vegetative cells	Spores
β-Hydroxybutyrate	28·5	0
Carbohydrate	18·2	4·84[a]
Protein (N × 6·25)	39·4	68
Calcium dipicolinate	Nil	15
Ash	9·1	8[b]

[a] Carbohydrate included glucose, galactose and traces of uronic acid.
[b] Less calcium (3%).

fractionation of spore layers by exosporium stripping (Matz and Gerhardt, 1964), fractional centrifugation and enzymic digestion coupled with electron microscopy (Warth *et al.*, 1963a, b), and X-ray and radioactive labelling studies. Unfortunately most of the studies are limited to *Bacillus* species.

A. STRUCTURES

1. *Exosporium*

A preparation of exosporium from only one species, *B. cereus*, has been analysed (Table 2). It was notable for its high protein and lipid contents. Glucosamine was the only amino sugar detected, and only 0·2% diaminopimelic acid was present. Organic phosphate was mainly present as teichoic acid (2·0%). The non-hydrolysable fraction was believed to consist mostly of degraded carbohydrates (Matz and Gerhardt, 1964).

The exosporium of *B. cereus* is a complex structure composed of two main layers. The outer layer is made up of a nap of hair-like projections, irregularly distributed and about 250 Å deep; these arose from an intermediate covering about 60 Å deep. The inner basal layer is about 190 Å thick, and has a hexagonally perforate surface pattern of holes, about 76 Å from centre to centre, made up of lamellae which could fragment into crystal-like elements. The crystal-like nature was confirmed by X-ray diffraction analysis (Gerhardt and Ribi, 1964). The exosporium of *B. anthracis* has a longer nap, 560–690 Å deep (Gerhardt and Ribi, 1964; Y. Hachisuka, unpublished observations).

TABLE 2. Composition of the Exosporium of *Bacillus cereus* T
(Data from Matz and Gerhardt, 1964)

Component	% dry wt.
Amino acids (15)	37
Ether-extractable lipid	17·9
Sugars (glucose, galactose, rhamnose, ribose)	10·4
Glucosamine	11·2
Organic phosphate	2·1
RNA	1·2
Non-hydrolysable fraction	18·0
	97·8

2. *Coats*

Coats are a major fraction of the spore. The volume occupied by the coats forms about 50% of the spore volume and yields of water-washed integument preparations approximate 40–60% of the dry wt. In some reports this has been shown to decrease to 20–30% with chemical or enzyme treatment and the removal of cortical and exosporial material (Kadota *et al.*, 1965; Kondo and Foster, 1965).

During disruption and centrifugation of the insoluble integuments, considerable contamination with other spore membranes occurs. This was not fully appreciated until recently, partly because preparations were not thin-sectioned for electron microscopy. Strange and Dark (1956) realized that hexosamine-containing peptides were associated with their coat preparations, and that they were not peptide-bonded to the coats as the "spore peptide" could be removed by lysozyme but not by proteinases. Warth *et al.* (1963a, b) showed that, in sectioned integument preparations, adhering cortical material was present which could be removed by lysozyme or autolytic enzymes with loss of essentially all the glycopeptide-containing diaminopimelic acid.

Most methods of coat preparation suffer from possible changes in composition during disruption and purification. Also, the coat analyses

(Tables 3 and 4) are not altogether satisfactory because (i) many coats are composed of two or three layers and these have not been analysed separately, (ii) some of the preparations contained residual cortical material, a conclusion based on the criteria that diaminopimelic acid and hexosamine were present, and (iii) some preparations may have contained exosporium material. The data for several species have been averaged as the variation, except in a few instances, is not very great. The coats are composed mainly of protein together with 1–3% ash, 0–2·8% phosphorus, and a variable amount of lipid (Table 3). The protein content varies from 35% in *B. megaterium* spore coats (Kondo and Foster, 1965) to 80% in those from spores of *B. licheniformis* (Bernlohr and Sievert, 1962). The nitrogen content increases with removal of cortical material (Table 5). Figures for hexosamine and carbohydrate contents have been omitted from Table 3, and those for alanine,

TABLE 3. Average Composition of the Spore Integuments of Six *Bacillus* Species

	% dry weight	
	Average	Range
Total nitrogen	11·7	9·7–13·2
Amino-nitrogen	8·0	5·6–12·8
Total phosphorus	1·25	0·3–2·8
Ash	2·16	1·4–2·8
Total lipid	2·84	0·9–7·2

Data from Strange and Dark, 1956; Salton and Marshall, 1959; Bernlohr and Sievert, 1962; Warth *et al.*, 1963a; and Yoshida *et al.*, 1957.

glutamic acid and diaminopimelic acid contents from Table 4 as the evidence of Warth *et al.* (1963a, b) suggests that, in many species, all the diaminopimelic acid and hexosamine and a large part of the alanine and glutamic acid derive from the cortical material. Alanine and glutamic acid, however, are certainly not considered to be absent from coats. The coats are composed of a large number of amino acids, with all except one species having glycine and lysine as the predominant amino acids. Coats of *B. licheniformis* were high in isoleucine, phenylalanine, tyrosine and aspartic acid (Bernlohr and Sievert, 1962; Snoke, 1964). *B. subtilis* coats were high in lysine, aspartic and histidine (Salton, 1964).

In *Bacillus* strain 636, which had a ridged coat structure, lysozyme removed nearly all of the DAP but only about half of the glucosamine. It was suggested that the glucosamine in the lysozyme-treated coats may occur in the ridge structure. These coats also contained greater amounts of glutamic acid and taurine. Taurine was not present in the soluble fraction of this spore or in whole spore hydrolysates of fourteen other *Bacillus* species (Warth *et al.*, 1963b).

The low nitrogen content (5·1%) of *B. megaterium* coats reported by

TABLE 4. Average Amino Acid Composition of Spore Coat Preparations of Five *Bacillus* Species and a Parasporal Crystal

Amino acid	Mole ratio Min.	Max.	Av.	*Bacillus thuringiensis* parasporal crystal[a]
Glycine			10	10
Lysine	2·3	19·7	9·3	4·9
Serine	1·9	8·0	4·1	10·3
Threonine	2·0	4·1	3·0	8·7
Valine	tr.	4·2	3·4	8·7
Leucine	3·2	10·0	5·8	11·8
Isoleucine	3·0	20·0	7·2	8·7
Phenylalanine	2·3	10·0	4·2	7·4
Tyrosine	tr.	11·5	6·0	5·9
Aspartic acid	2·0	10·5	6·5	16·5
Arginine	1·8	6·0	3·8	8·7
Histidine	1·5	10·4	5·2	2·6
Proline	4·0	5·3	4·9	7·2
Cystine	1·5	3·6	2·7	0·9
Methionine	1·5	3·6	2·7	1·1
Ornithine	0·8[b]	10[b]	—	—
Tryptophan				4·6
α,ε-Diaminopimelic			[c]	—
Alanine			[c]	7·9
Glutamic acid			[c]	15·8

Data for analyses of spore coat preparations are from Salton and Marshall (1959), Hunnell and Ordal (1961), Warth *et al.* (1963b), Bernlohr and Sievert (1962), Snoke (1964), and Salton (1964).

[a] Data for *B. thuringiensis* parasporal crystal are from Holmes and Monro (1965).

[b] Data for *B. licheniformis* only.

[c] Data omitted, see text.

Kondo and Foster (1965) is in part a reflection of the high ash (25·9%), phosphorus (9·3%) and hexosamine (14·5%) contents, and their data suggest that a different type of coat is present in this organism or that the preparation contains exosporia or cortical material. Some strains are known to be rich in phosphorus (see p. 167). Kondo and Foster (1965) found that the lysozyme-treated fraction was a phosphoglycopeptide containing muramic acid, glucosamine and amino acids in mole ratios of aspartic acid, 1·8; glutamic acid, 1·0; glycine, 1·3; serine, 1·0; and lysine, 1·6. Fluorodinitrobenzene (FDNB) reacted only with lysine although not with all of this amino acid. Partial hydrolysis and purification on an ion-exchange column gave a fraction which, on complete hydrolysis, yielded (in mole ratios) lysine, 1·0; muramic acid, 0·8; and phosphorus, 4·8. The lysine in this fraction did not react with FDNB.

TABLE 5. Analyses of (Cysteine + Cystine) Sulphur and Nitrogen Contents of *Bacillus subtilis* Cells and Spores (data from Kadota *et al.*, 1965)

Structure	Content of (Cysteine + Cystine) S (μg. S/mg. N)	Content of N (% dry wt.)
Vegetative cell	26	
Vegetative cell wall	10	
Intact spore	108	
Crude spore coats	190	11·2
Spore coats after lysozyme treatment	212	
Spore coats after treatment with lysozyme + ribonuclease	252	
Purified spore coats	292	14·5

The contents of (cysteine + cystine) sulphur were determined by the method of Kuratomi *et al.* (1957).

Spores contain about 26 μg. cystine sulphur/mg. protein nitrogen, compared with 7 μg. in vegetative cells; this occurs mainly as cystine in the coat fraction (Vinter, 1960). Most of the cysteine + cystine of the sporangium is incorporated into the spore integument preparation, and it is not degraded during disruption or germination (Vinter, 1959). In a mutant strain of *B. cereus*, grown in the presence of cystine, Bott and Lundgren (1964) found a disulphide content as high as 0·79% dry wt. Enzyme-purified coats of *B. subtilis* contained 292 μg. cysteine + cystine – S per mg. N, similar to wool or hair keratin (Table 5; Kadota *et al.*, 1965). Coats are resistant to many enzymes including proteolytic enzymes, lipases, lysozyme, phosphatase, bacterial proteinase (Strange and Dark, 1956; Salton and Marshall, 1959; Hunnell and Ordal, 1961; Warth *et al.*, 1963a, b; Kadota *et al.*, 1965). Whether some of these or other enzymes can remove small groups or split bonds has not been reported. Possibly the inner and outer coats and the laminae need to separate before breakdown occurs, e.g. by breaking disulphide bonds between and in the laminae to permit protein digestion (Gorini and Audrain, 1956).

The resistance of coats to many chemicals, such as 8 *M*-urea, 8 *M*-LiBr, 2 *M*-$CaCl_2$ (90°), formic acid (98%), performic acid (98%), *N*-NaOH (100°), trichloroacetic acid (5%), phenol (80%) and surface-active compounds, suggests the presence of relatively stable covalent crosslinking, and that their resistance does not entirely depend on disulphide crosslinking or hydrogen, ionic or hydrophobic bonding. Alkali- and surfactant resistance suggests that normal lipid and lipoprotein are not very important to structural integrity (Warth *et al.*, 1963b). Strong alkali causes coats to become gelatinous (Strange and Dark, 1956). Coats of *B. coagulans* resist 0·1 *N*-HCl at 121° for 2 hr. (Hunnell and Ordal, 1961).

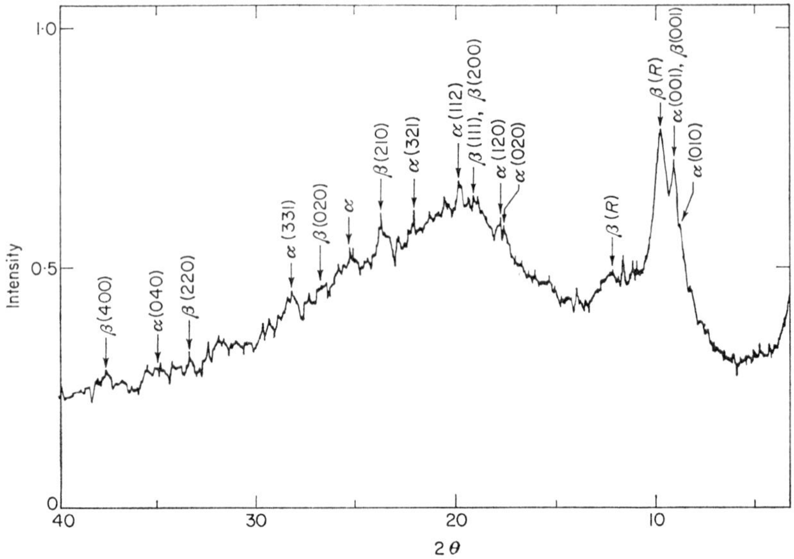

FIG. 6. Crystallite forms and Miller indices of the peaks in the X-ray diffraction pattern of purified spore coats. Peaks corresponding to those in wool keratin are indicated. Miller indices are indicated in parentheses. From Kadota *et al.*, 1965.

Coats are therefore composed of very chemically-stable disulphide-bonded proteins with 1–2% lipid, phosphorus, inorganic matter and, in some instances, glycopeptide or hexosamine. If the coats originate from protein-phospholipid basal membrane units, some chemical change would have to enhance their stability. Little information is available on their chemical structure. Knowledge of their biosynthesis helps a little (see p. 228). Bernlohr and Sievert (1962) suggest that coats are composed of polypeptide structures linked by phosphate to hexosamine (of the cortex presumably). Recent X-ray diffractometry indicates that enzyme-cleaned coats of *B. subtilis* contain periodic (crystalline) structures with a lattice distance of 9·8 Å, very similar to wool keratin (Kadota and Iijima, 1965; Kadota *et al.*, 1965). The diffraction patterns showed 9 and 9·6° peak structures, and eighteen peaks corresponded to those in wool keratin, nine with α- and nine with β-keratin (Fig. 6), suggesting that coats are composed of α and β forms like keratin.

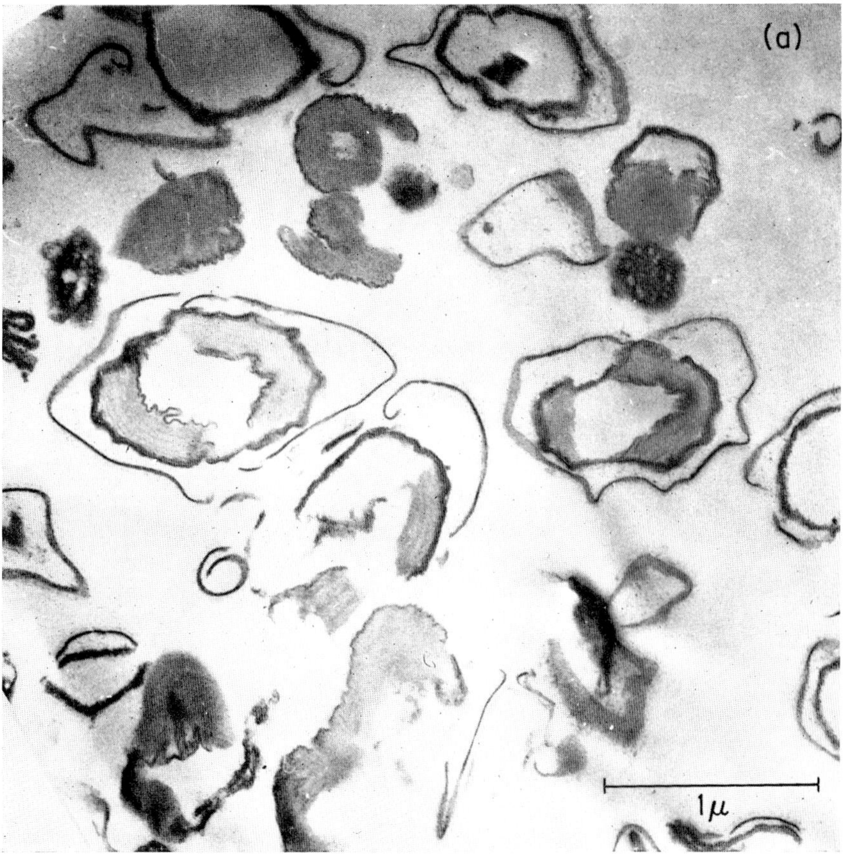

FIG. 7(a)

FIG. 7. Electron micrographs of (a) integument preparations of *Bacillus cereus* T; (b) an enlargement of the integuments of a single spore from preparation (a) showing the irregular involuted margin of the cortex after spore disruption; (c) an integument preparation after lysozyme treatment; note the absence of cortical material. ex, exosporium; cx, cortex; sc, spore coats; gcw, germ cell wall.

3. *Cortex*

Disruption of spores under conditions of inhibited lytic activity (pH 9–10, plus EDTA to bind Ca^{2+}), or after heating to inactivate the lytic enzymes, yields integument preparations containing nearly all of the structural material of the cortex including the germ cell wall (Fig. 7), and essentially all of the hexosamine and diaminopimelic acid of the spore. The insoluble cortical material can then be removed with lysozyme or the

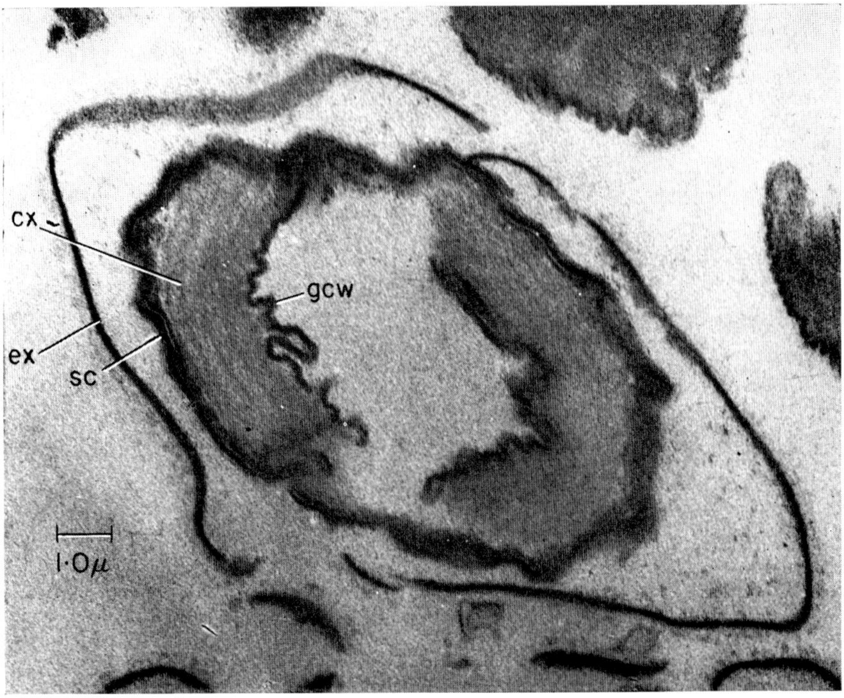

Fig. 7(b)

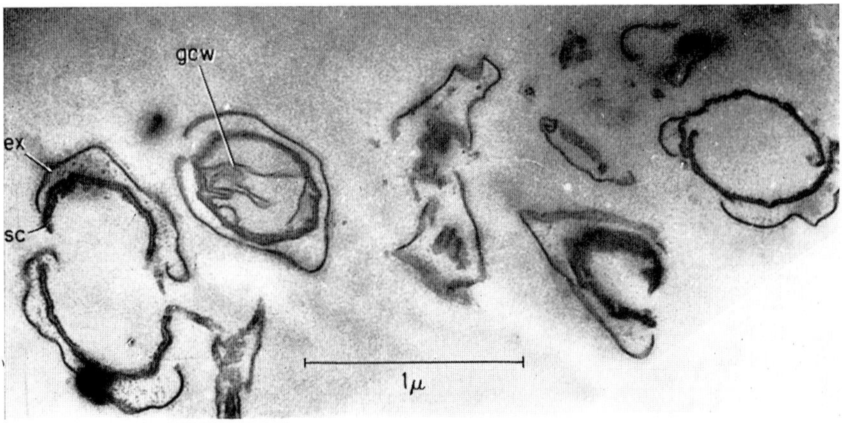

Fig. 7(c)

spore lytic enzymes, and analysed (Fig. 7c). This provides no information on substances possibly occurring within the cortex of the intact spore and which may be released during disruption (Warth *et al.*, 1963a, b).

Enzymic degradation of the cortical structure yields non-dialysable glycopeptide units and a small amount ($<4\%$) of low mol. wt. solutes. These include amino acids and peptides similar to those in the peptide side-chain of the glycopeptide; they may derive from this polymer or from the coats as a result of weak protease activity. Strange and Powell (1954) removed a small amount of protein from the glycopeptide of *B. megaterium* by trichloroacetic acid precipitation.

The glycopeptides in germination extracts and extracts from disrupted spores of several species were very similar (Strange and Powell, 1954). As the spore peptides are released during germination by the lytic enzyme(s), this is not surprising. The germination-exudate peptide had a mol. wt. of 15,300 (Record and Grinstead, 1954). Some of the glycopeptide from *B. megaterium* passed through the dialysis membrane, indicating a mol. wt. of about 10,000. Strange and Powell (1954) suggested that various preparations may contain molecules of different chain length; viscosity measurements supported this.

The non-dialysable glycopeptide from four species contained about 45% amino sugars, amino acids (7% total N; 3% α-amino-N) and small amounts of phosphorus and ash (mainly calcium) (Strange and Powell, 1954; Warth, 1965). The major constituents occurred in the mole ratios of α,ϵ-diaminopimelic acid (1), glutamic acid (1), alanine (3) and acetylhexosamine (6–8; Table 6). Lower hexosamine values occurred when muramic acid was not determined. Small amounts of other amino acids were present in the acid hydrolysates (Strange and Powell, 1954; Warth *et al.*, 1963b). This was more common when the spores were autoclaved before disruption, suggesting coagulation of included protein. A more detailed analysis of a preparation of *B. coagulans* gave glucosamine, muramic acid, L-alanine, D-alanine, D-glutamic and mesodiaminopimelic acid in the molecular proportions 2·8 : 3 : 1·6 : 1·0 : 1·0 : 1·0.

Uronic acids, purines, pyrimidines (Strange and Powell, 1954), hexoses, pentoses and methylpentoses (Warth, 1965) were not found in the glycopeptide. Traces only of phosphorus (Table 14) indicated the absence of significant amounts of teichoic acid. The ash (mainly Ca^{2+}) probably arises from metal contaminants trapped by the free carboxyl groups (see p. 171) of the polymer during disruption or germination of the spore.

Data from titration, dinitrophenylation and composition studies of the glycopeptide of *B. coagulans* suggest the structure shown in Fig. 8 (Warth, 1965). About 80% of the amino groups of diaminopimelic acid

TABLE 6. Composition of Non-Dialysable Glycopeptide of the Spore Cortices of *Bacillus* Species

Component	Contents (mole ratios)							
	Bacillus megaterium			*Bacillus subtilis*	*Bacillus* strain 645	*Bacillus coagulans*		*Bacillus stearothermophilus*
	Germination exudate	spore extracts						
	(*a*)	(*a*)	(*b*)	(*b*)	(*b*)	(*b*)	(*c*)	(*b*)
α,ε-Diaminopimelic acid	1·00	1·00	1·00	1·00	1·00	1·00	1·00	1·00
D-Glutamic acid	1·35	1·18	0·8	0·80	0·88	0·95	0·95	0·78
L-Glutamic acid			0·11	0·20	0·22	0·03	0·03	0·04
L-Alanine, D-Alanine	3·8	3·1	2·2	2·4	2·6	2·5	2·5	2·6
Hexosamine	8·4	8·2	2·1[d]	2·5[d]	2·4[d]	4·7	5·6	1·9[d]
Glycine			0·20	0·26	0·22	0·03	0·05	0·08
Aspartic acid			0·17	0·21	0·18	0·03	0·03	0·07
Acetyl groups							3·6	

[a] Strange and Powell (1954).
[b] Murrell and Warth (1965).
[c] Warth (1965).
[d] This value does not include muramic acid.

and 3·8 carboxyl groups per polymer unit were free; alanine in trace amounts (1 in 300 residues) was the only N-terminal amino acid detected. The results are consistent with the cortical structure being a loosely cross-linked matrix with a considerable excess of carboxyl groups. One-fifth of the diaminopimelic acid amino groups are probably peptide-bonded with carboxyl groups, possibly in another chain. One L-alanine residue possibly occurs in each peptide side-chain. About six units (mol. wt. 11,000) probably occur for each non-dialysable glycopeptide fragment released by lysozyme. However, both Strange and Powell (1954) and Warth (1965) found evidence of inhomogeneity and variable polymer length.

Warth (1965) suggests that the molecules of soluble glycopeptide are linked by peptide bonds to form the macroscopic structure of the cortex, and that these bonds involve the ϵ-amino group of diaminopimelic acid; about one cross-link occurs in every thirty amino sugar residues.

The cortical structure was resistant to a variety of enzymes (Strange and Dark, 1956; Warth *et al.*, 1963b) and to many chemical treatments. Formic acid, performic acid, *N*-NaOH, 80% phenol, cetyl trimethyl ammonium bromide and polysept 103 S (quaternary ammonium glycine) released less than 14% hexosamine; *N*-NaOH for 1 hr. at 100° released 28%, and 5% trichloroacetic acid (15 min., 90°) released 43% (Warth *et al.*, 1963b) of the hexosamine.

The cortical material is not physically held between the protoplast membrane and the outer membrane and coats, but remains attached to the coats on disruption (Fig. 7; Warth *et al.*, 1963b). The type of linkage involved in this attachment is not known, but apparently it is not by peptide bonding (see p. 146).

No information has been published on the composition of *Clostridium* spore cortices, but germination exudates resemble in composition those of *Bacillus* (Riemann, 1963), and they contain diaminopimelate and hexosamine (Tsuji and Perkins, 1962) suggesting that similar types of glycopeptide are present.

4. *Germ Cell Wall*

In integument preparations of *B. subtilis*, *Bacillus* strain 636, *B. coagulans* and *B. stearothermophilus*, the germ cell wall disappeared during digestion of the cortex by lysozyme, suggesting a composition similar to that of the cortex. On the other hand, the germ cell wall of *B. cereus* strain T persisted. The presence of diaminopimelic acid in the lysozyme-treated integuments (Warth *et al.*, 1963b) and the presence of traces only in exosporia (see p. 145) also suggests a glycopeptide composition for the germ cell wall. The difference in susceptibility to lysozyme could result from small differences in composition.

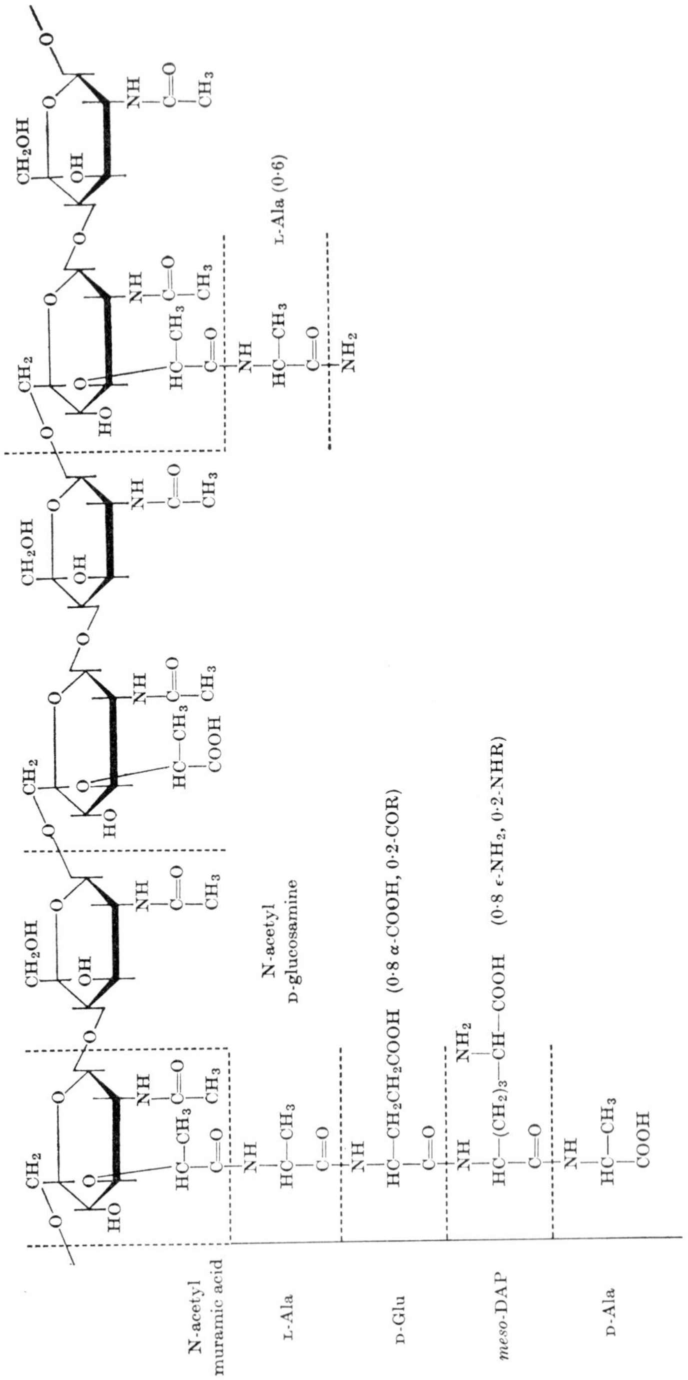

FIG. 8. Hypothetical average unit of the spore glycopeptide polymer. DAP indicates α,ϵ-diaminopimelic acid (after Warth, 1965).

5. *Protoplast*

The protoplast is the remaining part of the spore to be considered chemically. It is not yet possible to isolate and determine its composition directly. Disruption and removal of the insoluble integuments leaves a "soluble" fraction which includes the protoplast and any soluble material held in the other layers before disruption. The protoplast is probably the

TABLE 7. Chemical Analyses of Acid Hydrolysates of Soluble Extracts from Spores of *Bacillus* Species (data from Warth *et al.*, 1963b)

	Per cent dry wt.			
	Bacillus cereus	*Bacillus subtilis*	*Bacillus coagulans*	*Bacillus stearothermophilus*
Total nitrogen	9·1	9·7	—	9·2
α-Amino nitrogen	—	—	5·6	4·3
Total phosphorus	2·3	0·7	2·2	0·8
Hexosamine (as glucosamine)	5·5	7·0	2·8	2·6
Hexose (as glucose)	2·3	0·5	2·1	3·9
Methyl pentose (as rhamnose	1·09	0·75	0·0	0·0
Dipicolinic acid	9·4	12·3	17·5	26·4
Fraction (% whole spores)	52	48	34	40
Sugars:				
Glucose	+ +	+ +	+ + +	+ + +
Galactose	+	+	+ + +	+
Mannose	−	−	−	+ + + +
Rhamnose	+ + +	+ +	−	−
Ribose	+ +	+ +	+ + +	+ +

The soluble extracts were obtained as a supernatant after disruption of the spores and centrifugation at 10,000 *g* for 15 min.

major location of enzymes, ribosomes, RNA and DNA, and certain membranous elements that derive from the vegetative cell during septation or are synthesized in the spore protoplast before it is transformed into the dormant state (Fig. 5d; Black and Arredondo, 1966). Certain enzymes, however, are likely to be located externally to the protoplast.

a. *Soluble fraction*. The soluble fraction, after hydrolysis, from four species contained about 9% total nitrogen, 4–6% α-amino nitrogen, 1–2% phosphorus, 6–7% sugar and 10–26% DPA (Table 7). The content of DPA was greater in the more heat-resistant species, but this resulted mainly from a decrease in the soluble fraction from the more resistant spores, i.e. a similar quantity of DPA was extracted in a smaller soluble fraction. The soluble fraction also contained hexosamine derived from lytic breakdown of the cortex during disruption. The mesophilic species contained more hexosamine in the soluble fraction, as the lytic enzymes of these species are apparently more active at 1° than those from the thermophiles. The amino acid compositions were in general similar to

TABLE 8. Enzymic Activities of Vegetative Cells and Spores of *Bacillus subtilis*

Enzyme	Enzyme activity (moles substrate reacted/hr./mg. protein)		Reference
	Vegetative cell	Spore	
RNA polymerase	—	—	Balassa (1963c)
Ribonuclease	—	—	Balassa (1963a)
Proteolytic enzyme	—	—	Bishop and Doi (1966) Benassi and Bonanni (1965)
DNA polymerase	0·08	0·05	Falaschi *et al.* (1965)
Adenylate kinase	21·0	12·9	Falaschi *et al.* (1965)
Deoxyadenylate kinase	8·4	8·2	Falaschi *et al.* (1965)
Deoxycytidylate kinase	0·3	1·7	Falaschi *et al.* (1965)
Deoxythymidylate kinase	< 0·2	0·4	Falaschi *et al.* (1965)
Deoxyguanylate kinase	< 0·2	0·4	Falaschi *et al.* (1965)
Adenosine triphosphatase	—	1·0	Falaschi *et al.* (1965)
Deoxythymidine kinase	—	0·3	Falaschi *et al.* (1965)

those for the coats (Table 4), except that no cystine was detected. Glucose, galactose and ribose were the main sugars; in addition, rhamnose occurred in *B. cereus* and *B. subtilis* and mannose in *B. stearothermophilus*.

b. *Enzymes.* Numerous active enzymes have been isolated from disrupted spores, some in greater amounts than in vegetative cells (Halvorson, 1962). More recently, other important enzymes have been shown to be present (Table 8). Whether these occur in an active form in the intact resting spore is difficult to assess, as most disruption and extraction methods do not rule out changes due to mechanical germination (Rode and Foster, 1960a) which could activate the enzymes. The properties and heat stability of the extracted enzymes in a soluble state are in most cases similar to those of the vegetative cells with the following exceptions: alanine racemase (Stewart and Halvorson, 1954); catalase (Baillie and

TABLE 9. α-Amino Nitrogen Contents of Alcoholic Extracts of *Bacillus subtilis* (data from Pfennig, 1957)

	Vegetative cells		Spores	
	Before hydrolysis	After hydrolysis	Before hydrolysis	After hydrolysis
Amount (% dry wt.)	0·54	1·06	0·12	0·57
Number of amino acids detected	11	12	9	12

Norris, 1962, 1963); see also Halvorson (1965) and Halvorson *et al.* (1966). Thus, although certain enzymes bound to insoluble debris show greater heat resistance and differences in properties, the majority appear identical to those in the vegetative cells.

c. *Free solutes.* The content of unbound solutes provides valuable information about the chemical state of the resting spore. Free amino acids account for about 1% of the spore dry weight (Tables 9, 10). In *B. subtilis*, the amount of free amino acids in the unhydrolysed extract of spores was one-fifth of that in vegetative cells, but was half as much after hydrolysis, suggesting that there is more nitrogenous material in a peptide or polypeptide form in the unhydrolysed extract. The three additional amino acids detected were tyrosine, phenylalanine and threonine. The same twelve amino acids were present in the hydrolysed

TABLE 10. Free Amino Acids in Spores of *Bacillus megaterium* (data from Lee and Ordal, 1963)

	Content (mmoles/100 g. dry wt.)		
Amino acid	Boiled in water for 90 min.	Broken in 50% ethanol	Stored at 3° for 40 days, then boiled
Glutamic acid	7	5	5
Threonine		0·2	
Leucine, Isoleucine	0	0	3
Alanine	0	0	1·5
Glycine	0	0	1·5
Histidine, Arginine	3	3	5
Lysine	0	0·5	1·5
Total (% dry wt.)	1·53	1·31	2·41

extracts. Young (1958) extracted ten amino acids from *B. megaterium* spores, five of which were associated with DPA (see p. 180). Lee and Ordal (1963) found about 1% free amino acids, mostly glutamic acid, using two extraction procedures; additional amino acids became free during cold storage of intact spores (Table 10). Farkas and Kiss (1965) observed only two faint ninhydrin-positive spots on the chromatogram of a hot-water extract from *B. cereus*; X-irradiation (1000 Krad.) released twenty-one amino acids with the loss of only about 12% of the DPA.

Although disruption must allow some lytic action to occur unless the lytic system is inhibited, and hot extractions probably cause some hydrolysis, the above results suggest that free amino acids are present in only small amounts. An estimate of the free amino acids and other substances present in extracts from disrupted spores that had not been subjected to a heat treatment or hydrolysis of the extracts would be informative.

d. *DNA*. The amount of DNA in the spore is about 1% of the dry wt. as in vegetative cells (Table 11). On cytological evidence only about a quarter of the nuclear material in the axial cord at stage I enters the forespore (Ryter, 1965). The final amount of DNA-phosphorus in the spore compared with the DNA-phosphorus in the vegetative cell is somewhat less than half (Table 11). Fitz-James (1955, 1957) and Young and Fitz-James (1959a, b), however, showed that the amount in vegetative cells varied from about 1·5 to 3 times that of the average amount in spores, depending on the state of division, but that the amount per spore remained remarkably constant in each species. This was true for one species grown on two different media, and when the aeration was varied (Fitz-James and Young, 1959). The initial amount in the spore at nuclear division is not indicated by the above data.

The amount of DNA-phosphorus/spore in ten species (fourteen strains) varied four-fold in multiples of two, three and four times the minimum level, with *B. thuringiensis* being an exception. All strains appeared to contain one nucleoid, and the DNA content was reflected in the size of the nucleoid (Fitz-James and Young, 1959).

In *B. subtilis*, the helix-coil transition temperature was 88·1° indicating a base composition of 42·3 mole % guanine + cytosine, similar to that of vegetative cells; the denatured DNA showed renaturation and hyperchromic effects as expected of a double-stranded molecule (Mandel and Rowley, 1963). Masui *et al.* (1960), however, reported a difference in the guanylic acid + cytidylic acid content of spores and vegetative cells of *B. subtilis* (60·8 and 57·2%, respectively). Wake (1963) and Dennis and Wake (1966) estimated from the spore DNA content ($5{\cdot}0 \times 10^{-15}$ g.) that the *B. subtilis* genome had an upper limit of $3{\cdot}0 \times 10^{9}$ daltons (1700 μ of the B form). Autoradiographic determinations of the vegetative cell

TABLE 11. Nucleic Acid Composition of Spores and Vegetative Cells of Bacilli

Species	Amount (% dry wt.)							
	Vegetative cells			Spores				
	RNA	DNA	DNA/RNA ratio	RNA	DNA	DNA/RNA ratio	DNA-P/cell / DNA-P/spore	Reference
Bacillus megaterium	3·15	1·2	0·38	2·7	1·3	0·48		a
	9·35	0·50	0·05	4·57	0·89	0·20		b
Bacillus cereus	10·5	1·12	0·11	5·42	1·0	0·18		b
	14·3	1·0	0·07	5·0	1·0	0·20		c
	Nucleic acid phosphorus (10^{-16} g./cell)							
Bacillus megaterium	75·7	13·4	0·18	7·09	4·05	0·57	3·3	
Bacillus cereus	52·6	12·9	0·25	7·07	5·43	0·77	2·4	d
		16·4			9·25		1·8	b
		23·3			10·5		2·2[i]	e
Bacillus cereus var. *alesti*		39·5			20·8		1·9[i]	f
Bacillus species[h]						0·14–0·52		g

[a] Tinelli (1955a).
[b] Fitz-James (1955); average figures Tables 2 and 6.
[c] Stuy (1958); four strains.
[d] Hodson and Beck (1960).
[e] Young and Fitz-James (1959a).
[f] Young and Fitz-James (1959b).
[g] Fitz-James and Young (1959).
[h] Twenty-two strains of ten species.
[i] At time t_0.

chromosome length gave a value of 800–900 μ, suggesting that two chromosomes may occur in the spore. However, considering the fragility of the bacterial chromosome, they did not rule out the possibility of a longer chromosome only one of which occurred in the spore. From the number of γ-radiation hits required to inactivate spores of *Cl. botulinum* type A, Grecz *et al.* (1966) calculated about 10^7 nucleotides per spore nucleoid with a length of 1800 μ, occupying 0·5–1·0% of the spore volume. This agrees well with the chemical estimations above.

From their cytological and chemical studies on acid-treated spores, Fitz-James *et al.* (1954) suggested that some histone and a labile phosphorus-containing material were associated with the spore chromatin.

e. *RNA and microsomal particles.* RNA accounts for a considerable portion of the spore protoplast, and contains 50% of the total phosphorus of spores of *B. cereus* (25% in the high residue P strain of *B. megaterium*; Fitz-James, 1955). The nucleic acid content is lower, but the DNA/RNA ratio is higher in spores than in vegetative cells (Table 11). The smaller amount of nucleic acid in spores is due chiefly to the lower RNA content, but this does not necessarily mean that the concentration of RNA in the cytoplasm is lower than in vegetative cells; it probably reflects the smaller amount of cytoplasm relative to non-cytoplasmic material in the spore. In fact, as the protoplast accounts for only about a third or less of the spore volume, both nucleic acids must be present in much greater concentrations in the spore than in the vegetative cells. Fitz-James and Young (1959) observed that (i) the amount of RNA-phosphorus varied with spore volume, larger spores containing up to twice the amount of smaller ones, (ii) the media affected the spore size and the RNA-phosphorus content, (iii) the amount of RNA-phosphorus varied about eightfold in different *Bacillus* species, ranging from 12×10^{-16} g. RNA-phosphorus/spore in *B. apiarius* spores to 100×10^{-16} g./spore in *B. medusa* and *B. anthracis*, and (iv) the DNA-phosphorus/RNA-phosphorus ratio ranged from 0·14 to 0·5 (Table 11).

Most of the RNA in spores occurs in the form of r-RNA (65%) and t-RNA (35%) whereas, in vegetative cells, only half of this amount of t-RNA is found, and all the r-RNA occurs in the ribosomes (Balassa, 1966a). No m-RNA has been found in the mature (resting) spores (Woese *et al.*, 1960; Doi and Igarashi, 1964b). Balassa (1963a, 1966a) reported a polydisperse fraction with some properties of m-RNA or precursor r-RNA, but he now agrees that the functional test indicates that m-RNA is absent (Balassa and Contesse, 1965). Doi and Igarashi (1964b) initially found 8–10S particles as did Balassa (1963a), using 0·01 M-$MgCl_2$ during extraction, but not with 0·05 M-$MgCl_2$. These particles had an r-RNA base ratio, not that of m-RNA.

Woese *et al.* (1960) found only 68S and 50S particles in about equal quantities (about 4 mg./g. dry wt.) in the spore, whereas all the microsomal particles appeared during the initiation of germination and early outgrowth (Woese, 1961). The 30S and 100S particles, if present at all, were present at less than 0·04 mg./g. dry wt. The Fig. 2 of Woese *et al.* (1960) suggests that a non-microsomal soluble fraction, sedimenting with a coefficient less than 10S, amounted to as much as 1·2% of the dry weight. A major part of the spore-RNA, however, remained in the crushed spore debris; this was not released during germination and may be present as membrane-bound microsomes. Thus only a small part of the RNA of the spore and the early germinating form seemed to exist as microsomal RNA.

TABLE 12. Base Ratios of *Bacillus subtilis* Spore and Vegetative Cell RNAs (from Doi and Igarashi, 1964b)

Type of RNA	Sedimentation coefficient (S)	Base composition (moles %) cytidylate	adenylate	uridylate	guanylate	GC content (%)	Purine/Pyrimidine
Spore	4	29·4	18·9	18·2	33·5	63	1·10
	16	24·0	24·4	20·6	30·9	55	1·24
	23	22·6	26·2	19·8	31·4	54	1·36
Vegetative cell	4	28·5	19·1	19·4	32·9	61	1·09
	16	23·4	24·8	21·1	30·8	54	1·25
	23	22·8	26·4	20·7	30·0	53	1·30

Analyses were accurate to ±1·0%. The GC composition of *Bacillus subtilis* DNA is 43% (Marmur *et al.*, 1963).

The base ratios of spore RNA are similar to those of the vegetative cell RNA (Table 12); the base composition of the t-RNA appeared to be that of pure 4S suggesting that the t-RNA did not arise as a breakdown product (Doi and Igarashi, 1964b). The ratio of t-RNA/r-RNA is 30% as compared to 15% in vegetative cells (Doi and Igarashi, 1964b). Doi and Igarashi (1964b) suggest that this is related to the differences in biosynthetic activity.

The ribosomal state of *B. subtilis* spores was recently re-examined using [^{32}P]labelled spores, and employing vegetative cells to displace the spore particles adsorbed to the alumina used for grinding the cells (Bishop and Doi, 1966). Ribonuclease and protease activity were present in the spore extracts and were capable of degrading ribosomes and r-RNA. This suggested that previous figures for ribosome particles in spores are low not only because of such factors as incomplete spore breakage and failure to disrupt membranes but also because of enzymic degradation. Protoplast membranes and mesosomes are certainly likely to remain with the particulate debris (Warth *et al.*, 1963a). Bishop and Doi (1966) observed 70S, 50S and 30S particles. The 70S and 50S particles contained 40% protein and 60% RNA, as in vegetative cells; fractionation of the RNA from the 70S particles indicated the presence of 16S and 23S ribosomal components with a base composition essentially similar to that for vegetative cells (Table 12).

The resting spore of *B. subtilis* thus contains all the RNA fractions with the exception of m-RNA. Whether this applies to all species of spores has yet to be determined.

B. Chemical Fractions

1. *Lipid*

The many membranous structures in spores suggest that phospholipids are of considerable importance, but knowledge of the type and distribution of lipid substances in the spore is not very satisfactory. Analyses for lipid in spores vary from less than 1 to as much as 13% dry weight in *Bacillus*, and to 38% in *Clostridium* spores (Table 13). Surface contamination is of major importance in the preparation of spores for analysis (Long and Williams, 1960), and better extraction is obtained with disrupted spores (Fitz-James, 1955; Yoneda and Kondo, 1959). Many extraction procedures with intact spores have little or no effect on viability or heat resistance (Church *et al.*, 1956; Sugiyama, 1951). Poly-β-hydroxybutyrate (PHB), an important reserve material in many organisms at a stage preceding or during sporulation (see p. 202), has been reported by several workers not to occur in the mature spore (Table 13), but Yoneda and Kondo (1959) report as much as 4·1% PHB in spores of *B. subtilis*; much of this occurred in the coat and exosporium fraction. The lipid content of integument preparations is normally low (Table 3).

The lipid in *B. cereus* T and *B. polymyxa* spores contained mainly C_2–C_8 fatty acids, with large amounts of acetic acid, less butyric and propionic acids, and some non-saponifiable waxy residue with a low melting point. Small amounts of C_{14}, C_{16}, C_{18} and C_{20} acids were present in *B. polymyxa* spore lipid (Church *et al.*, 1956). Church *et al.* (1956) also made the interesting observation that "fattened" spores could be grown in the presence of 5% glycerol. When grown under these conditions the spores of *B. cereus* T and *B. polymyxa* increased in volume five- and eight-fold, and the lipid content 9 and 8·6 times, respectively. Spores of another strain of *B. cereus* and of *B. subtilis* var. *niger* showed no increase in size; their lipid contents were not reported. One wonders how much

TABLE 13. Lipid Contents of Spores

Species	Type of lipid	Amount (% dry wt.) Spores	Amount (% dry wt.) Insoluble fraction (integuments)	References
Bacillus megaterium	β-Hydroxybutyrate	0		Tinelli (1955a); Grelet (1957); Slepecky and Law (1961)
Bacillus cereus	C_2–C_8 Acids	4		Church *et al.* (1956)
	Acid hydrolysate		0·9	Strange and Dark (1956)
Disrupted spores	β-Hydroxybutyrate	4·1	5·3	Yoneda and Kondo (1959)
Disrupted spores	Lipid phosphorus	0·012		Fitz-James *et al.* (1954)
	Lipid phosphorus in acid hydrolysate	0·03		
Bacillus subtilis	Acid hydrolysate		1·1	Strange and Dark (1956);
	Acid hydrolysate		3·0 (1·4 before hydrolysis)	Salton and Marshall (1959)
Bacillus polymyxa	C_2–C_8 Acids	8·0		Church *et al.* (1956)
Bacillus licheniformis			about 2·0	Bernlohr and Sievert (1962)
Bacillus stearothermophilus	Surface	6·9–13·1		Long and Williams (1960)
	Total constituent	1·3–1·9		
	Phospholipid (22·5% of total constituent lipid)	0·3–0·4		
Clostridium perfringens	C_{12}–C_{22} Acids (behenic, stearic, palmitic, linoleic, myristic and lauric acids)	38·0		Meisel-Mikolajczyk (1965)

surface lipids contributed to the "fattening" process, although the spores were washed with water and methanol before extraction.

Matches *et al.* (1964) identified the following phospholipids in both spores and vegetative cells of *B. polymyxa*: phosphatidylethanolamine, lysophosphatidylethanolamine, lysophosphatidylserine, lysolecithin, phosphatidic acid, phosphatidylglycerol. Phosphatidylethanolamine was the major fraction in both. Quantitative estimations were not made.

2. *Phosphorus*

Phosphorus has many important functions in cells and, in addition to its commoner metabolic functions, it may be of considerable importance in the laminated layers of the spore integuments. It occurs in all fractions

TABLE 14. Phosphorus Contents of Spores and Spore Fractions

Fraction	Phosphorus content (% dry wt. of fraction)	Reference
(*a*) *Morphological*		
Exosporium	2	Matz and Gerhardt (1964)
Coats	0·3–2·8	Table 3 (p. 147)
Insoluble cortical material	0·04	Warth (1965); Strange and Powell (1954)
Whole spores (12 species)	0·2–2·7	Curran *et al.* (1943)

(*b*) *Chemical* (Fitz-James, 1955; data from Table 7)

	Phosphorus content			
	(% total P fraction)		(% spore dry wt.)	
	Bacillus cereus	*Bacillus megaterium*	*Bacillus cereus*	*Bacillus megaterium*
Cold acid-soluble	6	4	0·05	0·07
Lipid phosphorus	7	3	0·06	0·06
Hot 5% TCA-soluble				
Total phosphorus			0·57	0·51
RNA-phosphorus	50	20	0·45	0·35
DNA-phosphorus	10	4	0·09	0·07
Labile phosphorus (7 min. in N-HCl at 100°)	9	—	0·08	—
TCA residue (empty coats, lysozyme-resistant)				
Total phosphorus	22	60	0·20	1·1
Labile phosphorus	14	—	0·12	—
Sum total phosphorus			0·9	1·7

of the spore (Table 14), and its distribution shows considerable variation in the different species. The lipid and cold acid-soluble phosphorus (0·2 N-$HClO_4$ or 10% trichloroacetic acid (TCA), 0–2°) contents could be properly estimated only with disrupted spores. Chemical and cytological studies with acid-treated spores confirmed that most of the RNA-phosphorus and DNA-phosphorus was located in the protoplast (Fitz-James *et al.*, 1954; Mayall and Robinow, 1957; Robinow, 1960). After removal of the nucleic acid-phosphorus with hot 5% TCA, some phosphorus-containing compounds remained. The larger part of this was alkali-soluble and acid-labile, and resembled polymetaphosphate. The smaller part was stable to ribonuclease and was acid- and alkali-insoluble; this was associated with the coats. In some strains of *B. megaterium* spores, with twice the phosphorus content of *B. cereus* spores, the acid- and alkali-insoluble residual phosphorus comprised 60% of the total phosphorus of the spore, whereas this was only 4% of the total in *B. cereus* (Fitz-James, 1955). In twenty-two strains (ten species and subspecies) of *Bacillus*, the amount of this fraction varied 90-fold (Fitz-James and Young, 1959).

C. Inorganic Composition

1. *Metal and Dipicolinic Acid Composition*

Spores accumulate many metals, and these metals play important functions in the formation and final properties of the spores. Ca^{2+} is particularly important, both quantitatively and qualitatively. Any excess of one metal affects the content of others, and this is especially true of Ca^{2+} and other divalent metals (Slepecky and Foster, 1959). Further, any treatment that affects the Ca^{2+} content usually affects DPA content and heat resistance (Black *et al.*, 1960). Similarly, in reverse, any condition affecting the DPA content will probably affect or perhaps determine the Ca^{2+} content, the contents of other metals, and the heat resistance.

Normally, spores of most species contain 2–3% Ca^{2+} and 5–15% dry weight DPA in about a 1:1 mole ratio. In many of the experiments in which the DPA content was varied by organic inhibitors, the Ca^{2+} and other metal contents were not determined (Table 18). However, the general instability (readiness to initiate germination) of spores low in DPA makes it unlikely that Ca^{2+} is retained. It is, however, likely that, in the presence of certain compounds such as penicillin and cycloserine (p. 221), when DPA formation is probably normal although the acid is lost because the spores are unstable, Ca^{2+} uptake is also normal. When the Ca^{2+} content of the medium is decreased to such an extent that the DPA content is affected, there is a direct relation between Ca^{2+} availability and the Ca^{2+} and DPA contents (Figs. 9a and b). With these points in mind, the inorganic composition of the spore will be examined and then the Ca-DPA status of the spore discussed.

2. *Inorganic Constituents*

Spores produced in most media with a reasonable metal balance (Table 1; Murrell, 1961b) contain a wide variety of inorganic elements (Curran *et al.*, 1943; Table 15). The major elements are Ca, K, P, Mg, Mn and occasionally Cu and Si in some crops. The last of these elements probably results from added antifoam or from the glassware. In addition, a large number of minor or trace elements occur in spores (Table 15).

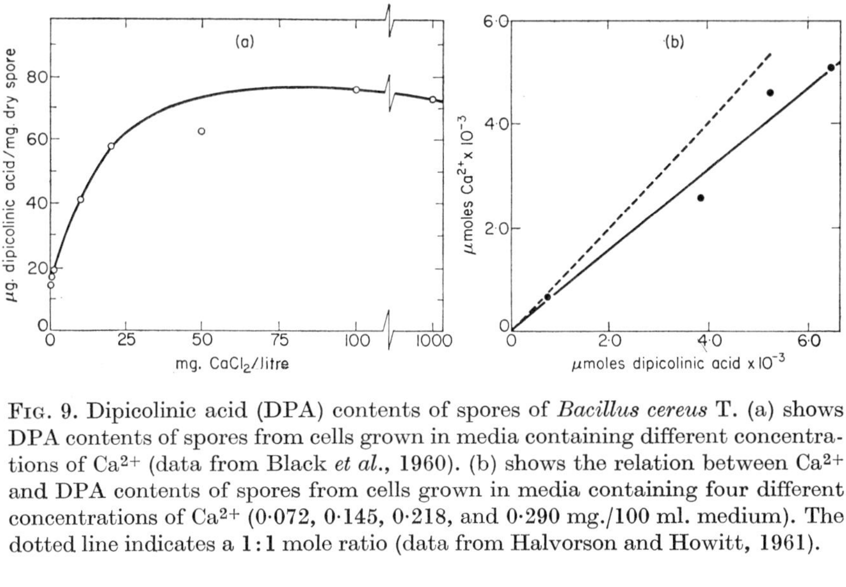

FIG. 9. Dipicolinic acid (DPA) contents of spores of *Bacillus cereus* T. (a) shows DPA contents of spores from cells grown in media containing different concentrations of Ca^{2+} (data from Black *et al.*, 1960). (b) shows the relation between Ca^{2+} and DPA contents of spores from cells grown in media containing four different concentrations of Ca^{2+} (0·072, 0·145, 0·218, and 0·290 mg./100 ml. medium). The dotted line indicates a 1:1 mole ratio (data from Halvorson and Howitt, 1961).

Copper and some of the others are probably of functional importance; the remainder are probably simply adsorbed by the integuments of the spore. Copper and manganese occurred in greater amounts in the spore than in vegetative cells (Curran *et al.*, 1943). Aluminium, copper and iron were usually present at higher concentrations in the spore than in the medium, indicating some degree of accumulation.

Spores are particularly rich in Ca^{2+}, and the more heat-resistant species usually contain more Ca^{2+} (Walker *et al.*, 1961; Murrell and Warth, 1965). Its rôle in spore development is discussed later.

The potassium content varied considerably; the two most heat-resistant species had low potassium contents, while four others had exceptionally high contents (W. G. Murrell, unpublished observations).

The two species (*B. apiarius* and *Bacillus* strain 636) with most potassium had additional ridge-like integuments, and strain 636 also contained taurine and additional glutamic acid and hexosamine, possibly in these layers (Warth *et al.*, 1963b). The Mg^{2+} content and the Mg/Ca ratio (Fig. 29) decreased significantly with increasing heat resistance (Walker *et al.*, 1961; Murrell and Warth, 1965).

TABLE 15. Contents of Inorganic Elements in Spores

	% dry wt.: data of				
	Murrell *et al.*[a]		Curran *et al.*, 1943[b]		Thomas, 1964[c]
	Range	Mean	Range	Mean	
Major elements					
Ca	1·2–4·48	2·16	1·0–2·4	1·78	2·4
K	0·03–1·47	0·44	0·1–0·9	0·34	3·4
Mg	0·10–0·41	0·20	0·3–1·0	0·52	0·26
Na	0·03–0·22	0·08	—		< 0·1
Mn	0·007–0·25	0·18	0·003–0·011	0·0065	1·5
Fe	0·03–0·6	0·12	0·008–0·038	0·017	0·02
Si	0·3–1·2	0·53	—		—
P	0·45–1·6	0·78	0·6–2·7	1·5	1·45
Al			0·009–0·066	0·026	
	μg./g. dry wt.				
Trace elements	Range	Mean			
Cu	1–9000	140 (1030)[d]		200	
Ba	30–450	120			
Cr	2–150	36			
Ti	< 10–400	55			
Ni	1–200	40			
B	< 2–70	10	1–20	9	
Sr	< 25–100	60			
Pb	6–500	120			
V	1–80	15			
Zr	< 6–100	22			
Sn	2·5–80	20			
W	< 20–100	30			
Mo	1–30	6			
Co	< 0·9–2·5	1			
Ag	0·2–1	0·5			
Ga	< 0·1–0·2	< 0·1			

[a] Twenty strains of sixteen species; Ca, Mg, K, Na, Zn were determined by atomic absorption spectroscopy, others by spectrography (Clarke and Swaine, 1962). Unpublished observations.

[b] Twelve species; spectrographic method.

[c] *Bacillus megaterium.*

[d] Average with two high values included.

The inorganic requirements for spore formation are qualitatively and quantitatively different from those for vegetative growth. K^+, Ca^{2+}, Mn^{2+}, Mg^{2+}, Fe^{3+}, Zn^{2+}, Cu^{2+}, Mo^{2+}, Co^{2+}, PO_4^{3-}, and SO_4^{2-} appear to be required for specific rôles in sporulation, for the production of sporangia able to sporulate satisfactorily, and for the production of fully heat-resistant spores (Curran, 1957; Murrell, 1961b; Halvorson, 1962; Kolodziej and Slepecky, 1964).

The possible rôles of the various inorganic ions has led to much speculation (Foster and Heiligman, 1949; Curran, 1957; Halvorson, 1962). The rôles of Ca^{2+}, PO_4^{3-} and Mg^{2+} are discussed elsewhere. The potassium ion was usually present in much lower amounts in spores than in vegetative cells, having a relationship in the two types of cells which is the reverse of that of Ca^{2+} (Curran *et al.*, 1943). This ion is necessary in high concentration for good spore formation (Foster and Heiligman, 1949) and is probably involved in similar functions to that in vegetative cells, namely in protein synthesis (Lubin and Ennis, 1965) and in the regulation of the ionic environment. It may also be involved in Ca^{2+} uptake.

Manganese ions are essential and probably have several functions in sporulation (Charney *et al.*, 1951; Benger, 1962; Weinberg, 1964; Weinberg and Tonnis, 1966). A number of lytic enzymes require, or are stimulated by, Co^{2+}, Mg^{2+}, Cu^{2+}, Ni^{2+} and Mn^{2+} (Strange and Dark, 1957a, b), and these enzymes degrade the spore cortex, enabling swelling and development of the germ cell during outgrowth with production of the germination exudate (Powell and Strange, 1953). Manganous ions are involved in the activation of a proteolytic enzyme and of a pyrophosphatase possibly involved in the initiation of germination (Levinson and Sevag, 1954; Levinson *et al.*, 1958; Murrell, 1961b). Benger (1962) pointed out that Mn^{2+} are required for activation of a number of TCA cycle and related enzymes (e.g. oxalosuccinate decarboxylase and malate dehydrogenase); an active TCA cycle is essential for sporogenesis in *B. cereus* (see p. 190). Benger (1962) found that Mn^{2+} were required for the oxidation of glycollic acid to glyoxylic acid which may interconnect with the TCA cycle and also provide C_2 compounds for DPA synthesis (see p. 218). When glyoxylate is provided, *B. megaterium* sporulates well (57%) without exogenous Mn^{2+}. Better sporulation (88%) was obtained when glutamate was provided as well. Essentially no sporulation occurred in the absence of Mn^{2+} and glyoxylate (Benger, 1962). An involvement in the activation of the TCA cycle and in DPA synthesis would account for the critical rôle of Mn^{2+} in sporulation. Deprivation of Mn^{2+} decreased the amount of sporulation and production of the associated protease activity in *B. megaterium*; the Mn^{2+} requirement could be replaced by an amino acid mixture added at about stage III (Gruft *et al.*, 1965).

3. *Variation of Metal Content*

The metal content of spores is affected greatly by the variety of preparative and cultural procedures used. Spores, or some of the peripheral layers, behave essentially as an ion-exchange system (Alderton and Snell, 1963), and so can lose and gain many metals during harvesting and cleaning procedures. Losses also occur from initiation of germination. Metal-containing precipitates derived from the medium are very difficult

TABLE 16. Metal Contents of Spores Formed in Media Containing, Respectively, Minimum and Maximum Concentrations of Individual Metals (data from Slepecky and Foster, 1959)

Metal ion	Metal ion concentration in medium (μg./ml.)	Ca^{2+} in in medium (μg./ml.)	Ca^{2+} in spores (% dry wt.)	Metal ion in spores (% dry wt.)
Ca^{2+}				
Min.	—	0·4	0·69	—
Max.	—	722	3·04	—
Zn^{2+}				
Min.	0·2	1·8	3·60	0·08
Max.	454	1·8	0·62	0·46
Ni^{2+}				
Min.	0	1·8	3·60	$<0·003$
Max.	11·2	1·8	0·62	1·81
Cu^{2+}				
Min.	0	1·8	0·65	$<0·0001$
Max.	0·16	1·8	0·77	0·01
Co^{2+}				
Min.	0	1·8	0·65	0·04
Max.	40	1·8	0·78	0·14
Mn^{2+}				
Min.	0·22	1·8	0·47	0·10
Max.	2·28	1·8	0·10	0·78

to remove from spores unless a procedure such as the two-phase system of Sacks and Alderton (1961) is used (Murrell and Warth, 1965). Acid-washing procedures may remove some precipitates (Fitz-James and Young, 1959; Slepecky and Foster, 1959); however, care needs to be exercised, especially with freeze-dried spores, as constituent Ca^{2+} may be removed (Murrell and Warth, 1965). Some of the "acid-released" Ca^{2+} and PO_4^{3-} of spores in crops grown in the presence of high concentrations of Ca^{2+}, phosphate and amino acids (Levinson and Hyatt, 1964) was very likely derived from inorganic precipitates in the crops; the very high Ca/DPA ratios (2–5) emphasize this possibility.

The metal content of spores can be varied to a large extent by altering the availability of the metals during growth, the temperature of growth, and chemical treatments such as base exchange. Slepecky and Foster (1959) grew apparently normal spore crops when the metal concentrations were varied by as much as 2000-fold (Ca^{2+} and Zn^{2+}), although these concentrations resulted in 3–600-fold variations in metal content (Table 16). More metal was incorporated the higher the concentration in the medium. A significant concentration mechanism was involved in the uptake of Ca^{2+}, Ni^{2+}, Zn^{2+}, and Mn^{2+}. At high concentrations, Zn^{2+}, Ni^{2+} and Mn^{2+} competed for Ca^{2+} sites in the spore (Slepecky and Foster, 1959; Levinson and Hyatt, 1964). The increased content of the substituent metal was less than the decrease in the amount of Ca^{2+}, i.e. it was not a quantitative substitution (Table 16). The low content of Ca^{2+} following the Co^{2+}, Cu^{2+} and Mn^{2+} treatments resulted possibly from interference by other metals in the medium (Mn^{2+}, Fe^{2+}, Zn^{2+}; Slepecky and Foster, 1959); these metals were not assayed in these crops. A Ni^{2+} content of 1·8% was attained when the concentration of Ni^{2+} in the medium (toxic limit) was 1/95 that of Ca^{2+} in the calcium treatment, indicating a much stronger adsorption of Ni^{2+} than Ca^{2+}.

Spores of *B. subtilis* and *B. coagulans*, formed in cultures grown at higher temperatures, contained more calcium, magnesium and manganese (Lechowich and Ordal, 1962). Spores of *B. cereus* T, formed at various temperatures after growth at 30° to stage II, had a calcium content which increased with the temperature from 15° to 37°; at 41°, a super-optimum temperature, the content decreased even though the DPA content was higher (Murrell and Warth, 1965).

Alderton and Snell (1963) studied the base exchange system of *B. megaterium* spores by adjusting spores to the H^+-form at pH 4 (no DPA leakage occurred at this pH value) and adding back a variety of bases. These changes affected the heat resistance considerably; Ca^{2+}, K^+ and Na^+ resulted in the highest pH values and the most resistant spores (it should be noted that the spores were heated in an unbuffered system). Data on the metal content of "stripped" (pH 4) spores and the base-adjusted spores were not given. Acid-washing treatments (see p. 171) suggest that titration to pH 4 removes only readily exchangeable metals in the spore, possibly in the coats. Further, if Ca^{2+} and DPA exist in the chelate form in the spore, this chelate is not fully dissociated until a pH of about 1 is attained (Tichane and Bennett, 1957). At this pH value, explosive disruption, the acid reaction of spores, occurs with loss of Ca^{2+} and DPA (Robinow, 1953; Mayall and Robinow, 1957; Fitz-James *et al.*, 1954). Presumably, at pH 4, there is insufficient change in the permeability of coats to allow any dissociated DPA to leach out; and there is insufficient dissociation to disrupt the spore.

4. *Location of Metals*

The location of Ca^{2+} in spores is particularly important; its site probably is also the site of DPA, and their location must be explained or fit in with any mechanism for the resting state of the spore and heat resistance. Several approaches have yielded interesting results on metal distribution,

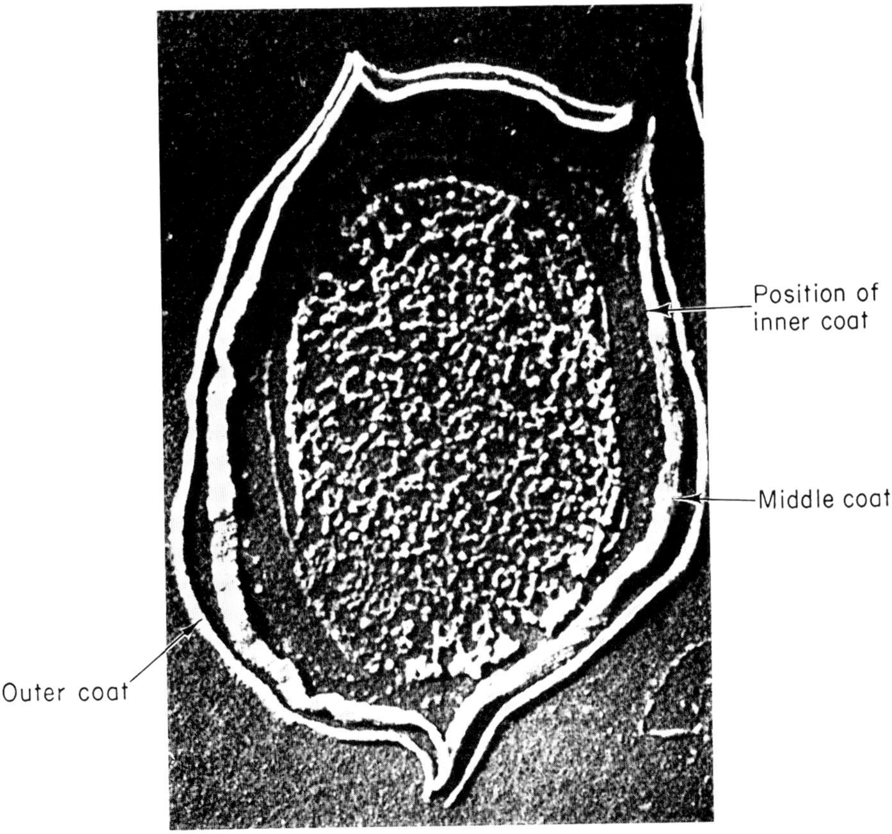

FIG. 10. Electron micrograph of a thin-sectioned *Bacillus megaterium* spore preparation, from which the methacrylate was removed, and which was then incinerated at low temperature by electrically excited oxygen. The preparation was then shadowed with uranium. × 50,000. Where the ashed middle layer of the coat has collapsed onto the support film, it shows the 100 Å fibrillar fine structure. Otherwise this layer and the ashed outer layer stand neatly erect off the support film. Note also that the ash of the protoplast has a granulated, reticular texture (from Thomas, 1964).

but the interpretation is difficult. The spodography experiments (micro-incineration for 30 min. at 500–525°) of Knaysi (1965), using the light microscope, suggest that the mineral ash is concentrated in the protoplast. Thomas (1964) studied low temperature-ashed sections; the electron micrographs show clearly the distribution of ash in different spore layers (Fig. 10). A considerable amount of ash occurs in the protoplast region and in the two coat layers, described by Thomas as the middle and outer coats; these appear to correspond to the inner and outer coats described elsewhere (Robinow, 1960; Ohye and Murrell, 1962). The inner coat layer and cortex are missing as a result of formalin fixation and other preparative procedures. How much of the mineral matter is lost with the cortex, and during sectioning onto water, is not revealed; the residue ash in Fig. 10 may perhaps represent only a minor part of the mineral content of the spore. If, on the other hand, it represents the major part then the density of the ash pattern and the relative proportions of the spore parts suggest that the major part is localized in the two coat layers. The ash pattern of the middle layer shows a remarkably fine structural pattern of minute fibrils lying parallel in regular arrays with a periodic spacing of 100 Å. The outer coat layer has a loosely knit meshwork pattern.

A large proportion of the ash revealed in this pattern may be due to phosphorus, particularly as some *B. megaterium* strains have a high content of coat-phosphorus (see p. 167). It is notable that *B. cereus*, which has a lower phosphorus content, had a less dense ash pattern (Thomas, 1964).

D. Dipicolinic Acid and Calcium

Dipicolinic acid has been found in spores of all species of *Bacillus* and *Clostridium* analysed and also in spores of *Sporosarcina ureae* (Thompson and Leadbetter, 1963), and always in about a 1:1 mole ratio with Ca^{2+}.

The DPA content of spores of different species ranges from about 5 to 15% with 1–3% Ca^{2+}. This is true of spores produced on media suitable for growing good crops of spores typical and characteristic of the species (Table 17). Although, in any one species, the DPA content can be varied widely, rather drastic conditions have to be applied to achieve this. Normal media usually produce spores with very similar composition and with similar properties and levels of heat resistance (Tinelli, 1955a; Benger, 1962; Hermier, 1964; Table 7 and 8, Murrell and Warth, 1965). In the data in Table 17, there is a significant relation between the Ca^{2+} content and heat resistance of the different species, but not between the DPA content and heat resistance; the more resistant spores, however, tended to have more DPA. Thus the DPA content or the Ca/DPA ratio, although important in the stability of the mature spore of each species, does not appear to be of direct importance in the 700-fold difference in heat resistance of this group of species.

Some of the conditions and factors that affect the DPA content of spores within a single species are listed in Table 18. Ca^{2+} availability has

TABLE 17. Heat Resistance and Dipicolinic Acid Contents of Spores of *Bacillus* Species (from Warth and Murrell, 1965)

Organism	D_{100} (min.)	DPA (% dry wt.)	Ca-DPA ratio
Bacillus cereus T	0·83	9·42	0·96
Bacillus 668	0·99	8·78	0·86
Bacillus 653	1·39	6·90	0·88
Bacillus subtilis var. *niger* (1)	1·67	7·35	1·07
Bacillus 652	2·00	7·14	0·95
Bacillus megaterium	2·10	8·80	0·76
Bacillus 645	2·38	9·34	0·89
Bacillus apiarius	5·00	5·06	1·02
Bacillus subtilis var. *niger* (2)	6·67	9·30	1·00
Bacillus 611	8·33	6·60	1·12
Bacillus cereus var. *mycoides*	10·00	7·36	1·28
Bacillus cereus	14·2	6·14	1·47
Bacillus licheniformis	24·1	6·99	1·16
Bacillus 636	35·2	7·06	1·12
Bacillus 669	43·8	11·28	0·94
Bacillus 670	68·5	5·74	3·25
Bacillus 671	81·3	9·25	1·14
Bacillus coagulans	270	10·42	0·94
Bacillus stearothermophilus (1)	459	9·77	0·86
Bacillus stearothermophilus (2)	714	13·55	0·96

The decimal reduction time (D_{100}) is the time required for a 90% decrease in the number of viable spores at 100°.

the greatest effect, and several of the other treatments may directly or indirectly interfere with Ca^{2+} availability, e.g. chelation by phenylalanine, precipitation by oxamic acid (formed by hydrolysis of ethyl oxamate). Cysteine and tetracyclines possibly interfere with Ca^{2+} adsorption sites (Vinter, 1957a, b, 1962a, b).

The Ca/DPA ratio is not always close to unity. Whether the divergence results from partial replacement of Ca^{2+} by other metals, analytical errors, or unknown factors is not known. A plot of all published analyses (Fig. 11) suggests that there does not exist a true 1 · 1 stoicheiometrical relationship between the two constituents, unless the spread of the points results from extraneous Ca^{2+} and assay errors. The spread seems too

TABLE 18. Factors Affecting the Dipicolinic Acid Contents of *Bacillus* Spores

Factor	Species	Effect	Reference
Nutritional			
Yeast extract (0–2 mg./ml.)	*Bacillus cereus* T	Dependent on concentration	Church and Halvorson (1959)
Pantothenic acid	*Bacillus cereus* T	Increases	
Thiamine	*Bacillus cereus* T	Increases	
Calcium (0–36 μg./ml.)	*Bacillus cereus* T	Determines content (see Fig. 9a)	Black *et al.* (1960)
Physical			
Temperature	*Bacillus subtilis*	Increase with temperature to near maximum growth temperature	Lechowich and Ordal (1962)
	Bacillus coagulans		
	Bacillus cereus T		Murrell and Warth (1965)
Stimulatory			
Glycollic acid	*Bacillus megaterium*	84% increase	Benger (1962)
Glyoxylic acid	*Bacillus megaterium*	12% increase	
Inhibitory			
Cystine	*Bacillus megaterium*	Interferes with Ca^{2+} incorporation	Vinter (1957a)
Phenylalanine	*Bacillus cereus* T	Decreases	Church and Halvorson (1959)
α-Picolinic acid	*Bacillus cereus* T	Decreases	Gollakota and Halvorson (1960)
Tetracyclines	*Bacillus cereus*	Inhibits Ca^{2+} incorporation	Vinter (1962a, b)
Ethyl oxamate, Ethyl malonate, Ethyl pyruvate, Ethyl pimelate, Ethyl succinate	*Bacillus cereus* T	Inhibits DPA formation indirectly	Benger (1962) Gollakota and Halvorson (1963); Kominek (1964)

great for this, and the lack of an exact relationship is not surprising if the specificity of Ca^{2+} is not absolute.

1. *Specificity of Calcium*

Refractile, non-staining spores can be grown in media with quite low concentrations of Ca^{2+} (1 μg./ml.); the spores have low DPA contents

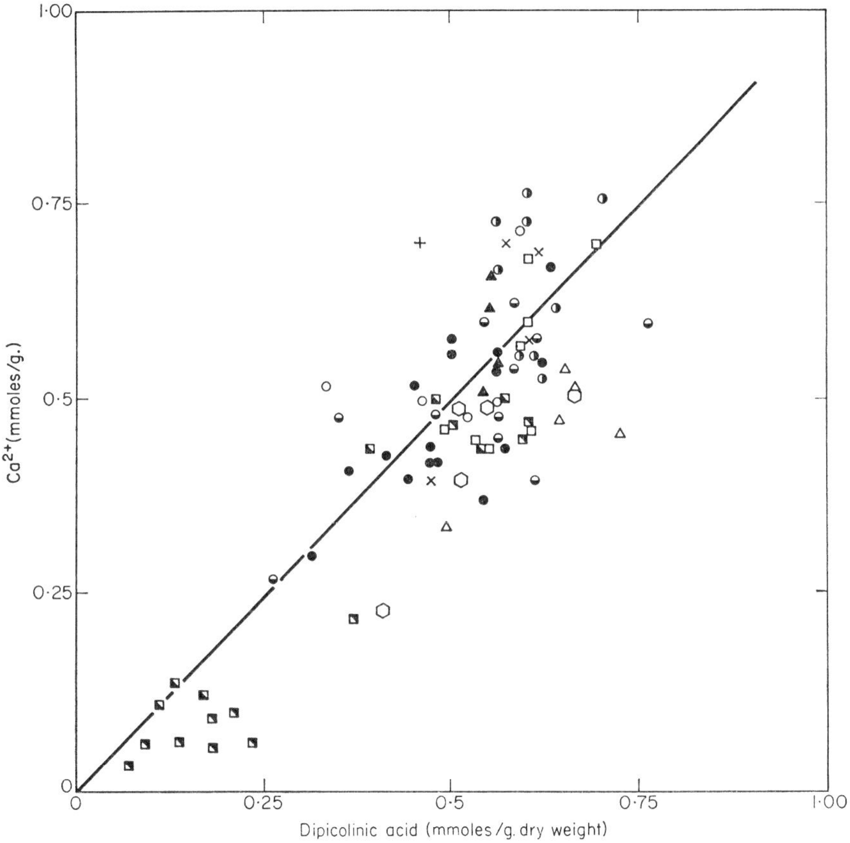

FIG. 11. Plot of Ca^{2+} and dipicolinic acid (DPA) determinations in the literature for *Bacillus* species. Line drawn at 45° slope. Data from El-Bisi *et al.*, 1962; Lechowich and Ordal, 1962; Pelcher *et al.*, 1963; Fleming and Ordal, 1964; Levinson *et al.*, 1961; Murrell and Warth, 1965; Walker and Matches, 1965.

and are unstable, becoming non-refractile and stainable during storage (Keynan *et al.*, 1962; Young and Fitz-James, 1962). Slepecky and Foster (1959) grew crops of spores in media containing 1·8 μg. Ca^{2+}/ml. in the presence of low and high concentrations of other ions (Table 16). At the high concentration, the Ca^{2+} content was decreased and the other ions (Zn^{2+}, Ni^{2+}, Mn^{2+}) accumulated in the spores instead of Ca^{2+}; but the replacement was not quantitative. The spores had less than 1% Ca^{2+}, but their DPA content was not determined. The spores were not stable to 10 min. at 60°, and so were probably low in Ca^{2+} and DPA. The high concentrations of Zn^{2+}, Ni^{2+} and Mn^{2+}, although apparently displacing Ca^{2+}, either were not as satisfactory as Ca^{2+} or were held at another site. On disruption of the spores, they were present in the extracts as the metal-DPA chelate (Slepecky, 1961).

TABLE 19. Replacement of Calcium by Other Metals in *Bacillus* Spores

Organism	Replacement metal (present as the chloride; m*M*)		DPA (% dry wt.)	Metal/DPA ratio	Heat resistance	Reference
Bacillus cereus T						
stage I[a]	Ca^{2+}	0·9	8·8		30	Black *et al.* (1960)
	Sr^{2+}	0·9	7·5		30	
			μg./spore (× 10^{-8})			
stage II[b]	Ca^{2+}	0·9	1·55			Halvorson and Howitt (1961)
	Sr^{2+}	0·6	1·60			
	Ba^{2+}	0·5	0·32	0·92		
			% dry wt.	Ca/DPA		
Bacillus cereus[a]						
	Ca^{2+}	0	3·0	0·50	< 5	Pelcher *et al.* (1963)
		0·1	6·2	0·59	5	
		1·0	9·9	0·96	60	
		10	8·4	0·93	60	
	Sr^{2+}	0·1	6·1		22	
	Ba^{2+}	0	2·9	0·71	< 5	
		0·1	1·9	0·94	8	
		1·0	2·2	1·00	23	
		1·0	3·0	0·33	29	
	Mg^{2+}	0·1	3·8	0·26	5	
	Zn^{2+}	0·1	1·5	0·67	< 10	
	Co^{2+}	0·1	1·1	0·43	< 5	
	Ni^{2+}	0·1	2·4	0·41	23	

[a] Cells grown to stage I and then placed in the metal-containing solutions.
[b] Spores formed in the glucose–yeast extract–salts medium of Stewart and Halvorson (1953) containing different metal ions.
Heat resistances of the *B. cereus* T spores are given as the percentage of survivors after 30 min. at 80°, and those of the *B. cereus* strain as the time (min.) for 99% death at 85°.

The results of replacement sporulation studies with Sr^{2+} and Ba^{2+} are summarized in Table 19. Sr^{2+} appears to be functionally equivalent to Ca^{2+}, whereas Ba^{2+} and the other divalent metals appear quantitatively and qualitatively less efficient. However, all these studies fail to indicate either the effect of the replacing metal on the DPA content, or the amount of Ca^{2+} and other metals in the spore. Is the Sr/DPA ratio 1, or are some of the other cations, e.g. Mg^{2+} and Mn^{2+}, replacing Sr^{2+} in the spore? If Ca^{2+} has two functions, namely in DPA biosynthesis and in the chelation of DPA, can Sr^{2+} replace Ca^{2+} equally well in both functions, and can Ba^{2+} and some of the other divalent metals participate in the second, perhaps less specific, function? Several of these questions are answered in the recent paper by Foerster and Foster (1966).

2. *Chemical State of DPA in the Spore*

DPA is apparently nearly always extracted from spores in the form of the Ca-DPA chelate (Powell and Strange, 1956), but sometimes as a chelate with other divalent metals (Slepecky, 1961) and perhaps as a DPA-Ca–amino acid complex. Evidence of a precursor form is lacking. Perry and Foster (1956) have isolated and identified monoethyl-DPA in spore extracts of *B. cereus* var. *mycoides* and *B. megaterium*; this ester accounted for only 1% of the DPA. Enough ethyl groups occurred, however, in spores to account for the possibility of the occurrence of DPA in the esterified form in the spore. The acid extraction method caused hydrolysis, and other mild extraction methods failed to give higher yields. Perry and Foster (1956) suggested, therefore, that DPA may exist in the spore in a bound form, and that conditions causing its release also cause hydrolysis of the ester linkage. Methyl dipicolinate has also been isolated in trace quantities from spores of *B. cereus* var. *globigii* (Hodson and Foster, 1965). Spectrometry studies suggest that DPA occurs in the chelate form. The infrared difference spectra between dry intact spores and germinated spores resembled those for DPA, but were too imprecise to differentiate DPA esters and chelates (Norris and Greenstreet, 1958). Electron paramagnetic resonance (EPR) spectra of spores produced on a Mn^{2+}-enriched medium strongly suggested the existence of a Mn-chelate in spores, while the spectra for Cu (II), on the other hand, indicated the presence of a Cu-protein-complex (Windle and Sacks, 1963). Ultraviolet spectra of dry spores embedded in potassium bromide were very suggestive of the presence of calcium and manganese chelates; other calcium and manganese complexes did not show a peak at 280 mμ. The spores were damaged during embedding, but water was "substantially" absent from the spores in the operations preceding u.v.-spectroscopy (Bailey *et al.*, 1964).

DPA also occurs in extracts in combination with amino acids (Young, 1958). Chromatograms of the hot- and cold-water extracts gave a DPA

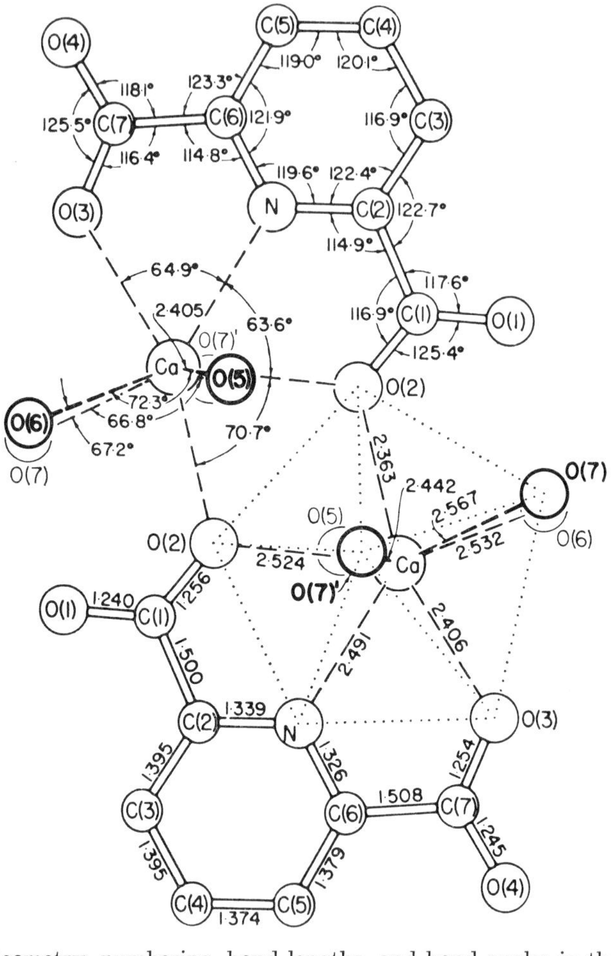

FIG. 12. Geometry, numbering, bond lengths, and bond angles in the Ca-DPA-$3H_2O$ dimer. Hydrogen atoms are omitted for clarity. (From Strahs and Dickerson (1965).

spot with considerable tailing. This spot contained DPA and six amino acids, tentatively identified as α,ϵ-diaminopimelate, glutamate, tyrosine, valine and isoleucine. The difference in tailing between the above combination and a synthetic mixture suggested a difference in the degree of polymerization. Ca^{2+} was required for stability on the paper. Young, therefore, suggested that DPA occurred in spores as a DPA-amino acid or -peptide complex with calcium, from which amino acids were released by hydrolysis or Ca^{2+} removal. The amino groups of the amino acids, whilst combined with Ca^{2+} and DPA, were masked (ninhydrin-negative) suggesting a combination in which DPA provided protective stabilization. Lund (1961) reported a greater solubility of spore Ca-DPA than synthetic Ca-DPA, suggesting that the spore Ca-DPA was in a form other than the simple chelate.

The order of chelate stability is Cu > Ni > Zn > Co > Cd > Ca > Mn > Sr > Ba > Mg, the stability constant ($\log K$) decreasing from 10 for Cu to 4·4 for Ca to 2·4 for Mg (Riemann, 1963; Fleming, 1964). The $\log K_{MA}$ values for the Ca-DPA chelate were decreased in the presence of higher ammonium acetate concentrations, by increasing the temperature, and by decreasing the pH value (Fleming, 1964). Strahs and Dickerson (1965) crystallized Ca-DPA in two forms, as a sesquihydrate and a trihydrate. The X-ray diffraction pattern of the trihydrate indicated a dimer of planar units related by a centre of symmetry half-way between the calcium atoms (Fig. 12). They considered that the dimer could be quite stable because the Ca—O (2′) bond length was the shortest Ca—O bond in the structure. The crystal is held together by several intermolecular hydrogen bonds. Oxygen atoms 1 and 4 both accept two hydrogen bonds from water molecules, while O (3) accepts one hydrogen bond. The polymeric Ca—O chains and the extensive hydrogen bonding may contribute to spore stability (Strahs and Dickerson, 1965).

Treatments which bring about the release of Ca-DPA are varied and include moist heat (Table 20), germination (Powell, 1953, 1957), irradiation (Rode and Foster, 1960b; Falcone and Cavallo, 1958; Farkas and Kiss, 1965), sonication (Berger and Marr, 1960), mechanical breakage (Young, 1958; Rode and Foster, 1960a, b), 80% acetone, methanol and ethanol at 56°, phenol, hydrogen peroxide and acid (Rode and Foster, 1960b), 1% thioglycollic acid in 8 *M*-urea with lysozyme (Gould and Hitchins, 1963a, b), long-chain alkylamines, and a variety of cationic surface-active compounds (Rode and Foster, 1960b, c, d). Most of these reagents probably do little to break bonds, and it is more likely that they bring about permeability changes or disrupt coats, thereby allowing leakage of DPA. The action of moist heat is complex; high temperatures increase the rate of release; the amount released by a given heat treatment is related to the spore's heat resistance (Table 20; Lund, 1958); release always lags behind loss in viability; and excess time of heating may decrease the rate of release (Lund, 1959). The last effect may result from precipitation reactions within the spore. Disruption of spores in cold water (2–4°) releases all of the DPA (at least from *B. megaterium*; Young, 1958); this indicates that the DPA is probably not peptide-linked to spore matter unless mechanical germination (Rode and Foster, 1960a) activates enzymes causing rapid hydrolysis. DPA was always released more slowly than Ca^{2+} from spores forming in the presence of penicillin (Vinter, 1964).

A number of other treatments do not release DPA; these treatments include electrodialysis (Rode and Foster, 1960b; Harper *et al.*, 1964),

TABLE 20. Release of Dipicolinic Acid from Spores by Moist Heat

Species	Conditions	Amount released (% total)	Reference
Bacillus megaterium	4°; 6 months	0	Rode and Foster (1960b)
	70°; 20 min.	10	
	80°; 20 min.	50	
	90°; 20 min.	90	
	100°; 2 min.	83	
	100°; 20 min.	100	
Bacillus cereus (in 5 m*M* buffer; pH 7)	86°; 10 min.	10	El-Bisi *et al.* (1962)
	86°; 20 min.	30	
	86°; 40 min.	40	
	86°; 60 min.	40	
Bacillus subtilis (in 2·5 m*M* phosphate buffer; pH 7)	80°; 8·5 hr.	6	El-Bisi *et al.* (1962)
	85°; 6 hr.	25	
	85°; 12 hr.	48	
	85°; 18 hr.	72	
	90°; 3 hr.	34	
	90°; 6 hr.	70	
Bacillus coagulans (in 2·5 m*M* phosphate buffer; pH 7)	45°; 10 hr.	3	El-Bisi *et al.* (1962)
	45°; 30 hr.	4	
	45°; 216 hr.	7	
	80°; 3 hr.	8	
	80°; 8·5 hr.	8·5	
	90°; 1·5 hr.	7	
	90°; 3 hr.	17	

repeated freezing and thawing, many solvents at 25°, NaOH (pH 12·2; Rode and Foster, 1960b), and many anionic and non-ionic surface-active compounds (Rode and Foster, 1960c).

Although Ca^{2+} and DPA are not strongly bonded to spore material, it is evident that more data on the rates of release under various conditions, and on the complexes containing DPA, are required. If DPA is in a combination with α,ϵ-diaminopimelic acid, a DPA location in the cortex is indicated.

3. *Location of DPA*

The presence of other u.v.-absorbing materials has so far prevented the location of DPA by u.v. microscopy (Hashimoto and Gerhardt, 1960). The rapid release of Ca^{2+} and DPA under a wide variety of treatments, the cytological disappearance of the cortex during germination (Mayall and Robinow, 1957), and Ca^{2+} uptake and the formation of DPA at the time of cortex formation (Section IV, p. 216) were suggested as evidence favouring the cortex location (Murrell, 1961b). Hachisuka and Kuno (1963) further studied this possibility by examining the electron microscopic and stained appearance of spores of *B. subtilis* before and after DPA extraction by the Janssen *et al.* (1958) method. The spores (73 mg.) lost 11·2 mg. of dry matter, of which 10·3 mg. was DPA, and the only obvious change was the apparent disappearance of the cortex. Carbol fuchsin filled the translucent cortical areas formed by the almost specific loss of DPA. The inability of osmium tetroxide to resolve the cortex in near-mature and intact-mature spores was overcome by uranyl acetate staining of the thin sections. This may have removed cortical glycopeptide, as the heat treatment does not (Fig. 7; Warth *et al.*, 1963b). On the other hand, the insoluble cortical material may not have been resolved.

The observation of Donnellan and Setlow (1965) of similar unidentified photoproducts in hydrolysed DNA from u.v.-irradiated spores, and in irradiated DNA previously dried in the presence of Ca-DPA or salts, suggests a protoplast location for DPA.

The approximate 1:1 mole ratio of Ca^{2+} and DPA in spores, and their close quantitative relationship during sporulation, could be interpreted as indicating a molecular association and support their occurrence in the chelate form. If DPA is not associated with Ca^{2+} in the spore, then other cations must be present to neutralize the DPA. In nearly all species, the mole sum of all other inorganic cations determined was much less (average $< 0{\cdot}5$; Table 15, p. 169); hence, unless a very high concentration of organic cations is present, much of the DPA must be associated with Ca^{2+}. If calcium, the major ash component, was not lost with the cortex, the micro-incineration studies of Thomas (1964) would rule out the cortex as the site for Ca^{2+} and hence DPA, and suggest the coats or the protoplast as the site for these constituents (Fig. 10, p. 173).

4. *Rôle of DPA*

Halvorson and Howitt (1961) reviewed a variety of rôles that DPA may play in spores. They grouped them as non-specific (overall heat resistance) or specific (related to a given reaction or process). Specific rôles that have been observed include stimulation of electron transport (Doi and Halvorson, 1961), enzyme inhibition (Hachisuka *et al.*, 1965) and enzyme stabilization (Doi and Halvorson, 1961; Y. Hachisuka, unpublished observations; Powell, 1957) and initiation of germination (Riemann and Ordal, 1961). For further discussion, these papers should be consulted. DPA is clearly involved in the production of the heat-resistant, resting state of spores (Section IV, p. 216), but it does not appear to be a determining factor in the degree of heat resistance of each species (see p. 240; Table 17, p. 175).

IV. Biochemistry of Spore Formation

It is apparent from the cytological development and composition of the spore, already described, that the development and biosyntheses occurring during this metamorphosis involve a series of sequential integrated biochemical events and reactions. These have been studied rather indiscriminately, but considerable clarification of many of the reactions has occurred in recent years. There are still, however, many events which have not been studied or are unknown; also most studies have concerned only a few *Bacillus* species. Spore formation, occurring after vegetative division has stopped, is a relatively long process (taking 6–8 hr.) compared with an average generation time for vegetative bacilli of about 40 min. The most useful results on the biochemical changes involved have been obtained with well-synchronized cultures, in which structural changes have been followed by thin sectioning and electron microscopy. Studies on spore formation have used sporulating cultures as such, or washed cells at stage I or II resuspended in water or buffers ("endotrophic sporulation") or in a suitable simple medium ("replacement sporulation"; Hardwick and Foster, 1952).

Ideally, the cells should be in perfect synchrony, but even if they are well synchronized at the beginning of the process, inevitably some cells appear to get out of phase during the long process. In pairs of cells arising from the same division, one cell will be seen to attain refractility before its partner, and sometimes one of such a pair divides and gives rise to two spores while the other of similar size and identical age gives rise to one large spore. In many cases, however, such examples are infrequent and the cultural change from the first few refractile immature spores to refractility in nearly all of the maturing spores is obtained in 1–2 hr.

Despite imperfect synchrony and the lack of details in many studies of developmental changes and the degree of synchrony, an attempt will be made to relate all of the biochemical events to the development time-scale of the seven stages described earlier (Figs. 2 and 3). Stage I begins at t_0, the end of exponential growth, and stage VI ends about 6–8 hr.

later (t_{6-8}). Not all species are likely to pass through these stages at the same rate. However, each of those species commonly studied (*B. cereus*, *B. subtilis* and *B. megaterium*), appear to take about 6–8 hr. from stage I to VI under near optimum conditions. Under adverse conditions the process is slowed considerably; for example, at 15°, *B. cereus* strain T takes about 96 hr. compared to 5 hr. at 41° (Murrell and Warth, 1965).

During the exponential phase of synchronized growth (three to four divisions with the usual 10% inoculum) the nutrients are rapidly utilized and a rapid accumulation of metabolic products occurs. These products probably do not accumulate in sufficient concentration in the first half of the exponential phase to affect growth or metabolism. However, during the life time of the organisms of the last and perhaps penultimate generation (75–80% of the aggregate life times of the culture), the medium becomes depleted (75–80%) of the main energy and nitrogen sources and a major accumulation of metabolic products occurs. Faced with a practically complete depletion of original substrates, the cells adapt to at least some of the metabolic products. It is under these conditions that sporogenesis takes place; there is an overall change in the metabolic pattern of the cells associated with the synthesis of many new enzymes and eventually the new integuments of the spore protoplast. In non-synchronized cultures, the individual cell micro-environment probably changes similarly, and each cell undergoes independently similar metabolic changes.

A. Metabolic Changes during Spore Formation in *Bacillus* Species

1. *Carbohydrate Metabolism*

In most studies on spore formation the carbon source has been glucose and, with *Bacillus* species, glucose is usually converted first to organic acids with a marked decrease in the pH value of the culture fluid. The extent of the fall in the pH value depends on the buffering capacity of the medium and the initial carbohydrate concentration. If the latter is too high and the pH value falls too low (<5) the culture may fail to increase in pH value and successfully sporulate (Knaysi, 1945; Bernlohr and Novelli, 1960a; Bergère and Hermier, 1964). This is probably because the low pH value is unfavourable for adaptive enzyme formation and for the action of TCA cycle enzymes. Depletion of the carbohydrate substrate ends exponential growth and, with utilization of the acids, the pH value rises and the cells sporulate. This pattern of events seems fairly general among *Bacillus* species (Table 21), but it may not be without exception, especially in media lacking fermentable carbohydrate.

TABLE 21. A Summary of the Extent of Carbohydrate Metabolism in *Bacillus* Species

Species	Substrate	Typical pH fall	Organic acids formed	α-Picolinic acid inhibition	Acetate-oxidizing system in vegetative cells[a]	References
Bacillus cereus var. *mycoides*	Glucose	To around pH 5	Lactic			Knaysi, 1945; Hardwick and Foster, 1952
Bacillus cereus var. *lacticola*	Glucose + glutamate	+				Hardwick and Foster, 1952; Lundgren and Beskid, 1960
Bacillus cereus var. *albolactis*	Glucose	+		+		Nakata, 1963
Bacillus cereus T	Glucose	+	Pyruvic, acetic	+	–	Nakata, 1963; Nakata and Halvorson, 1960
Bacillus cereus T	Glucose	Buffered	Acetic, lactic			Nakata, 1963
Bacillus cereus T	Glucose	+	Acetic, pyruvic, acetoin			Nakata, 1963
Bacillus cereus NCTC 232	Glucose	+			–	Hanson *et al.*, 1963a, b
Bacillus cereus NCTC 6222	Glucose	+			–	Hanson *et al.*, 1963a, b
Bacillus cereus ATCC 4342	Glucose + glutamate + glycine	+	Acetic, pyruvic, α-ketoglutaric[b]			Buono *et al.*, 1966
Bacillus megaterium	Glucose	+		+		Hanson *et al.*, 1963a, b

Bacillus subtilis	Glucose	+	Acids and Acetoin	+	–	Hanson *et al.*, 1963a, b; Szulmajster and Hanson, 1965
(i)	Glucose	+				Bergère and Hermier, 1964
(ii)	Amino acid	–				Bergère and Hermier, 1964
Bacillus coagulans	Glucose	+		+		Hanson *et al.*, 1963a, b
Bacillus brevis	Glucose	+			–	Hanson *et al.*, 1963a, b
Bacillus licheniformis	Glucose + lactate	+	Lactic			Bernlohr and Novelli, 1959, 1960a
Bacillus stearo-thermophilus	Glucose	+				

[a] Acetate-oxidizing activity was measured by determining the extent of oxidation of [6-^{14}C]glucose to $^{14}CO_2$.
[b] Glycollic, glyoxylic and succinic acids have also been identified (Malveaux and Cooney, 1964).

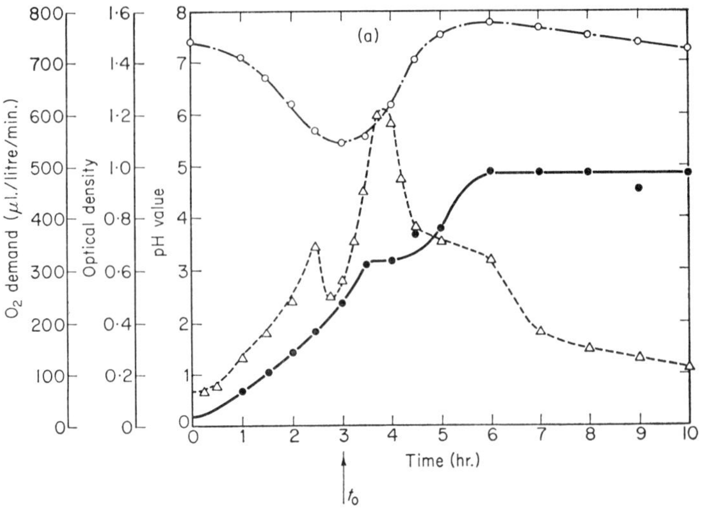

FIG. 13a

FIG. 13. Metabolism of *Bacillus cereus* T during sporulation. (a) The relationship of oxygen demand (△) to pH value (○) and optical density (●). From Halvorson, 1957. (b) Production of pyruvate and acetate from glucose during growth and subsequent utilization of the acids during sporulation at 30°. Cells characteristic of the prespore stage appeared shortly after 4 hr. From Nakata and Halvorson, 1960. (c) Ability of cells harvested at different ages to oxidize [1-^{14}C] acetate to CO_2. From Hanson *et al.*, 1963a. (d) Changes in the pH value, turbidity, and poly-β-hydroxybutyrate (PHB) concentration during growth and sporulation. From Kominek and Halvorson, 1965.

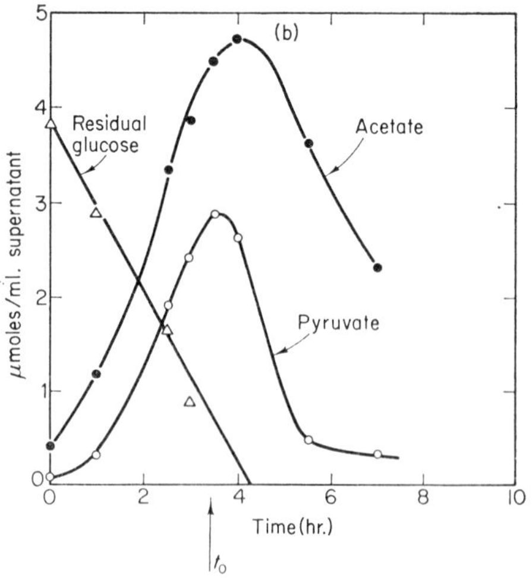

FIG. 13b

a. Bacillus cereus. The biochemical changes associated with this cycle of events in *B. cereus* strain T have been studied extensively by Orin Halvorson and his colleagues (Fig. 13). Under standardized conditions

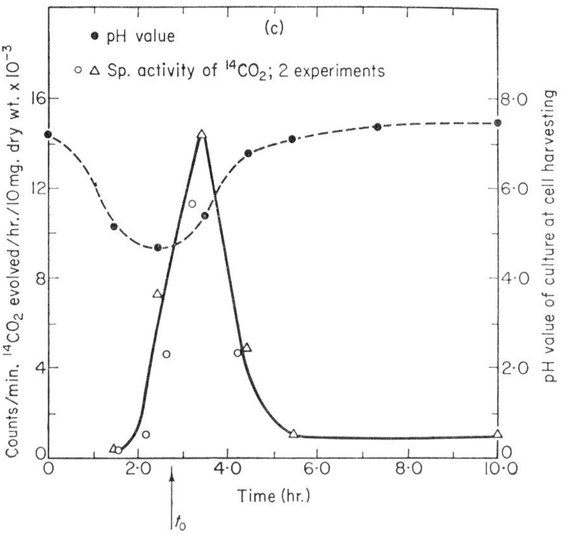

FIG. 13c

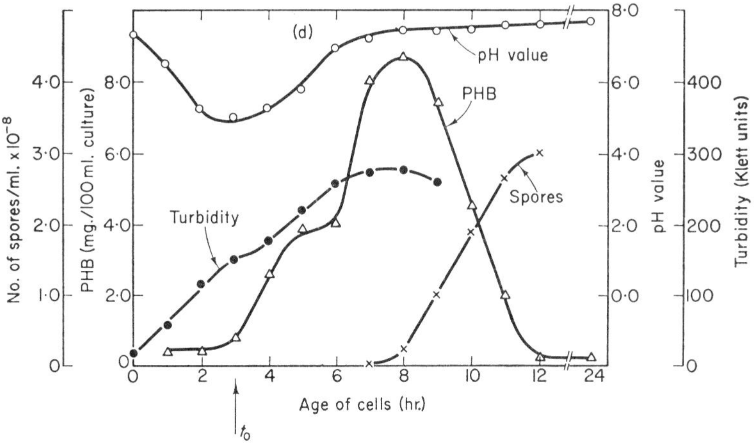

FIG. 13d

with well-synchronized cultures, the pH value falls to near 5, reaching a minimum at about the same time as the minimum in the oxygen demand curve occurs (Fig. 13a). As soon as the oxygen demand curve rises again, the pH value begins to rise, reaching about 7·7 about 1·5 hr. after the second peak in the curve is reached. The pH value from then on shows a slow but steady decrease. The latter part of the second rise in the oxygen demand curve is considered not to be due to a rise in cell population, but to be the result of an actual increase in oxygen demand/cell (Halvorson, 1957).

The pH minimum occurs at about time t_0 and results mainly from the accumulation of acetic and pyruvic acids (Nakata and Halvorson, 1960). Glucose is all used by the time the pH value begins to rise (about time t_2). Acetic acid accounts for 60–65% of the initial glucose carbon. Maximum accumulation of pyruvate occurs about 30 min. before maximum acetate accumulation (Fig. 13b). The vegetative cells occurring before the pH minimum lack a pathway for the complete oxidation of glucose. Utilization of the organic acids, co-incidental with the second increased oxygen demand, is essential for sporulation. The enzymes required for this utilization are adaptively formed. Alpha-picolinic acid specifically inhibits the formation of the enzymes for acetate utilization, and so prevents sporulation (Gollakota and Halvorson, 1960; Nakata and Halvorson, 1960).

These observed changes in the metabolic pattern occurred together with the appearance of cells characteristic of the prespore stage, presumably about stages I and II. The induction of the new enzyme system for the rapid utilization of acetate occurred during the transition from vegetative growth to sporulation and the induction appears associated with the first recognizable step in sporogenesis (Hanson *et al.*, 1961, 1963a).

Martin and Foster (1958) suggested the involvement of the TCA cycle in sporulation. Evidence for this was provided by Hanson *et al.* (1963a, b) who found that, during the period of minimum pH value, only the TCA cycle operated and acetate was converted to CO_2 and poly-β-hydroxybutyrate. Before or after this period fluoroacetic acid did not inhibit growth and sporulation. It failed to inhibit the vegetative cells because they lacked a condensing enzyme to convert acetate to fluorocitrate and afterwards because the TCA cycle was apparently not operating. Addition of fluoroacetate during the period of minimum pH value blocked citrate utilization and prevented sporulation. The acetate-oxidizing system reached peak activity in the latter half of the pH-minimum period (Fig. 13c), that is, after the maximum population was reached; it then declined rapidly. The decrease was considered to be the result of the cells becoming impermeable to acetate rather than to the disappearance of the enzyme system.

There has been no investigation to determine whether these changes in the enzyme systems occur in the forespore or in the mother cell.

b. Bacillus subtilis. Although this organism possibly prefers an amino acid-type medium for growth and spore formation (Bergère and Hermier, 1964), there is evidence of an involvement of, and need for, the TCA cycle for the physiological expression of the sporulation genome (Szulmajster and Schaeffer, 1961a, b; Hanson *et al.*, 1964a, b; Szulmajster and Hanson, 1965). In a glucose-containing medium, the Marburg strain showed a typical fall in pH value followed by a rise with acid utilization (Fig. 1; Szulmajster and Hanson, 1965). The final pH value and extent of spore formation were higher, however, on various amino acid-containing media (Bergère and Hermier, 1964). When the pH value was buffered by the amino acids, the value continued to rise from 14 hr. onward; whether a slight fall occurred before this was not reported. With media containing NH_4Cl and glucose, the final pH value was lower than with media lacking NH_4Cl.

The evidence for the involvement of the TCA cycle in sporogenesis is as follows. A sporogenic strain (Sp^+) showed increased activity of a particulate NADH oxidase during sporogenesis while in several asporogenic mutants (Sp^-) the activity remained low and constant during the corresponding period. Transformation of the Sp^- mutant with wild-type DNA led to sporulation and the appearance of $NADH_2$ oxidase activity similar to that in the Sp^+ strain (Szulmajster and Schaeffer, 1961a). The Sp^- mutants showed genetic differences in reciprocal transformation behaviour; as a result, Schaeffer and Ionesco (1960) suggested that the Sp^- mutation may affect different enzymes in the complex $NADH_2$ oxidase system. This was investigated with cells at stages IV–V by Szulmajster and Schaeffer (1961b). The Sp^+ strains had high $NADH_2$ oxidase, $NADH_2$ dehydrogenase, lipoamide dehydrogenase, cytochrome *c* reductase and cytochrome *c* peroxidase activities. In the Sp^- mutants these enzymes, with the exception of cytochrome *c* peroxidase, were low in activity or absent. Both Sp^+ and Sp^- mutants lacked a functional cytochrome *c* oxidase. The oligosporogenous mutants (Osp, or low frequency sporulating strains) had slightly higher $NADH_2$ oxidase activity than the Sp^- strains, trace activities of $NADH_2$–cytochrome *c*–reductase and succinate–cytochrome *c*–reductase, and active cytochrome *c*–peroxidase and cytochrome *c*–oxidase activities.

Further, an aconitate hydratase-negative mutant (large albino colonies), requiring glutamate for growth, was Sp^-. Even when grown in mixed culture with the wild type (small brown colonies) to eliminate environmental effects, it formed no spores while the wild-type colonies sporulated normally (Hanson, Blicharska and Szulmajster, 1964a).

The activities of several enzymes of the TCA cycle are repressed during exponential growth in media containing glucose alone or with glutamate, arginine and yeast extract. The enzyme activities increase at times t_1

and t_3 in a glucose-containing medium following exhaustion of glucose at time t_0 (Hanson *et al.*, 1964b; Szulmajster and Hanson, 1965). In the presence of acetate and lactate, glutamate did not cause repression of aconitate hydratase. Glucose on its own was believed to act by leading to a large glutamate pool (see later) which would thus repress.

The involvement of the TCA cycle is therefore supported by evidence of an increased activity of TCA cycle enzymes at the time of sporogenesis, the derepression of these enzymes when glucose is exhausted, and the occurrence of asporogenic mutants lacking these activities. Transformation of the Sp^- mutants leads to the recovery of both enzyme activity and ability to sporulate.

c. *Other* Bacillus *species.* A number of other *Bacillus* spp. were examined by Hanson *et al.* (1963a). These species (*B. subtilis, B. coagulans, B. cereus* var. *albolactis, B. megaterium*) all showed inhibition of the rise in pH value and of sporulation by α-picolinic acid, suggesting an interference in the acetate-oxidizing system as in *B. cereus* T (Table 21). Several other strains of *B. cereus, B. coagulans* and *B. brevis* and another strain of *B. subtilis* all showed a pH minimum at 2·75–4 hr. and negligible oxidation of [6-^{14}C]glucose to $^{14}CO_2$ before the pH minimum, indicative of lack of an acetate-oxidizing system in the vegetative cell stage (Table 21). Similar changes in pH value were observed with *B. cereus* var. *lacticola* on a semisynthetic medium containing an excess of glutamate over glucose (Lundgren and Beskid, 1960). *B. licheniformis* showed a typical pH value profile, with the accumulated lactate being used up during the period of rise in pH value (Bernlohr and Novelli, 1959, 1960a).

Contrary to the above results with *B. cereus* T, Hardwick and Foster (1953) observed considerably less oxidative activity (malate, succinate, α-ketoglutarate and pyruvate dehydrogenases) in *B. cereus* var. *mycoides* and three other species, in cells at about stage I and II, as compared with vegetative cells.

d. *Metabolism in media lacking fermentable carbohydrate.* Many *Bacillus* species grow and sporulate well in media lacking added carbohydrate. Is the development of a TCA cycle essential for sporogenesis under these conditions? With *B. cereus* T no fall in pH value was observed if glucose was omitted from the glucose–yeast extract–salts medium of Stewart and Halvorson (1953), or if the media had a 15-fold excess of phosphate buffer; in the latter case, the acetate oxidation system was present as before suggesting that a change in pH value was not responsible for the metabolic shift (Hanson *et al.*, 1963a; Nakata, 1963).

In *B. licheniformis* grown on glutamate as the carbon and nitrogen source, synthesis of all of the enzymes that are low in activity or not found in vegetative cells growing on glucose was derepressed, and the activities were as high or higher than in sporulating cells grown in the presence of glucose. This is true for the four enzymes involved in arginine metabolism (see Fig. 15, p. 199), for ornithine carbamoyltransferase and the protease that becomes active during sporulation (R. W. Bernlohr, private communication; Laishley and Bernlohr, 1966). Glycerol also causes repression of arginase synthesis; pyruvate and acetate do not. Bernlohr suggests, therefore, that many of the enzymic activities associated with sporulation (see above, also Table 25) are really enzymes under glucose catabolite repression, and are not necessarily involved in sporulation metabolism. However, the fact that such enzymic activities can be induced in vegetative cells by derepression does not necessarily mean that their activities are not important during sporulation. Derepression in the vegetative cells may not cause vegetative cells to become sporangia. The enzymes, which are active while the cell is undergoing morphogenesis in the absence of the original substrate, are likely to be essential or of considerable importance to the sporulating cell.

e. *Rôle of oxygen in* Bacillus *metabolism.* The supply of oxygen for sporulation of *Bacillus* species is often critical, and is perhaps necessary for the oxidation of the accumulated organic acids by the TCA cycle (Knaysi, 1945). Tinelli (1955b, c) observed lysis and lack of spore formation with insufficient aeration of *B. megaterium*; with better aeration an increase in oxygen consumption occurred just before the appearance of spores, although oxygen consumption then declined. Halvorson (1957) confirmed this marked requirement for oxygen with *B. cereus*, observing two peak demand periods (Fig. 13a). The oxygen demand, however, may be less important in other species. Grelet (1952) observed a decreased requirement in a strain of *B. megaterium*, and Bergère and Hermier (1964) produced more homogeneous spore crops with less aeration and lower final culture pH values. In the latter case, with bacteria growing in an amino acid-containing medium, fewer organic acids were probably present. The need for high aeration rates and for the operation of the TCA cycle in sporogenesis may largely depend on the amounts of organic acids in the medium, although, in the following experiments, cell population level was probably also involved. At higher glucose concentrations, better aeration was needed (Halvorson, 1957). With *B. licheniformis* no effect of aeration rate was observed when the glucose concentration in the medium was lowered from 227 to 27 μmoles/ml.; at the higher concentration, however, the aeration rate could still be too high, with the pH value falling too low and failing to rise; no spore formation or bacitracin formation occurred under these conditions (Bernlohr and Novelli, 1960a). The authors suggested an inactivation of TCA cycle enzymes by oxygen, or an inhibition of their adaptive synthesis by the low pH value and high aeration rate.

2. *Nitrogenous Metabolism*

As in carbohydrate metabolism there is evidence of a considerable shift in the pattern of metabolism of nitrogenous compounds from that of log-phase cells to that of sporulating cells. Although not many species have been studied, the metabolism of glutamate appears to play a central rôle in the biochemistry of the sporulating cell.

With *B. cereus* var. *mycoides*, Grelet (1955) observed a requirement for a high concentration of glutamate (50 m*M*). Most of this glutamate was not used during growth and spore formation but it seemed to play a rôle in annulling the effects of high concentrations of other amino acids. These amino acids were used but their metabolism during spore formation was not studied.

In *B. megaterium* Millet and Aubert (1960) found that, although glucose was completely utilized by about time t_0, the glutamate concentration in the cells continued to increase till about time t_2; it then decreased (see Fig. 16b, p. 201). At time t_0 the cells contained about 0·4%, and at time t_1 1·2% glutamate. This increase was not accompanied by an increase in the total radioactivity of the cells, i.e. no growth occurred on the products of the degradation of the radioactive glucose. The glutamate in the medium apparently did not derive from the hydrolysis of the proteins or glutamyl peptide, as no other free amino acids appeared. It was considered to derive from the metabolism of glycoside and lipid reserves. After time t_1, the free glutamate in the medium decreased until the release of free spores. The amounts in spores were about equivalent to those in vegetative cells at the end of growth but, relative to dry mass, were twice the amount in vegetative cells. When ^{14}C-labelled glutamate was added at various times during sporulation, relatively large amounts were incorporated into the spore from time t_0 to t_5. The label was distributed in the alcoholic extract (free glutamate and DPA), the hot TCA extract (nucleic acids and DPA) and in proteins. About 30% of the radioactivity of the proteins occurred in glutamate; aspartate, alanine and proline were also labelled. About two-thirds of the glutamate carbon was lost as CO_2. The later the ^{14}C-labelled glutamate was added, the less isotopic dilution occurred. Exogenous glutamate was thus mixing with pool glutamate, and was metabolized into CO_2, DPA, amino acids and protein during stages V and VI.

When *B. cereus* was grown in a medium containing glucose and glutamate, 95% of the glucose was used by time t_0 but only 10% of the glutamate (Buono *et al.*, 1966). During spore formation, another 10% of the glutamate was used. Halving the concentration of glutamate from 70 m*M* to 35 m*M* decreased the growth rate and the glucose oxidation rate, and delayed spore formation. At a concentration of 14 m*M*, the

TABLE 22. Amino Acid Requirements for Spore Formation in *Bacillus cereus* (from Buono *et al.*, 1966)

Compounds	Ability to replace glutamate (70 m*M*)	
	In culture	With washed cells at stage III
With 14 m*M* *glutamate*		
$(NH_4)_2SO_4$ (at 56 m*M*)	+	
Aspartic acid	+	
Alanine	+	
Glutamine	−	
Without glutamate		
Arginine (at 70 m*M*)	+	+ (all at 7 m*M*)
Ornithine	+	+ (60% effective)
Glutamine	−	+
Aspartic acid	−	+
Alanine	−	+
Histidine	−	+ (60% effective)
Citrulline	−	+ (60% effective)
Proline	−	+ (30% effective)
Hydroxyproline	−	+ (30% effective)
$(NH_4)_2SO_4$	−	
α-Ketoglutarate + $(NH_4)_2SO_4$	−	+
Pyruvate + $(NH_4)_2SO_4$	−	+
Fumarate + $(NH_4)_2SO_4$		+
Oxaloacetate + $(NH_4)_2SO_4$		+
Effective after 2–4 *hr. lag*		
Acetate + $(NH_4)_2SO_4$		+
Ineffective with $(NH_4)_2SO_4$		
Citrate, malate, succinate, isocitrate, *cis*-aconitate		−
Ineffective		
Gly, ser, leu, ile, lys, cyS, cySH, met, asp(NH_2), val, trp, phe, tyr, thre		−

effects were more pronounced. Thus again, a high glutamate requirement existed even though little (14 mM) was used. In media containing 14mM glutamate, the additional glutamate requirement for normal sporulation could be replaced by $(NH_4)_2SO_4$, aspartate or alanine (56 mM), but not by glutamine. Glutamine supported growth but not sporulation. Arginine and ornithine (at 70 mM) each replaced glutamate. Other amino acids

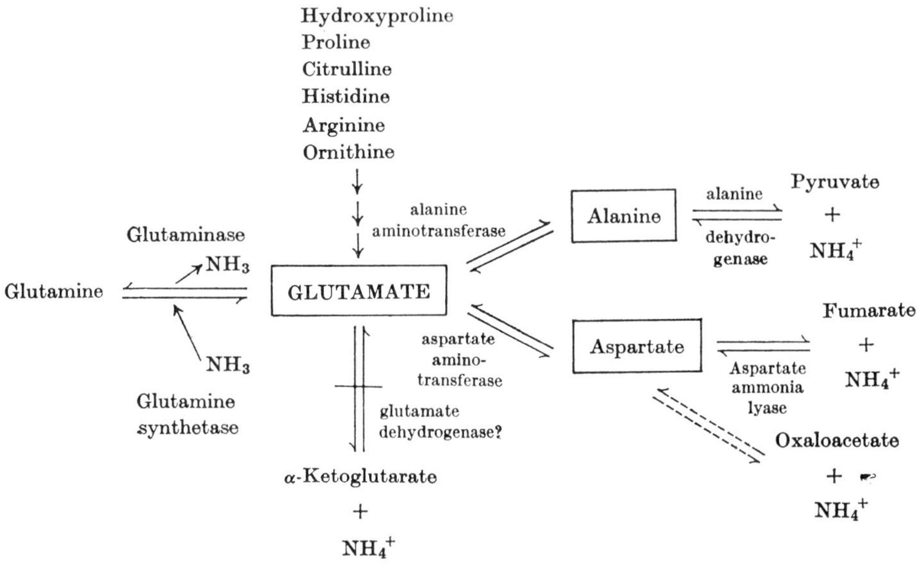

FIG. 14. Diagram showing the interrelationships of glutamate with other substrates that support sporulation of *Bacillus cereus* in replacement-salts medium. From Buono *et al.*, 1966.

and organic acids in the presence of ammonia would not (Table 22). The amino-acid requirement for sporulation of washed cells was studied in a replacement medium of salts. Without added amino acids, cells at $t_{0\cdot5\text{-}1\cdot5}$ gave less than 5% spores in 8 hr., those at $t_{2\cdot5}$, $t_{3\cdot5}$ and $t_{4\cdot5}$ giving 40–50, 70–75 and 90–95% spore formation respectively. With glutamate (7 mM) even t_0 cells gave 95% sporulation in 8 hr. Using $t_{3\cdot5}$ cells (stage III) glutamate could be replaced as shown in Table 22.

This metabolic pattern, together with the detection of alanine aminotransferase, aspartate aminotransferase and alanine dehydrogenase, but not glutamate and α-ketoglutarate dehydrogenases, led Buono *et al.* (1966) to suggest that glutamate metabolism occurred as indicated in

Fig. 14. Glutamate, because of its importance in the solute pool (see p. 202) and in sporulation, was given the central rôle. All the amino acids supporting spore formation in the replacement medium lead metabolically into glutamate. Alanine and aspartate aminotransferase activities have been demonstrated. Glutaminase was not assayed, but a glutamine synthetase has been reported in *B. cereus* T (Krask, 1953). The mechanism of NH_4^+ entry into the amino-nitrogen pool in *B. cereus* is not known (Buono *et al.*, 1966).

Mayer and Beers (1964) reported a pronounced change in the activity of enzymes metabolizing glutamate in *B. cereus*; in early exponential-phase cells, it was metabolized by a pathway involving carboxylation of α-ketoglutarate formed via glutamate dehydrogenase; later (stages I-II cells) is was metabolized through the TCA cycle.

The nitrogen metabolism in sporulating cells of *B. licheniformis* has been studied extensively by Bernlohr and colleagues. In most of these studies the cells were grown on a defined medium containing glucose, salts and often lactate. Growth and sporulation were much slower than in studies with *B. cereus* and *B. megaterium*. Stage I began at 6–10 hr. and stage VI at about 25 hr. The time from a low per cent of mature included spores (stage VI) to 75% or more at stage VI took at least 5 hr. or more. This suggests asynchrony but the time span is probably not unusual with slow differentiation. The changes in metabolic patterns relate clearly to different developmental stages. When 0·4% acid-hydrolysed casein was substituted for glucose and inorganic nitrogen in the medium, the time for the entire sporulation process was halved (Bernlohr, 1965a).

Cells in stage I or later, grown without exogenously supplied amino acids, were unable to concentrate amino acids or exchange externally added amino acids even when these were present at 1000 times the internal concentration. Cells grown with 0·2% casein hydrolysate had a larger amino acid pool, and probably possessed an amino-acid permease as rapid exchange occurred between the external source and the intracellular pool in both growing and sporulating cells. Low concentrations of labelled amino acids were incorporated into the hot trichloroacetic acid-insoluble fraction and at least 90% of the radio-activity remained in the added amino acid. Various amounts of the labelled amino acid were conserved during spore formation, becoming fixed in the free spore. Seventy-five per cent of the vegetative cell phenylalanine, 25% of alanine and 50% of a ^{14}C-algal hydrolysate were conserved (Bernlohr, 1965a).

Analysis for sixteen amino acids during sporulation showed that synthesis, turnover and oxidation were occurring (Table 23). The contents of histidine and tyrosine increased significantly, indicating synthesis of these amino acids. As no detectable quantities of polysaccharide or lipid were present at time t_0, they were probably synthesized from the carbon skeletons of other amino acids. Alanine and isoleucine decreased in amounts and were oxidized. Arginine and glutamate remained constant in amount, although they were being metabolized. Aspartate increased in amount despite oxidation. Oxidation was evident from the accumulation of ammonium ions (causing an increase in pH value), stimulation of endogenous oxygen uptake, and from the evolution of $^{14}CO_2$. Leucine disappeared but only a small amount was oxidized. Alanine, glutamate,

TABLE 23. Changes in the Amino Acid Contents of Cells of *Bacillus licheniformis* during Sporulation (from Bernlohr, 1965a)

Amino acid	Change (%)	Reason for change: Synthesis	Oxidation	Turnover	Time of change (sporulation stage)
Histidine	+115	++	−		
Lysine	+31	+	−		
Phenylalanine	+11	+	−		
Tyrosine	+150	+	++		
Leucine	−44		−	+	
Alanine	−31		++		I–III
Arginine	−6		++	+	I–II
Glutamic acid	−11		++	+	I–III
Isoleucine	−44		++		I–II; VI
Valine	−18		++		I–II; VI
Ornithine			++		I–III
Aspartic acid	+24		++		I–III

aspartate and ornithine were oxidized at significant rates during stages I–III. Isoleucine and valine were oxidized in two stages, the first during stages I–II and the second during stage VI (Fig. 6, Bernlohr, 1965a). The results suggested that, in addition to amino acid turnover and oxidation, interconversion was occurring.

All of the amino acids were oxidized at the time of addition, but a burst in oxidative activity occurred during stages I and II with L-glutamate and arginine. At the time of addition, most of the label entered the trichloroacetic acid-insoluble fraction. The increased activity in stages I–II was the result of turnover of vegetative cell protein. Only about one-fifth of the radio-active arginine was evolved as $^{14}CO_2$; as with glutamate, this probably resulted from differences in the pool size (Ramaley and Bernlohr, 1966).

During exponential growth the cells synthesize arginine in significant amounts from glutamate, but no arginine is converted to glutamate. On the other hand, sporulating cells do the reverse (Bernlohr, 1965a; Ramaley and Bernlohr, 1966). This metabolic shift occurs even though the free intracellular pool of glutamate is fifty times that of arginine in both vegetative cells and sporulating cells, a change in metabolism occurring soon after the cessation of growth (Ramaley and Bernlohr, 1965). Synthesis of arginase, protease ornithine transaminase and pyrroline-5-carboxylate reductase is induced at time t_0; ornithine racemase and α-transaminase activities could not be detected. Arginase activity

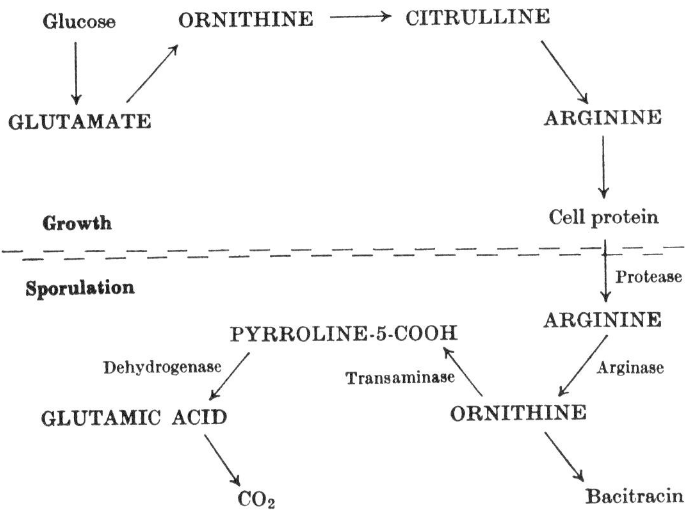

FIG. 15. A proposed scheme for the reversal of arginine metabolism in sporulating *Bacillus licheniformis*. The upper half describes vegetative cell metabolism and the lower half the metabolism of sporulating cells. From Ramaley and Bernlohr, 1966.

increased 165-fold during sporulation, while the enzymes involved in arginine biosynthesis (NADP-oxidase and ornithine carbamoyltransferase) became less active. The data are consistent with the scheme in Fig. 15. L-Ornithine is the precursor of D-ornithine in bacitracin (Ramaley and Bernlohr, 1963, 1966), but the conversion of L-ornithine into D-ornithine has not been demonstrated in sporulating bacilli. During phase I, the arginine catabolic system is developed, arginine is converted back to glutamate and some to bacitracin (L-ornithine, which arises from L-arginine being incorporated into the polymer) which is later released into the medium. Several factors are probably involved in the induction of the arginine catabolic system: (i) the increased size of the intracellular pool, resulting probably from the degradation of vegetative proteins by protease, and (ii) derepression of the synthesis of several enzymes by glucose exhaustion (Laishley and Bernlohr, 1966).

3. *Metabolism of Intracellular Reserves*

During sporulation, which occurs after the depletion of the main growth substrates, the mother cell and its spore protoplast both continue active metabolism, synthesizing new materials. This synthesis depends on residual nutrients and metabolic products and probably largely on

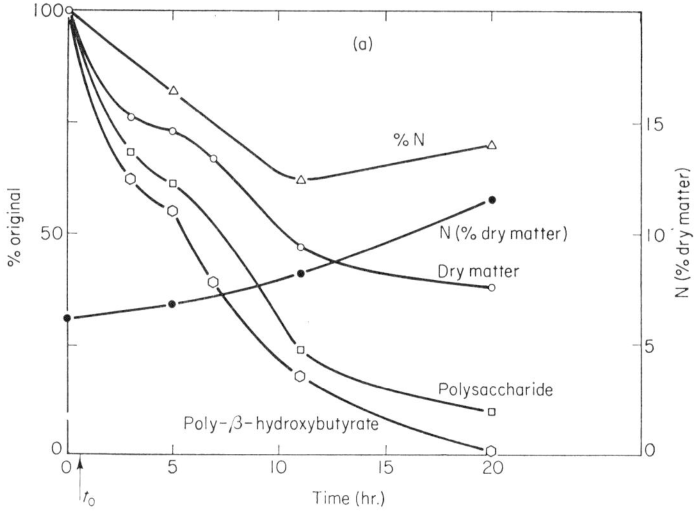

FIG. 16. Metabolism of sporulating *Bacillus megaterium*. (a) Changes in the gross composition of the sporangia during spore formation ("replacement"). Data from Tinelli, 1955a. (b) Glucose utilization and changes in the free glutamate and dipicolinic acid contents of cells during spore formation in a synthetic medium. Data from Millet and Aubert, 1960.

the intracellular reserve materials. That the cells can sporulate successfully on their own reserves is evident from the washed-cell experiments of Hardwick and Foster (1952), Perry and Foster (1954), and Black *et al.* (1960). Not all species and strains can be prepared in a suitable metabolic state to sporulate in this way, probably because of solute loss during washing, lack of sufficient endogenous reserves and minerals at harvest, or other special requirements (Powell and Hunter, 1953; Black *et al.*, 1960). Washed cells of many species sporulate successfully in minimal replacement media (Tinelli, 1955a; Black *et al.*, 1960; Pelcher *et al.*, 1963) in which the main supply of carbon and energy are intracellular reserves.

These reserves comprise (a) mother cell polysaccharides, proteins, ribosomal particles, (b) low mol. wt. solutes, and (c) special reserve materials such as poly-β-hydroxybutyrate (PHB).

a. *Non-specific reserve materials.* Hardwick and Foster (1952) showed that protein degradation and utilization of low mol. wt. solutes occurred during sporulation. Much of the protein was believed to be in the form of enzymes no longer required by the mother cell. Cell dry matter, polysaccharide and nitrogenous material decreased in amount throughout sporulation (Fig. 16a). Both structural and membrane materials are probably utilized. Considerable protein turnover occurs (Monro, 1961; Section IV D, p. 225) and proteolytic enzymes become active (Bernlohr,

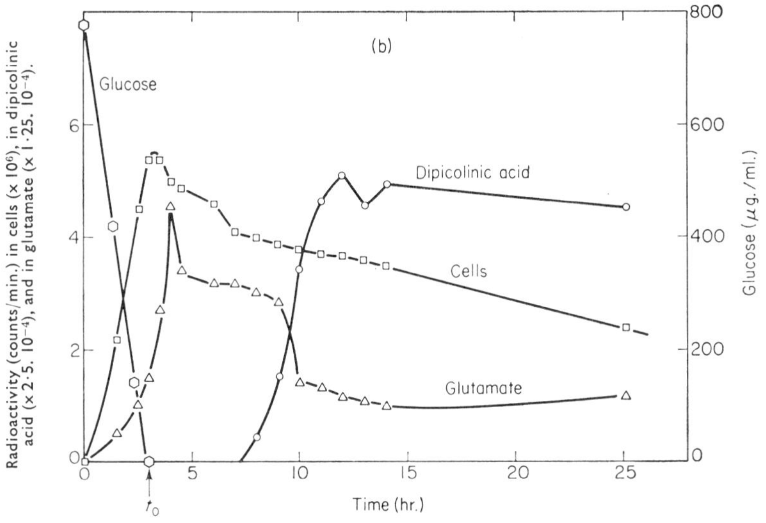

FIG. 16b

1964; Spizizen, 1965) and these are associated genetically with the physiological expression of the sporogenic character. RNA also undergoes turnover, and is re-utilized in the synthesis of spore RNA (see p. 211). Degradation of mother cell material cannot be too drastic however, as it is possible for the mother cell, before the end of stage III, to regenerate into an active vegetative cell again (Fitz-James, 1963).

b. *Utilization of solutes in the intracellular pool.* The main solutes built up in the free pool just before or during sporogenesis are amino acids, purines, pyrimidines and nucleotides. Metabolite analogue inhibition and reversal studies indicated that these preformed metabolites were essential for sporulation (Hardwick and Foster, 1952).

The main amino acids in the pool of *Bacillus* species appear to be glutamate and alanine. The increase in the concentration of glutamate from time t_0 to t_1 in *B. megaterium*, and its utilization throughout sporulation have been described (Fig. 16a). In *B. cereus* and *B. cereus* var. *alesti*, glutamate and alanine were the only amino acids present in significant concentration in the pool. The free alanine concentration remained roughly constant in both organisms from stages I to VI. In *B. cereus* var. *alesti*, glutamate was used steadily throughout this period but, in *B. cereus*, a sharp increase in glutamate concentration occurred during the forespore stage; thereafter the concentration decreased (Young and Fitz-James, 1959b). In *B. licheniformis*, glutamate and alanine made up 90% of the amino acid pool; these amino acids increased in amount during sporogenesis and were utilized thereafter. The rate and quantity of turnover of mother cell protein, and of the amino acid pool, indicated extensive re-organization of cell protein (Bernlohr, 1965b).

The amount of acid-soluble nucleotides is 75% higher in stage III than in exponential-phase cells of *B. licheniformis*; the nucleotides decreased in concentration during the later stages. The relative proportion of each nucleotide remained fairly constant during sporulation. Monophosphate exceeded di- and tri-nucleosides in amounts except in the case of uridine nucleotide; with these nucleotides, the amount of diphosphate exceeded the amount of monophosphate. ATP, GTP and UTP were detected only at stages II and III. The mole per cent of guanine + cytosine in the mononucleotide pool was 43%, compared with 46–48% in the DNA of *B. licheniformis*. Intracellular ribonuclease activity, and a decrease in the total nucleic acid content, indicated that the increased pool constituents probably arose from polymer breakdown. Not all the nucleotide was accounted for; the total nucleic acid decreased by 40 μmoles/10 g., the pool increased by 5 μmoles/10 g., while the extracellular nucleotide production was 60–80 μmoles/10 g. wet wt. cells (Leitzmann and Bernlohr, 1965).

c. *Poly β-hydroxybutyrate* (*PHB*). Stationary-phase cells of many *Bacillus* species contain reserve granules of PHB which disappear during sporulation (Fig. 5, p. 140). Cells of *B. megaterium* at time t_0 contain 28% dry wt. PHB; this decreased to less than 20% by the free spore stage (Fig. 16a). PHB or its degradation products were not found in the medium, except under conditions of partial spore formation during which lysis occurred (Tinelli, 1955a, b, c).

The amount, time of synthesis and use of PHB in relation to sporulation differ somewhat for different strains and environmental conditions such as the degree of aeration, amount of glucose and acetate in the medium and the differential effect of pH value on the rate of synthesis and utilization. Synthesis usually begins before or by the time t_0, with the peak content occurring 3–5 hr. later, and depletion by the end of

TABLE 24. Poly-β-hydroxybutyrate Synthesis and Utilization in Relation to Spore Formation

Species	Beginning of synthesis (hr. from t_0)	Peak cell content		Time of about 90% utilization		Conditions		Reference
		Time (hr. from t_0)	Concentration	Time (hr. from t_0)	Degree of sporulation (%)	Aeration	pH value	
Bacillus megaterium		0 or earlier		20	Free spores	Shaken		Tinelli, 1955a
	0	5	4·6% dry wt. cells	> 9	100	Shaken		Yoneda and Kondo, 1959
	about – 8	– 5	1·8 μg./ml.	0	0	Shaken		Slepecky and Law, 1961
	– 5	1–2	250 μg./ml.	10	0	Forced aeration		Slepecky and Law, 1961
Bacillus cereus	– 6	0	23% dry wt. cells	6	20	Forced aeration		Alper *et al.*, 1963
	– 4	0	—	—	—	Forced aeration		Lundgren and Bott, 1963
Bacillus cereus T	0	4	5 μmoles crotonate/ml.	8	100 (immature)	Shaken	Unbuffered	Nakata, 1963
	0	3·5	6 μmoles crotonate/ml.	8	100(immature)	Shaken	Buffered 6·4	Nakata, 1963
	0	3	3 μmoles crotonate/ml.	7	100 (immature)	Shaken	Buffered 7·0	Nakata, 1963
	0	3	1 μmole crotonate/ml.	7	100 (immature)	Shaken	Buffered 7·4	Nakata, 1963
	0	5	86 μg./ml.	9	100 (immature)			Kominek, 1964

sporulation (Table 24). Where considerable quantities of PHB are built up, under particular medium and aeration conditions, PHB is not exhausted for many hours and sporulation is delayed until the period of its rapid depletion (Slepecky and Law, 1961). From a study of seven strains, Stevenson *et al.* (1962) concluded that PHB formation is enhanced in media containing high concentrations of glucose and acetate, and that poor aeration and high PHB content are associated with low spore formation. Cultures grown with mild aeration gave three times and, with rapid shaking, twelve times as much PHB as is present in a stationary culture. With the best aeration, there was better utilization of PHB and increased sporulation.

Slepecky and Law (1961) and Nakata (1963) concluded that PHB production was not a mandatory requirement for sporulation; good sporulation was obtained with or without its accumulation. Under some conditions (e.g. high pH values), the utilization of organic acids in spore synthesis may be so rapid that little PHB is formed. Accumulation and utilization of PHB were optimum at a pH value of 6–6·4, but sporulation may not have been optimum under these conditions (Nakata, 1963; Kominek and Halvorson, 1965). The number of spores formed endotrophically from stage I cells containing small and large amounts of PHB was similar, but the spores formed from cells containing small amounts of PHB were much smaller and rounder (Kominek, 1964).

A large proportion (65–70% at pH 6·4) of the label of [2-^{14}C]acetate was incorporated into PHB and 20–25% into other constituents. Virtually no label was lost as $^{14}CO_2$ from time t_0 to t_{5-6}; thereafter evolution co-incident with PHB degradation occurred until time t_9. By time t_9 the PHB content was essentially depleted and immature spores were visible. Only 20–25% of the label was lost as CO_2; the major part of the products from the polymer were used in spore synthesis, 17% being incorporated into DPA, 50% into proteins, 20–25% into the hot 5% trichloroacetic acid-soluble fraction, 4–5% into the lipid fraction and 15–20% into the cold acid-soluble fraction. Nakata (1966) thus confirmed the hypothesis that the major rôle of acetate and subsequently of PHB in *B. cereus* strain T was the provision of carbon precursors and energy for sporulation.

Biosynthesis of PHB occurs with the adaptive formation of acetoacetyl CoA reductase at time t_0. PHB accumulation was almost completely inhibited by fluoroacetate, α-picolinic acid and chloramphenicol, indicating the involvement of acetate metabolism and the TCA cycle and interference in induced enzyme synthesis (Kominek, 1964; Kominek and Halvorson, 1965). α-Picolinate and chloramphenicol added before the pH minimum prevented rapid synthesis of acetoacetyl CoA reductase.

After all the acetate was incorporated, acetoin incorporation increased giving rise to the secondary increase in PHB production (Fig. 13d). During incorporation, acetoin was oxidized to CO_2 via the 2,3-butanediol and TCA cycles, thereby providing an intracellular source of acetate for the synthesis of further PHB or spore material. This enables the supposed acetate permeability barrier to be overcome (Fig. 17), and so would explain the cessation, during initial stages of sporulation, of the ability of cells to oxidize acetate, although the TCA cycle was still functional (Kominek and Halvorson, 1965).

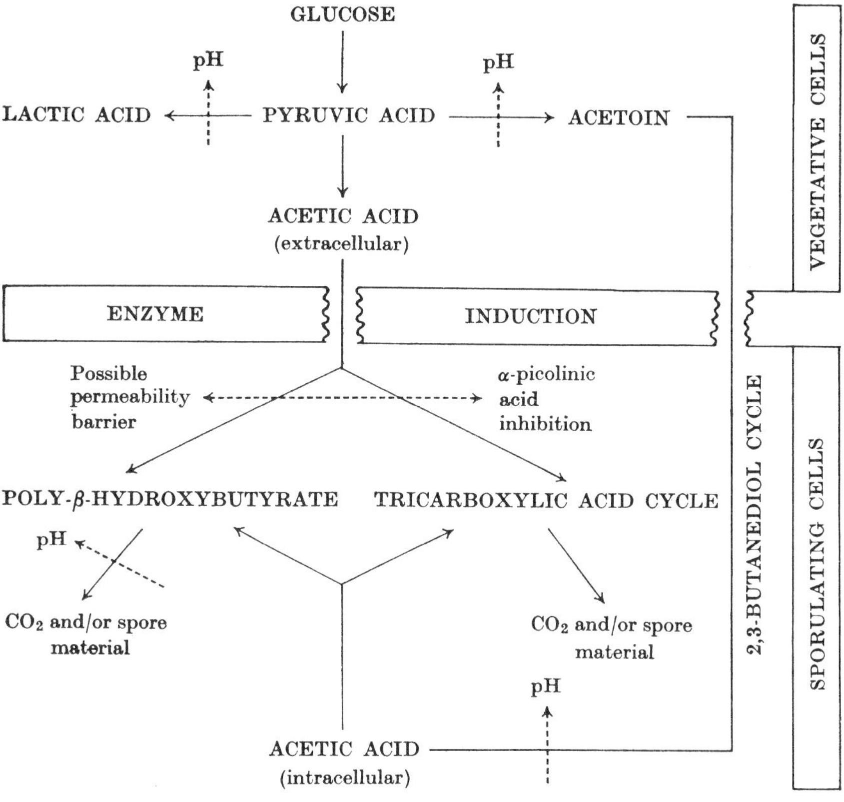

FIG. 17. A schematic summary of some of the metabolic transitions occurring during growth and sporulation of *Bacillus cereus* T in relation to poly-β-hydroxybutyrate production. From Kominek and Halvorson, 1965.

B. Metabolic Changes during Sporulation in *Clostridium* Species

1. *Carbohydrate Metabolism*

As in bacilli, carbohydrate fermentation in clostridia usually results in

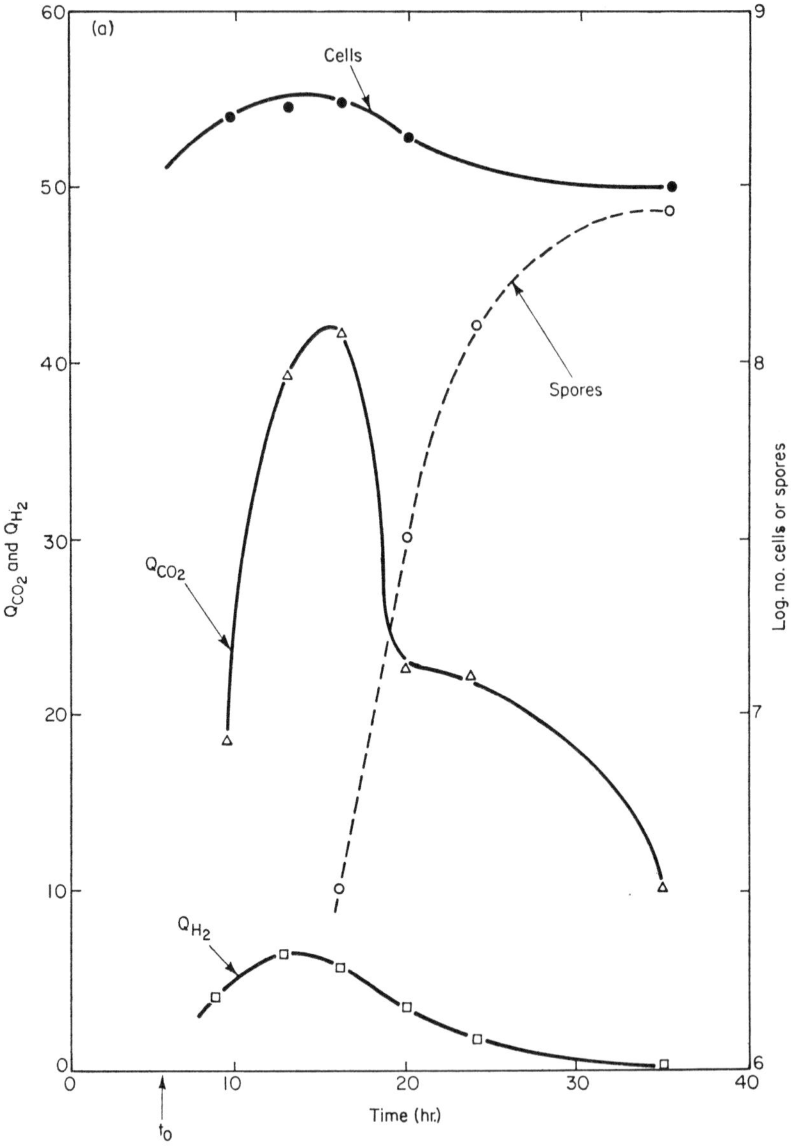

FIG. 18a

FIG. 18. Metabolism of sporulating *Clostridium botulinum*. (a) Catabolic activity of cells harvested at various stages of growth and sporulation; the substrates were L-alanine and L-proline. (b) Volatile acid production and pH changes during sporulation. From Day and Costilow, 1964a.

the accumulation of organic acids (Day and Costilow, 1964a; Bergère and Hermier, 1965b), which may decrease the pH value to 5·5 or less (Bergère and Hermier, 1965a). *Cl. botulinum* utilizes some of the acetate, at least during the early part of the period of rapid increase in number of refrac-

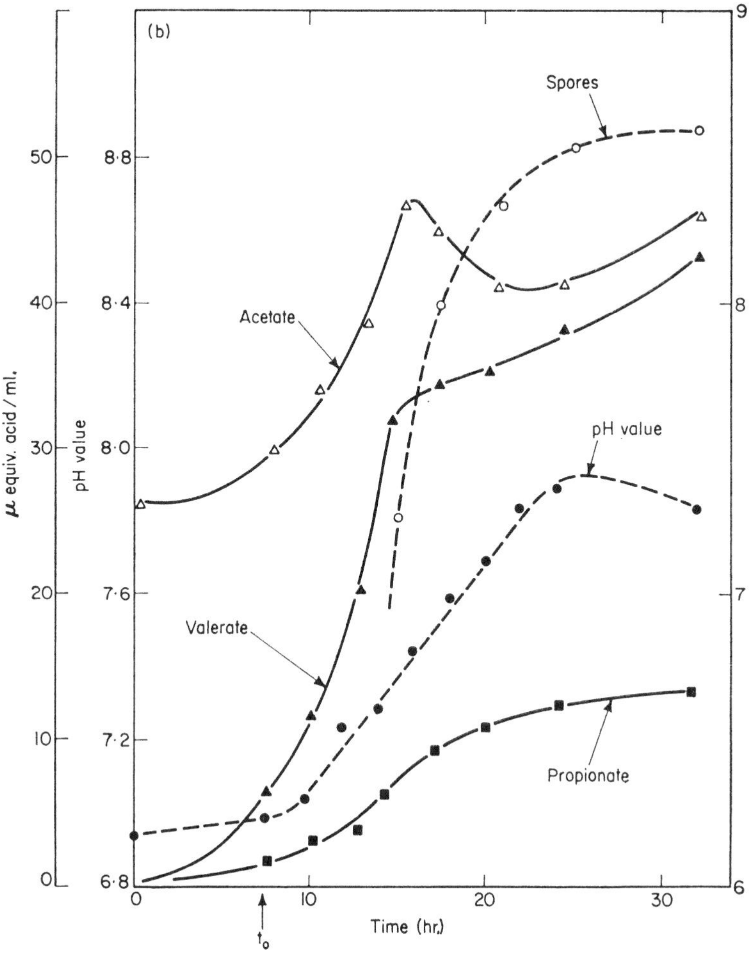

FIG. 18b

tile spores (Fig. 18b). Unlike *Bacillus* spp., *Cl. butyricum* in the presence of high concentrations of glucose (> 5 g./l.) does not continue vegetative growth until glucose is exhausted; the higher concentrations had little effect on the amount of growth, the number of spores produced, the final pH value or the amount of glucose fermented. Only 15% of the glucose was used in the exponential phase, and 55% after time t_0; 1·5 g./l. still remained at 21 hr. A decrease in the initial glucose concentration to < 5 g./l. reduced the population and the number of spores formed; all of the glucose was used. Bergère and Hermier (1965a) therefore suggested that glucose provided energy for growth and sporulation, and that carbon metabolism may be similar in the sporangium to that in the vegetative cell. The amount of growth in the presence of high concentrations of glucose was not limited by other nutrients, as increased concentrations of these did not increase the number of cells or spores formed. Growth was retarded by a heat-stable factor produced in the culture at about 12 hr. (t_1?); this factor induced sporulation (Section V; p. 234; Bergère and Hermier, 1965b).

2. *Nitrogen Metabolism*

Cl. botulinum 62A required arginine for sporulation; the onset of sporulation co-incided with the end of rapid arginine consumption (Perkins and Tsuji, 1962). Citrulline replaced arginine for growth and gave 3–5% sporulation; ornithine, urea or creatine would not support growth. Nearly all of the arginine consumed could be accounted for as citrulline or ornithine. High concentrations of arginine did not result in increased growth or increased incorporation into the cells, suggesting that the rôle of arginine in sporulation was to provide energy for spore synthesis. This was probably from ATP formed during citrulline breakdown (Perkins and Tsuji, 1962).

Cells of *Cl. botulinum* 62A at the forespore stage (about t_{2-3}) in replacement studies were able to obtain energy only from the mixtures L-alanine + L-proline, L-isoleucine + L-proline, or L-alanine + L-arginine, by the Stickland reaction. L-Ornithine failed to substitute for proline (Day and Costilow, 1964b). The amino acids were fermented most actively at about time t_{3-5} when 90% of the sporangia were swollen (Fig. 18a); the activity fell to the vegetative cell level in immature spores (Day and Costilow, 1964a). Acetate accumulated in these experiments during maturation of the spores, but was not apparently incorporated into the spores or sporangia. Nearly all of the label from [1-^{14}C]alanine appeared as CO_2, suggesting that the amino acids were primarily an energy source (Day and Costilow, 1964b).

The requirement for an exogenous energy source contrasts with the situation in *Bacillus* species; it possibly reflects lack of intracellular reserves in this organism grown in a minimal replacement medium. No information is available on the solute pool in clostridia in relation to sporulation. Day and Costilow (1964a) did not observe storage granules or detect PHB in clostridia. Bergère and Hermier (1965a), however, observed granulation in *Cl. butyricum* at time t_{1-2}.

Somewhat similar metabolic trends to those in bacilli are evident in the clostridia, but the information available is insufficient for a general description of metabolism in relation to sporulation. The exogenous energy requirement is apparently provided by glucose in *Cl. butyricum*, while *Cl. botulinum* may use part of the organic acids or amino acids. Although induction of sporulation was associated with substrate depletion in *Cl. botulinum*, in *Cl. butyricum* an unknown growth retardation factor was involved (Section V, p. 234).

C. Nucleic Acid Changes during Sporulation

1. *Deoxyribonucleic Acid*

According to the studies of Young and Fitz-James (1959a, b), the spore's DNA content is determined by division of the sporangial DNA at the moment that the axial thread divides and one part becomes enclosed by the spore septum. Growth and net synthesis of nucleic acids had occurred by this time (t_1–t_2) in *B. cereus* (Fig. 19). A cessation of DNA synthesis occurred during the axial filament stage (around time $t_{1\cdot0}$). In this organism, DNA synthesis then recommenced at a linear rate during development of the forespore. The amount of secondary synthesis was equal to the amount eventually found in the spores, and ceased when each sporangium contained three times the "spore" amount of DNA. The DNA content of the cells remained constant at this level and then decreased as lysis of the mother cell began (dotted line, Fig. 19). Removal of the sporulating cells from $^{32}PO_4^{3-}$ at times $t_{1\cdot0}$ or $t_{2\cdot25}$ caused no loss in the specific activity of the spore DNA-P. Removal before time $t_{1\cdot0}$ decreased the specific activity of the spore DNA-P. Removal before time $t_{1\cdot0}$ decreased the specific activity of the spores. Young and Fitz-James, therefore, concluded that the further synthesis occurred in the mother cell, and that the total spore DNA was partitioned from vegetative cell chromatin. Addition of $^{32}PO_4^{3-}$ to sporulating cultures at t_0–t_2 resulted in only slight labelling of the spores—this was attributed to spore-coat labelling in the residue-P fraction (Section IIIB, p. 166, Young and Fitz-James, 1959a). Canfield and Szulmajster (1964), using the ^{32}P-suicide technique with *B. subtilis*, concluded that synthesis of spore DNA was complete by time t_2.

Both of the above studies depend on cells from stage III onwards being permeable to phosphate in order to dilute an already labelled DNA or to label spore DNA. If the cells at stage IV or later become impermeable to phosphate then these experiments cannot reveal further DNA synthesis in the spore protoplast. Other experiments suggest the development of impermeability to several low-molecular weight substances, including 8-azaguanine and 2,6-diaminopurine (Young and Fitz-James, 1959c), phosphate (Fitz-James, 1963), and 5-fluorouracil (Aronson and Del Valle, 1964). However, *B. cereus* var.

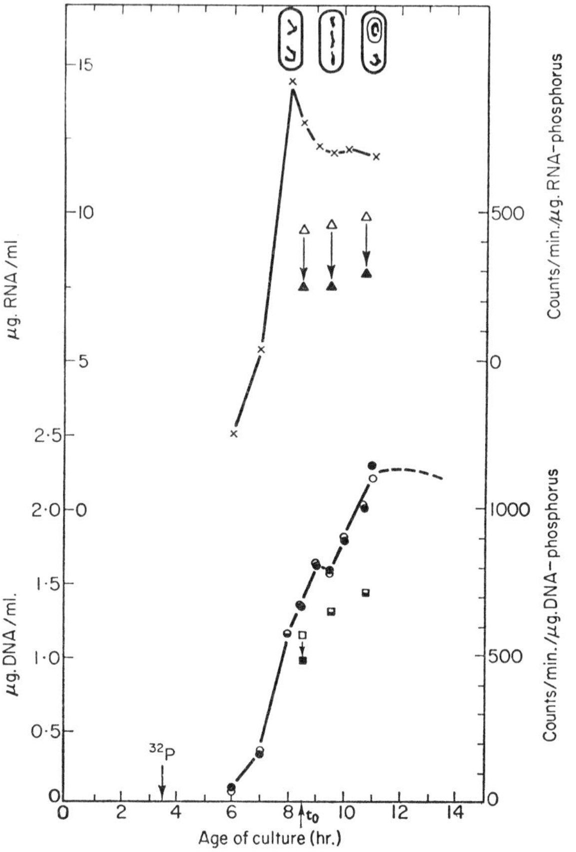

FIG. 19. Changes in the DNA-phosphorus (○—○, by Dische reaction; ●—●, by total-phosphorus) and in the RNA-phosphorus (×—×) contents of a culture of *Bacillus cereus* during growth and sporulation. Radioactive phosphorus was added (↓) to portions of the culture and the cells were transferred to non-labelled medium at 8·5, 9·5 and 10·75 hr. The specific activities of the DNA-phosphorus (□) and RNA-phosphorus (△) of the cells at the time of transfer are compared with the specific activities of the DNA-phosphorus (■) and RNA-phosphorus (▲) of the cells resulting from the transfer. From Young and Fitz-James, 1959a.

mycoides, *B. cereus* var. *alesti* and *B. cereus* var. *thuringiensis* did not synthesize this further amount of DNA. Perhaps, therefore, the spore DNA complement may be only one strand of DNA where further synthesis occurs, a further strand being synthesized in the forespore.

In those organisms in which no further synthesis occurs, the second strand may be completed at the same time or before septation. These possibilities appear unlikely but it seems surprising that (i) the spore should receive only a small part of the nuclear material (Section II, p. 137), and (ii) the mother cell should double its nuclear material shortly before its pending autolysis. These are possibly the reasons which led Fitz-James (1963) to suggest that in some cases this increase may result from a small number of asporogenous cells or cells in a state of unbalanced growth wherein DNA synthesis continues. Dennis and Wake (1966) from their autoradiography studies suggest that the *B. subtilis* spore may contain two chromosomes.

2. *Ribonucleic Acid*

Changes in the amounts and types of RNA fractions and in the messages carried are involved in the direction of the process of spore formation. Considerable attention has been focused on these in the last 3–4 years.

a. *Total RNA.* Net RNA synthesis in *B. cereus* strains ceases sharply within an hour of time t_0, sometimes before and apparently sometimes afterwards (Young and Fitz-James, 1959a, b; Fitz-James, 1963). In *B. subtilis*, RNA synthesis ceased almost exactly at time t_0 (Spotts and Szulmajster, 1962). The total RNA content then falls, despite recommencement of DNA synthesis, either abruptly (Young and Fitz-James, 1959a, b), or gradually (Spotts and Szulmajster, 1962). This is a natural "step-down" type of RNA metabolism associated with the end of exponential growth and the nuclear re-arrangement accompanying sporogenesis. The large decrease in total RNA during spore formation results probably largely from breakdown of polyribosomes (100S) and m-RNA in the mother cell (Fig. 20).

Although total content of RNA falls, active turnover and resynthesis of certain RNA fractions continue. Labelled RNA in *B. cereus* lost label when the organisms were placed in unlabelled medium at time t_0–t_2, and the ^{32}P taken up during this period persisted in the spore RNA (Young and Fitz-James, 1959a, b). Similar evidence in *B. subtilis* has been obtained with ^{14}C-uracil (Spotts and Szulmajster, 1962; Balassa, 1963c) and autoradiographically with ^{3}H-uracil (Ryter, Bloom and Aubert, 1966).

b. *Synthesis and degradation.* New RNA is synthesized from purines and pyrimidines in the medium (as shown by labelled incorporation studies) and in the free solute pool (Hardwick and Foster, 1952), and from degradation products of pre-existing RNA (Balassa, 1963b). The specific activity of RNA formed during the penultimate hour of growth was less than one-tenth that of the previous hour although RNA was being formed at a steady rate (Fitz-James, 1963). Fitz-James therefore suggested that an internal source of phosphate was being used in preference to externally supplied $^{32}PO_4^{3-}$. The relative rates of synthesis of

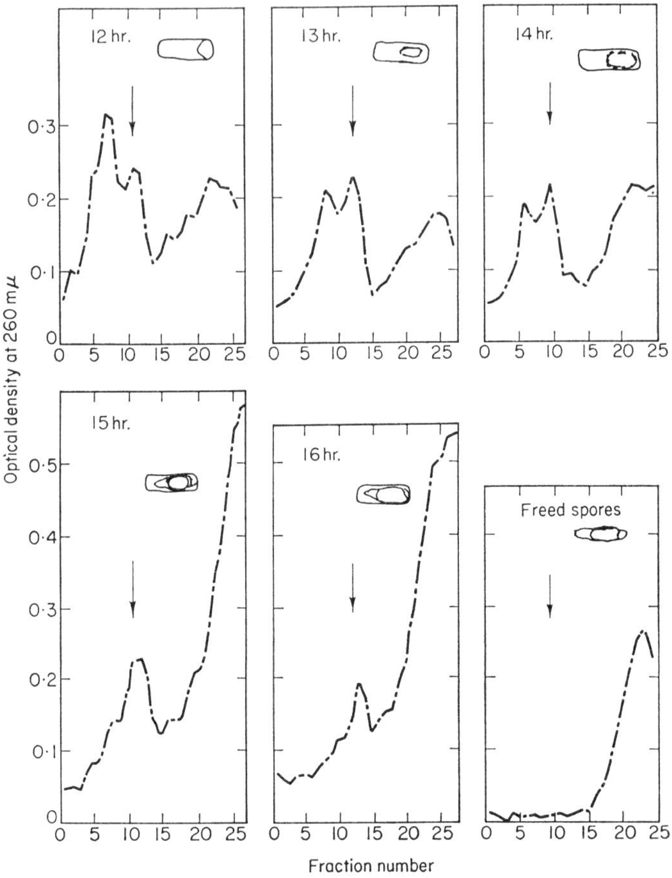

FIG. 20. Sedimentation analysis of the cell sap from *Bacillus cereus* var. *alesti* strain A⁻ during various stages of spore formation indicated by the number of hours of aeration and the line drawings. Time t_0 was about 10hr. The arrows indicate the approximate position of the 70S ribosomes. From Fitz-James, 1963.

the three RNA types are the same during sporulation and normal growth (Balassa, 1966b); the preferential degradation of r-RNA led to a relative enrichment in t-RNA, and an estimate of RNA turnover of 20% per hour. In wild-type *B. subtilis*, RNA synthesis during sporulation depended on the presence of amino acids, and rapid turnover was associated with proteolytic action (Balassa, 1966b). The rate of synthesis (estimated by ^{14}C-uracil incorporation) at time t_3 was two- to three times greater in the sporogenous (Sp^+) than in the asporogenous strain. Chloramphenicol increased incorporation eight times in the SP^+ strain but only slightly in the SP^- strain; this suggested that amino acid availability was limiting or controlling RNA synthesis. Chloramphenicol together with twenty amino acids gave similar incorporation in both strains (Balassa, 1964a, b). Messenger-RNA, formed in sporulating cells that had been labelled during growth with [^{14}C]uracil and chased with non-radioactive uracil, always contained part of the label. Degradation of stable RNA therefore provided precursors for the resynthesis of several RNA fractions, both stable and unstable. Balassa (1963b, 1964a, 1966b), therefore, proposed the following scheme interrelating RNA and

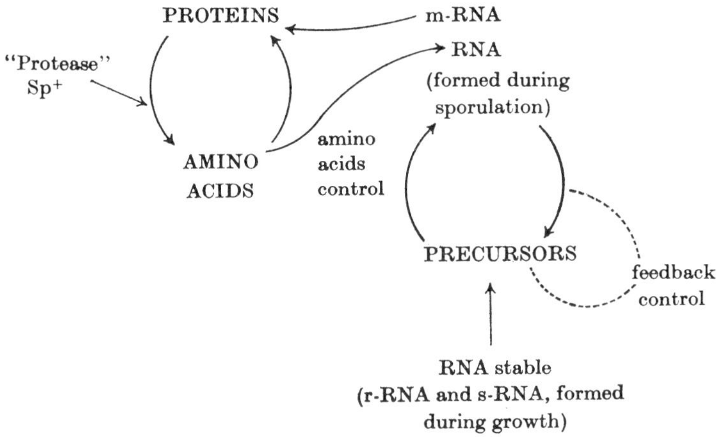

protein turnover. RNA synthesis would be controlled by the supply of amino acids. These would be provided by protein degradation which would be dependent on the induction of a specific intracellular protease produced during sporulation. The rate of protein synthesis would depend on the rate of m-RNA production which in turn would be proportional to the total RNA synthesis. Degradation of RNA might be controlled by a feedback inhibition mechanism exercised by the degradation products. Synthesis of the protease may control the rates of turnover of both RNA and protein. Product inhibition may control both the activity of the protease and degradation of r-RNA (Balassa, 1963b, 1964a, 1966b).

c. *m-RNA*. Labelling with [^{14}C]uracil for periods of 2 and 20 min. at time t_3 in *B. subtilis* indicated that rapid turnover was not limited to the one type of RNA, and provided further evidence for a difference in general metabolism associated with the increased rate of amino acid incorporation in this period (Spotts and Szulmajster, 1962). The major part of the activity with 2 min. labelling occurred in the 8S and 14S RNA fractions; it was very sensitive to the action of ribonuclease. Inhibition by actinomycin D, which inhibits the DNA-dependent RNA polymerase essential for the supply of m-RNA and continued protein synthesis, indicated the continued presence of labile m-RNA until time t_6 at least; incorporation of ^{14}C-uracil was totally inhibited as was that of ^{14}C-methionine within 5 min. of addition. During stage VI, inhibition was established more slowly but protein synthesis was still arrested (Balassa, 1963b; Szulmajster and Canfield, 1963; Szulmajster, 1964). The half-life was 1·5 min., similar to the figure obtained during exponential growth (Balassa, 1963b, 1966c). Del Valle and Aronson (1962) found insensitivity to actinomycin D and 8-azaguanine but sensitivity to chloramphenicol at a late stage in spore formation in *B. cereus* T, and therefore, considered that stable m-RNA was involved. Other investigators have found only actinomycin D-labile and short-lived m-RNA (Spotts and Szulmajster, 1962; Szulmajster, 1964; Balassa, 1963c; Fitz-James, 1963; Doi and Igarashi, 1964c; Leitzmann and Bernlohr, 1965). Further examination of the question using the technique of hybridization of RNA with DNA trapped in agar, showed that sporulation m-RNA was less heterogenous than that obtained from growing cells and that, at time t_{0+}, some m-RNA was synthesized which persisted throughout a large part of the sporulation process. This fraction of m-RNA was "spore message" *per se* as opposed to messages required for metabolic events occurring in the sporulating cell (Aronson, 1965a). Persistent m-RNA accounted for 10–20% of the initial amount. Both membrane-bound and soluble polysomes were found in exponentially growing and sporulating cells. Actinomycin D caused a loss of all soluble polysomes but not the membrane-bound ones. Membrane-bound polysomes from protoplasts of sporulating cells were more resistant to brief ribonuclease treatment and less readily removed from membranes by washing with buffers as compared with membrane-bound polysomes from exponential-phase cells. Sporulating cells treated with actinomycin D continued to synthesize proteins (both soluble and coat proteins) suggesting that polysomes surviving the treatment can make functional polypeptides (Aronson, 1965b).

d. *t-RNA*. This RNA fraction is present throughout sporulation and in the mature spore. The ratio t-RNA to r-RNA increased during sporulation from 17% in exponential-phase cells to 31% in spores (Doi, 1965). The absence of complete symmetry in elution profiles of t-RNA from vegetative cells and spores suggested a qualitative as well as a quantitative difference between the two populations of t-RNA.

e. *r-RNA*. The total ribosome population falls steadily throughout sporulation. The number of 100S particles decreases considerably but the 70S particles persist to the end of sporulation (Fig. 20; Fitz-James, 1963). The cells begin spore formation fully equipped with ribosomes, turnover of RNA being largely restricted to m-RNA. Nascent protein is formed on the ribosomes during the late stages of coat formation, showing that the ribosomes are actively functioning in protein synthesis during sporulation (Fitz-James, 1963). As the spore protoplast volume increases about six-fold (*B. coagulans*) from septation to the end of cortex formation (W. G. Murrell, unpublished observations) it is likely that there is a similar increase in the number of ribosomes as the near-mature spore appears well packed with ribosomes. This increase, however, is probably overshadowed by the large decrease in ribosomes in the mother cell.

f. *Sporal m-RNA*. As m-RNA is synthesized continually during sporulation, and as it is probably the major mechanism for controlling differentiation, studies have been made to determine whether sporulation is controlled at the transcription level and whether the genome being transcribed differs from that transcribed during vegetative growth. Messages necessary for growth, however, may still be produced concurrently with sporulation messages (Balassa, 1963c). The induction of β-galactosidase during sporulation in *B. megaterium* (Aubert and Millet, 1963a, b) favours this view.

Hybrid pattern differences between DNA and RNA of *B. subtilis* during sporulation, germination and stepdown transition (the passage from rapid to slow growth) and hybrid competition experiments indicated that m-RNA was derived from distinct genetic loci. Elution profiles of sporulation and germination m-RNA, labelled with ^{3}H and ^{14}C respectively, were markedly different. These results indicated that the sporal m-RNA population was different from the germination m-RNA; it contained unique RNA molecules not found in exponential-phase or germination RNA populations. The sporal m-RNA, however, contained all of the RNA molecules found in exponential-phase and germination RNA. The results implied that similar genetic loci were transcribed during the cultural phases except for those involved in the sporulation process, i.e. there is a differential transcription of the genome during morphogenesis. New genetic sites were derepressed giving rise to spore-specific m-RNA; genes being expressed during exponential-phase growth were not necessarily repressed (Doi and Igarashi, 1964b, c; Doi, 1965).

Hybrid formation studies between *B. cereus* T exponential-phase and sporulation m-RNA and DNA from other *Bacillus* species, indicated a high degree of conservation of r-RNA cistrons among *Bacillus* species, but that it was unlikely that there was a conservation of genes for spore formation during evolution (Doi and Igarashi, 1965).

D. Synthesis of Spore Components

1. *DPA Biosynthesis*

a. *Time of synthesis.* Although perfect synchrony of sporulating cells will be required to define precisely the starting point of DPA synthesis in relation to cytological development and other biochemical changes, it is evident that DPA synthesis starts early in stage IV and continues through stage V and possibly VI in parallel with cortex formation and Ca^{2+} uptake. The beginning and completion of synthesis may vary in different strains. The rate of formation was soon observed to be related to spore formation (Perry and Foster, 1955; Powell and Strange, 1956). With *Cl. roseum*, the number of immature spores (partially refractile, stainable, non heat-resistant spore forms) increased from 0–100% in 1·5 hr. and about 1 hr. after DPA formation (Halvorson, 1957). A similar sequence was found in *Cl. botulinum* (Day and Costilow, 1964a). Hashimoto *et al.* (1960) could not detect DPA in the forespore (stage III) of *B. cereus* T; 0–50% DPA formation occurred in stage IV, and synthesis continued during stage V and possibly stage VI. During this period refractility increased both in the number of refractile cells and in individual maturing spores; stainability decreased. DPA formation occurred about 1 hr. ahead of heat resistance; 40% of the DPA was synthesized while fewer than 1% of the sporulating cells were resistant to a heat treatment for 30 min. at 80°. In *B. cereus* var. *alesti*, Young and Fitz-James (1962) found the onset of Ca^{2+} uptake and DPA formation occurred at the time that the forespore starts to "whiten" in phase contrast which is about the time that one-third to one half of the cortex has formed; the Ca^{2+} uptake and DPA formation curves are practically superimposed. In the A^+ mutant, under conditions of slower DPA formation, the onset of Ca^{2+} uptake preceded DPA synthesis by almost an hour. The addition of Ca^{2+} to the white phase (late stage IV) cells grown in a complex medium deficient in Ca^{2+} resulted in the immediate uptake of Ca^{2+} and DPA synthesis. If the addition of Ca^{2+} was delayed until maximum refractility in the Ca^{2+}-deficient system had been attained (stage VI?) no Ca^{2+} uptake or DPA formation occurred (Fig. 21; Young and Fitz-James, 1962).

b. *Biosynthesis.* Although 13 years have elapsed since the discovery of DPA in spores (Powell, 1953), its biosynthesis is still unsolved. Several pathways have been explored. The formation from diaminopimelic acid by ring closure and removal of ammonia is not considered to be a major pathway (Perry and Foster, 1955; Martin and Foster, 1958; Finlayson and Simpson, 1961). Extracts of sporulating cells or whole cells failed to synthesize DPA from diaminopimelic acid (Powell and Strange, 1956).

This is perhaps not surprising as most of the diaminopimelic acid is decarboxylated or, if this is prevented, incorporated directly into the cortex (see p. 220). Martin and Foster (1958) found glutamate, aspartate, alanine, proline and serine were the most efficient precursors and that some 6·5% of the carbon of DPA came from CO_2. Labelling of the pyridine and carboxyl groups suggested that the carbon skeleton of DPA was

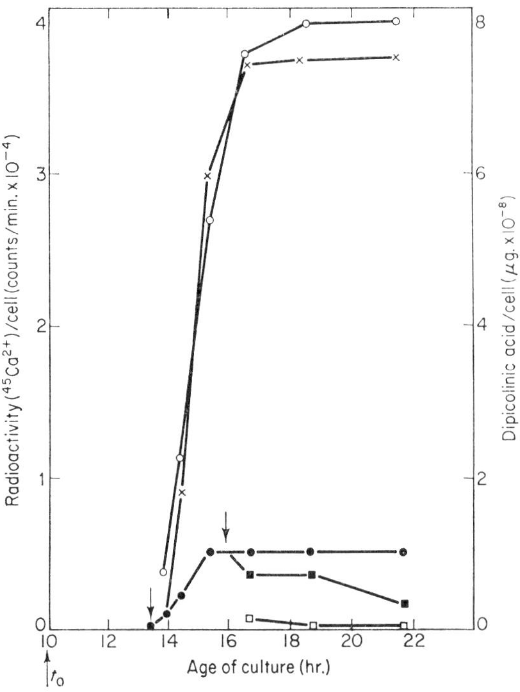

FIG. 21. Effect of the time of addition of Ca^{2+} on the subsequent intake of Ca^{2+} and synthesis of dipicolinic acid in cultures of *Bacillus cereus* var. *alesti* A^- sporulating in a Ca^{2+}-deficient medium. The graph shows synthesis of dipicolinic acid without added Ca^{2+} (●—●), with addition of Ca^{2+} ($10^{-3}M$) at 13·5 hr. (×—×) and at 16 hr. (■—■), and relative intake of Ca^{2+} following addition at 13·5 hr. (○—○) and 16 hr. (□—□). From Young and Fitz-James, 1962.

formed from a C_4+C_3 condensation of aspartate+pyruvate or their derivatives or alanine+oxaloacetate and their derivatives. DPA is also formed non-enzymically from α,ε-diketopimelic acid and ammonia in the presence of various non-specific bacterial homogenates (Powell and Strange, 1959); α,ε-diketopimelic acid was not detectable in cell extracts. It has, however, been isolated from *Penicillium citreo-viride* which forms DPA (Tanenbaum and Kaneko, 1964). Kondo *et al.* (1964) demonstrated diketopimelic acid oxidase activity in the mother cell cytoplasm (the extract from lysozyme-treated sporangia) but not in the forespores. No diketopimelic acid oxidase activity occurred in exponential-phase cells;

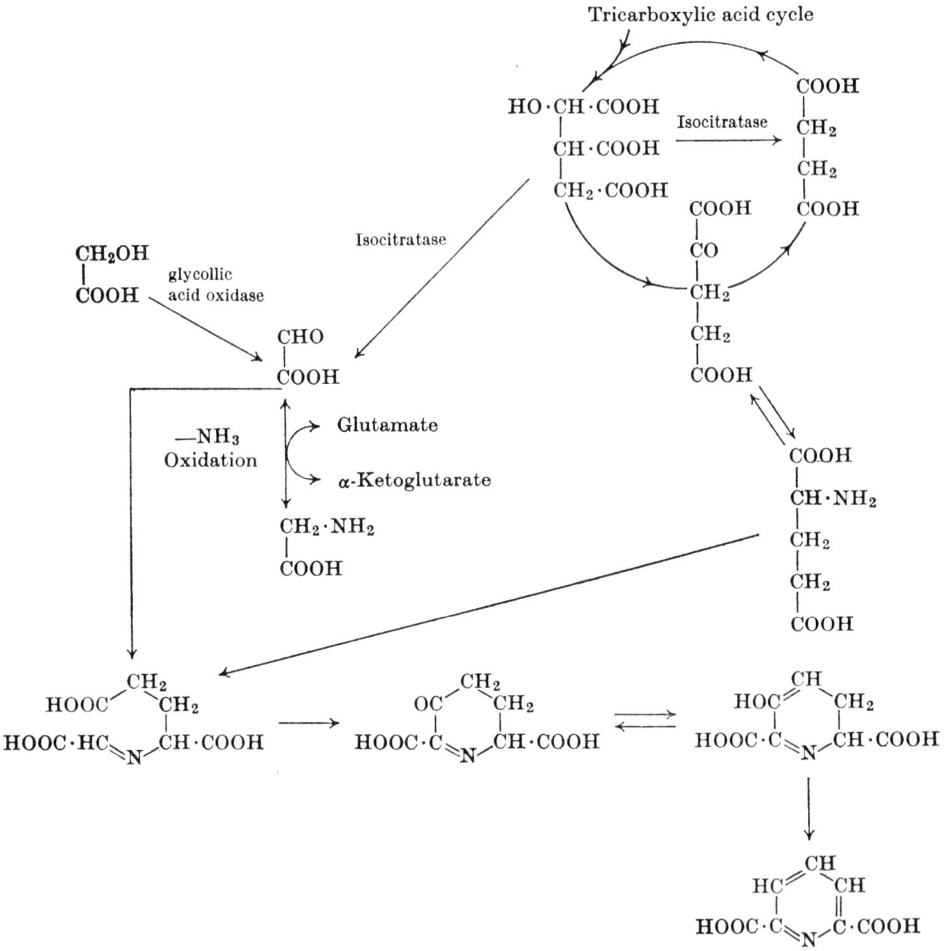

FIG. 22. Hypothetical scheme for the biosynthesis of dipicolinic acid. From Benger, 1962.

the activity developed suddenly at the onset of spore formation. Lysozyme, added at various stages during sporulation to lyse the mother cell but not the forespore, stopped further DPA synthesis.

The central rôle of glutamate in bacterial metabolism, the promotion of spore formation and hence DPA formation by C_2 compounds, the higher DPA content with glycollate- and glyoxylate-grown cells, the presence of enzymes of the TCA and glyoxylic acid cycles in sporulating cells, transamination between glyoxylic acid and glutamic acid, the requirement of Mn^{2+} for the oxidation of glycollic acid to glyoxylic acid and parallel biochemical interrelationships in other cells led Benger (1962, 1966) to postulate a $C_2 + C_5$ condensation as in the scheme outlined in Fig. 22. Glutamate is transaminated to azomethine which, with ring closure (e.g. as in orotic acid biosynthesis), would give 3-oxo-4,5-dihydropyridine-2,6-dicarboxylic acid, which by reduction and dehydrogenation gives DPA. However, the latter intermediate could not be detected.

DPA has, however, been synthesized by extracts of washed sporulating cells from pyruvate and aspartic β-semialdehyde (Bach and Gilvarg,

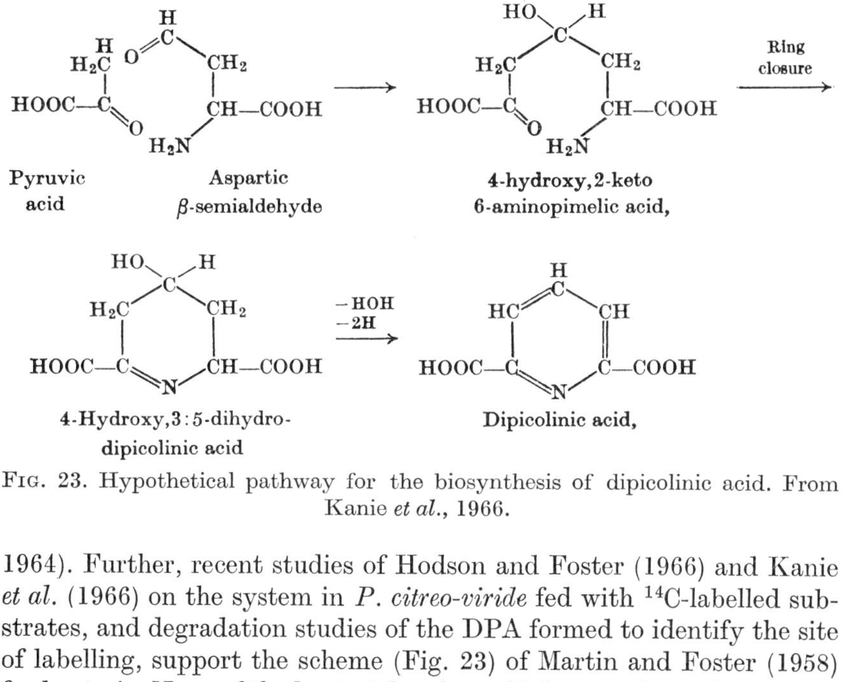

FIG. 23. Hypothetical pathway for the biosynthesis of dipicolinic acid. From Kanie *et al.*, 1966.

1964). Further, recent studies of Hodson and Foster (1966) and Kanie *et al.* (1966) on the system in *P. citreo-viride* fed with ^{14}C-labelled substrates, and degradation studies of the DPA formed to identify the site of labelling, support the scheme (Fig. 23) of Martin and Foster (1958) for bacteria. None of the bacterial and mould data are inconsistent with this, or with the formation from diketopimelic acid (Powell and Strange, 1959; Tanenbaum and Kaneko, 1964). The precursor origins of diketopimelate and ketoaminopimelate are very similar, and the distribution of radio-activity is consistent with the operation of respiratory cycles with the pyridine acid being formed from a C_3 primary building block with incorporation of one molecule of CO_2 (Kanie *et al.*, 1966). The C_7 straight-chain compound which undergoes cyclization has not been identified.

c. *Site of DPA synthesis.* DPA is generally assumed to be synthesized in the developing spore, as the compound has not been detected in the media or associated with vegetative proteins or debris during lysis of the mother cell (Halvorson and Howitt, 1961; Young and Fitz-James, 1962). The experiments of Kondo *et al.* (1964) with lysozyme-treated cells (see p. 218) showed that, in 16–18 hr. cells (time t_0–t_2?), DPA was only present in the supernatant of the lysate; in the same fraction from older cells, the amount of DPA had decreased while the amount in the immature spore fraction increased until, at 22 hr., no DPA was found in the mother cell lysate fraction. Similar results were obtained when

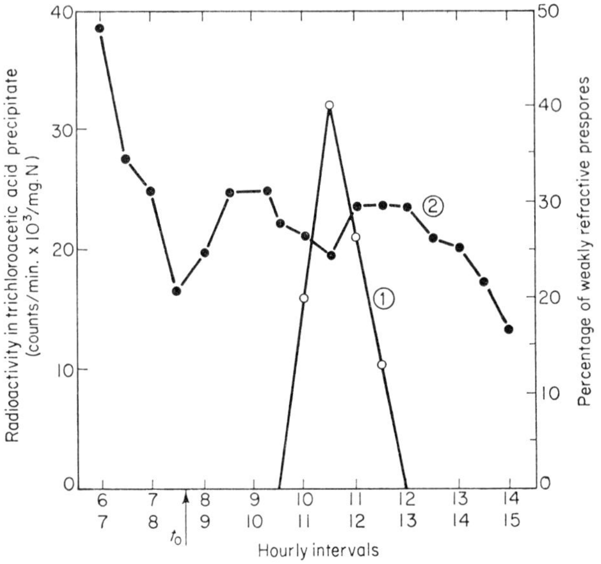

FIG. 24. Incorporation of [2-^{14}C]α,α′-diaminopimelic acid during individual phases of sporogenesis (excess ^{12}Clysine added). Curve 1 shows the increase in the percentage of weakly refractive prespores; curve 2, the radioactivity of the trichloroacetic acid precipitate. From Vinter, 1963.

the cells were lysed gently in hypertonic solutions of polyethylene glycol 5000. It was suggested, therefore, that in the early stages DPA was synthesized mainly in the mother cell or that the mother cell cytoplasm was required for the synthesis (Kondo *et al.*, 1964). A simple alternative explanation is that immature spores after liberation from the mother cell cannot retain DPA while the more mature spores can.

2. *Cortex Formation*

Cortex formation characterizes stage IV, but its synthesis may start earlier, and perhaps final crosslinking and molecular differentiation continue into stage V and even stage VI. Vinter (1963, 1964, 1965a, b) observed a marked incorporation of exogenous diaminopimelate into the hot trichloroacetic acid-insoluble fraction (i.e. spore integuments). As diaminopimelate is located mainly in the cortical glycopeptide (Warth *et al.*, 1963a, b), its incorporation into this fraction can be used to follow synthesis of cortical glycopeptide.

Two peak periods of diaminopimelic acid incorporation occurred in *B. cereus* (Fig. 24), one during septation (perhaps early stage II) and one during "whitening" or increasing refractivity of the prespores (stages IV–VI). Very little of the diaminopimelate of the mother cell was incorporated into the spore (Vinter, 1963). That most of the diaminopimelate was incorporated directly into the cortical glycopeptide was substantiated by studies on the distribution of labelled material in the spores and by inhibition studies. On disruption of the mature spores, most of the label was located in the soluble fraction after action of the autolytic enzymes on the spore integument fraction. Penicillin added during stages IV and V inhibited cortex formation (Fitz-James, 1963) and spore formation with the release of Ca^{2+} and DPA (Vinter, 1964). Inhibition of glycopeptide synthesis and glycopeptide polymer formation by penicillin suggests the necessity of the cortex for retention of Ca^{2+} and DPA, but does not necessarily mean that the cortex is the binding site of Ca^{2+} and DPA. When Ca^{2+} accumulation was complete, penicillin no longer prevented spore development (Vinter, 1964). This suggests either that side-chain addition to the glycopeptide (Fig. 8) and polymerization occurred simultaneously with Ca^{2+} uptake (and hence DPA synthesis), or that completion of the cortical polymeric matrix results in retention of all the Ca^{2+} accumulated in the protoplast or the cortex. Vinter (1964) found that Ca^{2+} release, caused by penicillin, was always accompanied by a slower release of DPA. In some cases further DPA was actually synthesized for a short time after penicillin addition (Vinter, 1962d). Penicillin did not inhibit ^{14}C-diaminopimelate incorporation, although loss of DPA occurred. After 2–3 hr. contact with penicillin, the cells lysed with the release of ^{14}C-diaminopimelic acid (Vinter, 1964); the obvious explanation of this incorporation in the presence of penicillin is that it becomes incorporated into the cortical glycopeptide units or precursors which fail to polymerize and later leak out.

Inhibition of ^{14}C-diaminopimelate incorporation by chloramphenicol (Vinter, 1964) suggests that some protein synthesis or continuing protein synthesis is essential for the synthesis of cortical glycopeptide, for its polymerization, or for its translocation from the site of synthesis (spore protoplast membrane?) to its location in the spore. Chloramphenicol does not affect the synthesis of vegetative cell walls (Hancock and Park, 1958; Mandelstam and Rogers, 1958, 1959). Szulmajster (1964) observed only partial inhibition by chloramphenicol of incorporation of ^{14}C-alanine and ^{14}C-glutamate in their racemic forms at time t_2 (stages II–III); almost immediate and total inhibition of incorporation of the amino acids found in spore envelopes occurred. Addition of ^{14}C-L-alanine led to negligible labelling of the spores. These data indicate that the D-alanine and D-glutamate are probably being added to the glycopeptide side-chain at this stage.

Ca^{2+} uptake and DPA formation slightly preceded diaminopimelate incorporation (Vinter, 1963), more incorporation occurring during and after maximum $^{45}Ca^{2+}$ intake and DPA formation.

The structure (possibly the germ cell wall) into which diaminopimelate was incorporated during stage II is relatively stable, and is not degraded during germination while the structure (probably the cortex proper) labelled during stages IV and V was readily degraded during germination (Vinter, 1965a). Evidence that the germ cell wall differed slightly in chemical structure in *B. cereus* T was indicated by its resistance to lysozyme (Fig. 7c; Warth *et al.*, 1963b).

The similarity in the composition of the glycopeptide in the cortex and vegetative cell (see p. 151), its cytological development, and the inhibition by penicillin of its synthesis, indicate a similar mode of synthesis of the spore cortex and vegetative cell glycopeptide. Further, if the enzymes involved in glycopeptide synthesis are of membrane origin (Chatterjie and Park, 1964), then the enzymes involved are probably located either in one or both membranes of the forespore (Section II, p. 139).

Cycloserine interferes with spore formation most probably during glycopeptide synthesis. Spores matured in the presence of cycloserine are initially refractile, but unstable, initiating germination in 1–2 days at 1°. The refractile spores are only one-fifteenth as heat resistant as control spores (Murrell and Warth, 1965). D-Alanine reverses the cycloserine effect as in vegetative cell-wall glycopeptide synthesis (Strominger, 1962). L-Alanine and other amino acids only slightly decrease the cycloserine effect. Ca^{2+} uptake and DPA formation are near normal, but Mg^{2+} is not excluded to the usual extent. Cycloserine probably prevents the addition of the terminal D-alanine residue to the peptide side-chain. The absence of D-alanine from isolated glycopeptide has not yet been shown. The rapid loss of Ca^{2+} and DPA from cycloserine-grown spores suggests that the terminal D-alanine residue plays a key rôle in the formation and maintenance of a stable cortical structure, essential for spore stability and Ca-DPA retention. Electron microscopy of thin sections during sporulation has shown no difference in cortex formation (R. A. Gordon, D. F. Ohye and W. G. Murrell, unpublished observations). This is contrary to the cycloserine-grown spores pictured

TABLE 25. Some Antibiotics, Enzymes and Proteins Synthesized during Sporulation

Substance	Species	Time of synthesis (sporulation stage)	References
Antibiotics			
Bacitracin	*Bacillus licheniformis*	I–VI	Bernlohr and Novelli, 1960a, 1963
Circulin	*Bacillus circulans*	I	Jann and Eichorn, 1964
Unknown	*Bacillus subtilis*	I	Schaeffer *et al.*, 1963
Toxins			
α-toxin	*Clostridium histolyticum*	IV	Sebald and Schaeffer, 1965
N-Succinylglutamate	*Bacillus megaterium*	I	Aubert *et al.*, 1961
Antigens	*Bacillus cereus*		Baillie and Norris, 1964; Norris and Baillie, 1962
	Bacillus subtilis		Cavallo *et al.*, 1963
Enzymes			
A. Not detectable in logarithmic-phase cells			
Glucose dehydrogenase	*Bacillus cereus*	I	Bach and Sadoff, 1962
Ribosidase	*Bacillus cereus*	IV	Powell and Strange, 1956
Protease	*Bacillus licheniformis*	I–VI	Bernlohr and Novelli, 1963; Bernlohr, 1964
Acetoacetyl-CoA reductase	*Bacillus cereus* T	I–III	Kominek and Halvorson, 1965
B. With greatly increased activity during sporulation			
Adenosine deaminase	*Bacillus cereus*	IV	Powell and Strange, 1956
Alanine racemase	*Bacillus cereus*	VI	Stewart and Halvorson, 1953
$NADH_2$ oxidase (particulate)	*Bacillus subtilis*	I–VI	Szulmajster and Schaeffer, 1961a
Citrate condensing enzyme	*Bacillus cereus* T	I	Hanson *et al.*, 1961
Aconitate hydratase	*Bacillus cereus* T	I–III	Hanson *et al.*, 1963b
Succinate dehydrogenase	*Bacillus cereus* T	→III	Hanson *et al.*, 1963b

TABLE 25—*continued*

Substance	Species	Time of synthesis (sporulation stage)	Reference
Malate dehydrogenase	*Bacillus cereus* T	I–III	Hanson *et al.* (1963b)
Butylene glycol dehydrogenase	*Bacillus cereus* T	I–IV	Kominek and Halvorson, 1965
Diacetyl reductase	*Bacillus cereus* T	I–IV	Kominek and Halvorson, 1965
Fumarate hydratase	*Bacillus cereus* T	→III	Hanson *et al.* (1963b)
Alkaline phosphatase	*Bacillus megaterium*	V–VII	Millet, 1963
Arginase	*Bacillus licheniformis*	I–VI	Ramaley and Bernlohr, 1966
L-Ornithine 2-oxo acid aminotransferase	*Bacillus licheniformis*	I–VI	Ramaley and Bernlohr, 1966
Pyrroline 5-carboxylate reductase	*Bacillus licheniformis*	I–VI	Ramaley and Bernlohr, 1966
$NADH_2$ dehydrogenase	*Bacillus licheniformis*	I–VI	Ramaley and Bernlohr, 1966
Amylase	*Bacillus licheniformis*	I–VI	Bernlohr and Novelli, 1963
C. Of similar activity but with different properties			
Alanine racemase	*Bacillus cereus* T		Stewart and Halvorson, 1954
Catalase	*Bacillus cereus*	I–III	Baillie and Norris, 1963; Lawrence and Halvorson, 1954; Sadoff, 1961
$NADH_2$ oxidase	*Clostridium botulinum*		Green and Sadoff, 1965

previously (Murrell and Warth, 1965); these apparently had already begun cortical breakdown.

3. *Protein Synthesis*

Considerable protein synthesis occurs during sporulation. From the size of the spore protoplast at the time of septation very few vegetative proteins are likely to be directly incorporated. More of these may be synthesized during protoplast enlargement. Most of them may be non-specific to the spore. Accompanying the major metabolic shifts described earlier, there occurs the synthesis of many new enzymes and products. Many of these products are spore-specific and are involved in the mechanisms and control of sporulation and the synthesis of new components and structures. Some of the proteins and enzymes occur in the mother cell and are probably not incorporated into the spore as, for example, are some of the TCA cycle enzymes (Szulmajster and Hanson, 1965). A major part of the protein synthesized is coat protein, but in addition many new enzymes, toxins, surface and spore-specific antigens and polypeptidic antibiotics are synthesized (Table 25). Halvorson (1965) estimates that the amount of *de novo* protein synthesis may be 80% and, with recycling and redistribution, near 100%.

In *B. cereus* var. *alesti*, a protein crystal-forming strain, there was a linear rate of [^{35}S]methionine incorporation from time t_{-2}–$t_{3 \cdot 5}$; then the rate increased greatly coincident with the beginning of coat synthesis (Young and Fitz-James, 1959b). Maximum rates of incorporation of labelled cysteine, lysine and phenylalanine occurred also at this stage, while alanine incorporation decreased (Table 26). The high concentration of alanine in the pool, and the considerable incorporation into the cortex somewhat earlier, possibly explain the lower or decreased incorporation during this period. Practically all of the cysteine + cystine of the mother cell and spore protoplast in *B. megaterium* are incorporated into the spore, mainly into the coats at the stage of coat formation (Vinter, 1959). This incorporation is accompanied by development of radiation resistance and this precedes calcium uptake (Vinter, 1961, 1962c).

When cells of *B. subtilis* are grown in media containing ^{14}C-labelled amino acids for various times, and are placed in equivalent unlabelled media to complete sporulation, the longer they are kept in the labelled medium the greater the amount of the label in the free spores. From times t_2 to t_5, a plateau occurred (Fig. 25), during which only 5% of the final label was incorporated, while after time t_5 (stage V) 25–30% of the final labelling occurred. Szulmajster and Canfield (1963) suggest that between times t_2 and t_5 protein turnover occurs during which enzymes required for sporulation are synthesized; at time t_5 synthesis of spore-

TABLE 26. Protein Synthesis During Spore Formation

Organism	Amino acid incorporated	Rate of incorporation			Reference
		Stage IV	Stage V	Stage VI	
Bacillus megaterium	[^{35}S]cysteine	Normal	Maximum		Vinter (1959)
	[^{35}S]methionine	Normal	Increased	Decreasing	Vinter (1960)
	[^{14}C]alanine	Normal	Decreasing		Vinter (1960)
Bacillus cereus	[^{35}S]cysteine		Maximum (t_3–t_6)		Vinter (1960)
	[^{35}S]methionine		Decreasing		Vinter (1960)
	[^{14}C]alanine		Decreasing		Vinter (1960)
	[^{14}C]lysine	Normal	3–4 × normal	2 × normal	Vinter (1963)
Bacillus subtilis					
Sp^{+} mutant	[^{14}C]phenylalanine	Constant	Much greater		Spotts and Szulmajster (1962)
Sp^{-} mutant	[^{14}C]phenylalanine	Greater than at stage V	Less than at t_2		Szulmajster and Canfield (1963)
	[^{14}C]alanine		Minimal		Szulmajster (1963)

specific proteins begins, probably being incorporated into spore coats. Maximum rate of incorporation also occurred after time t_5 during maximum synthesis (Table 26).

The sources of substrates for protein synthesis have been described earlier (Section IVA, p. 201). The type of protein synthesis appears characteristically normal in that it requires short-life m-RNA and ribosomes, and it is inhibited by chloramphenicol, puromycin, and

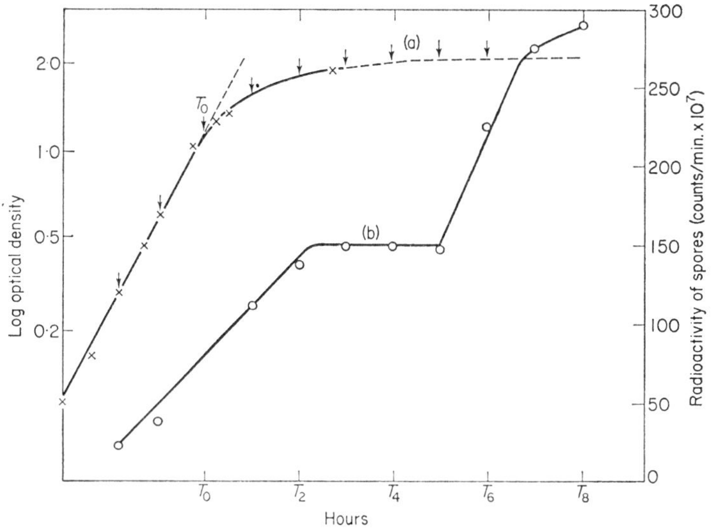

FIG. 25. Synthesis of specific spore protein by *Bacillus subtilis* growing in a medium containing [^{14}C]lysine, [^{14}C]phenylalanine, [^{14}C]proline, and [^{35}S]methionine. Curve (a) is the growth curve, and the arrows indicate the times at which cultures were transferred to non-radio-active media. Curve (b) relates the specific activity of the isolated spores (ordinate) to the time at which the culture was transferred to non-radio-active media (abscissa). From Canfield and Szulmajster, 1964.

actinomycin D to a very late stage (t_{6+}) in sporulation (Del Valle and Aronson, 1962; Fitz-James, 1963; Szulmajster and Canfield, 1963; Canfield and Szulmajster, 1964; Ryter and Szulmajster, 1965). Nascent protein was formed on ribosomes during late spore coat formation (stage V) which could be "chased" by a short exposure to unlabelled amino acids (Fitz-James, 1963). If, as is very likely, coat synthesis occurs in the mother cell (see p. 228), then the synthesis of m-RNA and ribosomal activity at this late stage are probably taking place in the mother cell cytoplasm.

4. *Coat Synthesis and Formation*

The biochemical reactions and mechanisms involved in coat formation are not known.

a. *Site of formation.* Coat formation occurs outside the two plasma membranes enclosing the forming cortex, and in species with an exosporium, within the surrounds of the exosporium (Fig. 5b, p. 141). The site of coat formation initially is thus about 0·1 μ from the spore protoplast, and in later stages when the cortex is about 0·1 μ thick this distance will be doubled; these facts, together with the difficulty of translocating proteins through the two plasma membranes and the cortex, favour the mother cell as the site of synthesis. Other evidence in support of this has been discussed previously (Ohye and Murrell, 1962). Briefly, this evidence is (i) the occurrence of laminated entities both within and outside an outer membrane and exosporium which are similar in many ways to the laminated coat (Tokuyasu and Yamada, 1959a, b), (ii) the presence of protein crystals and parasporal bodies free within a membrane enclosing the spore and sometimes contiguous with the laminated coat (Hannay, 1957, 1961), and (iii) the general chemical similarity of the above to coat proteins as indicated by staining and solubility properties (Hannay, 1953, 1956, 1957). The parasporal bodies and crystals are of protein or phosphoprotein nature (Hannay and Fitz-James, 1955; Fitz-James and Young, 1958) but a detailed chemical comparison is lacking. Analysis of the parasporal crystal of *B. cereus* var. *thuringiensis* shows a general similarity in amino acid composition to coat proteins (Table IV, p. 148), although more serine, leucine and aspartate were present. The continued synthesis of spore proteins at a very late stage, and the susceptibility of this synthesis to inhibition at a time when the inhibitors have little effect on the spores themselves, are observations compatible with such a site.

b. *Mechanism of formation.* In *Cl. pectinovorum* several single scalloped plates of coat material appear in sections outside the outer spore membrane, and these develop into a continuous layer of coat material (Fitz-James, 1962). Similar discrete laminated plates have been observed in *B. coagulans* (Ohye and Murrell, 1962) and in several *Clostridium* species (Takagi *et al.*, 1956, 1960). Fitz-James (1962) suggested that, as the coat plates appear in the narrow zone of cytoplasm between the outer membrane and the plasma membrane of the mother cell, these membranes or the perisporal mesosomes connected to the plasma membrane are intimately associated with coat formation.

In *B. cereus* T (see Fig. 5b, p. 141), the plate(s) of the basal laminae of the coat appears quite free in the cytoplasm on both sides of the oval forespore in a longitudinal cell-cut, within the developing exosporium.

Growth of the laminae, without noticeable thickening, then occurs towards each end of the spore as the cortex forms and the exosporium develops (D. F. Ohye and W. G. Murrell, in preparation for press). This type of development and the occurrence of the above laminated structures suggest a type of oriented deposition or crystal growth within the mother cell and within the exosporium of species with this membrane.

Significant quantities of mother-cell cytoplasm are trapped between the coat layer and the outer membrane of the forespore (Fig. 5b). This has also been observed by Tagaki *et al.* (1960) in *Cl. tetani*. This layer of material must become bonded to the coat and the outer spore membrane, as on disruption the cortex remains attached to the coats (Fig. 7). These data indicate that a chemical reaction between a plasma membrane (by origin) and the coat and cortex involving intermolecular bonding must occur during stage V or VI in the spore. The inclusion of cytoplasmic material in this region also means that various mother-cell proteins, including enzymes and ribosomes, are likely to become located in the innermost layer of the spore coat. These could be important in germination.

Bernlohr and Novelli (1960b, 1963) showed that a doubly-labelled polypeptide (bacitracin) was incorporated into spores without a change in isotope ratio. DL-Erythrochloramphenicol stopped sporulation but not bacitracin synthesis. D-Threochloramphenicol did not affect growth or bacitracin synthesis except at high concentrations. They therefore suggested that bacitracin is not synthesized by the same mechanism as in amino acid incorporation into proteins or in the synthesis of glutamyl-peptide (Hahn *et al.*, 1954). Bacitracin was incorporated into the coats (Bernlohr and Novelli, 1963), and the coats and bacitracin resembled each other in composition (Bernlohr and Sievert, 1962); bacitracin was also released into the medium during sporulation. Snoke (1964) obtained evidence for the presence of at most 0·7% bacitracin in lysozyme-treated spore coats; removal of cortical material by lysozyme digestion could cause a loss of physically incorporated bacitracin. Weinberg and Tonnis (1966) however obtained an inhibition of bacitracin synthesis by the three chloramphenicol isomers at concentrations lower than those required to inhibit sporulation. Further, Brenner *et al.* (1964) found no trace in spore hydrolysates of polymyxin B or of a fragment of this branched cyclic decapeptide containing six residues of L-α,γ-diaminobutyric acid, using a specific enzyme assay for this acid. However, it is quite likely that non-specific spore substances, such as bacitracin and other antibiotics, may be trapped in the mother-cell cytoplasm that is incorporated into the coats; this could easily amount to 2·4% of the dry weight as calculated by Bernlohr and Novelli (1963); such incorporation can be expected to be unaffected by inhibitors at concentrations permitting sporulation to continue, provided synthesis of the substance has already occurred.

In *B. cereus* var. *alesti* and *B. subtilis*, chloramphenicol prevented the development of the coat basal plates into a complete continuous layer (Fitz-James, 1963; Ryter and Szulmajster, 1965), presumably by the immediate stoppage of protein synthesis, rather than the mechanism involved in formation.

Although normal protein synthesis is required for spore coat formation (Section IV D, p. 225) it seems that several mechanisms may be involved in coat assembly such as deposition and crystal-like development, and even physical incorporation of mother cell material with tertiary bonding to hold the layers together; the last of these processes may involve phosphate or phospholipid material.

V. Genetic Control of Spore Morphogenesis and Induction of Spore Formation

In recent years sporulation has been shown, not unexpectedly, to be a complicated genetic exercise in unicellular differentiation (see

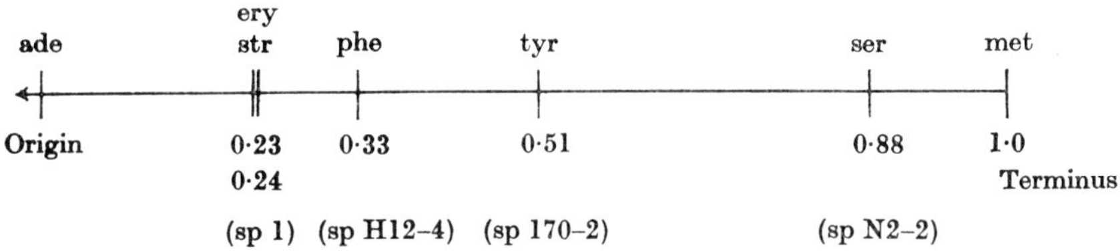

FIG. 26. Genetic map showing the location of amino-acid markers, antibiotic markers, and spore markers on the chromosome of *Bacillus subtilis*. ade, indicates the adenine locus; ery, erythromycin; str, streptomycin; phe, phenylalanine; tyr, tyrosine; ser, serine; met, methionine, sp, spore markers. From Takahashi, 1965b.

Halvorson, 1965). Definite genetic loci are involved, specific sporal m-RNA is formed and the process is under control at both the transcription and translation level.

a. *The spore genome.* Various mutational loci associated with sporulation occur in *Bacillus subtilis*. Schaeffer (1961) observed three unlinked loci with no associated auxotrophic markers. From both transduction and transformation studies, Takahashi (1965a, b) concluded that scores of genes were present at which sporulation mutations could occur; these were unlinked and some of the spore markers were closely linked to several auxotrophic and antibiotic resistance markers. From the mapping distance of these markers, he mapped approximately some of the genes controlling sporulation (Fig. 26). The variety of unlinked genetic sites was considered to be indicative of the many independent biochemical processes involved in sporulation, blockage of any one of which may prevent sporulation.

Spizizen's (1965) transformation studies with *B. subtilis* revealed a class of Sp^- mutants with a genetic defect in one locus; clustered with the gene controlling spore formation were genes for two proteolytic enzymes, an antibacterial substance and a competence factor; the separate gene elements for each were in a definite order. Future studies may confirm that the sporulation genome is comprised of many unlinked genes and several unlinked clusters of genes. Halvorson (1965) estimates that 100 or more structural and regulatory genes are involved, and are distributed in a number of operons that are transcribed in a specified order.

b. *Control of differentiation.* The presence of sporulation-specific m-RNA during sporulation (see p. 215) indicates regulation at the genetic transcription level. The m-RNA base composition, hybrid formation properties and competition between m-RNA populations for loci on DNA imply differential synthesis of m-RNA during morphogenesis, i.e. a controlled transcription of different genetic loci during "step-down" transition, spore formation and germination (Doi and Igarashi, 1964c, 1965). The "step-down" m-RNA was not complementary to DNA, and the authors therefore suggested that only one of the strands of DNA at the specific loci was being transcribed, leading to a sudden derepression of many genetic sites. Kaneko and Doi (1966) have also obtained some evidence of control at the translation level. Soluble RNA is believed to exert a regulatory function by controlling m-RNA translation (Ames and Hartman, 1963; Stent, 1964). The valyl-s-RNA elution profile pattern (obtained with a methylated-albumin-keiselguhr column) of RNA from cells at an early stage of sporulation was definitely altered; it returned to the characteristic pattern of vegetative cells during the late stages of sporulation. The change was a function of the s-RNA and not the aminoacyl-s-RNA synthetase activity, suggesting that a differential synthesis of proteins during morphogenesis may depend on the functional concentration of specific aminoacyl-s-RNA. The pattern for fifteen other amino acids was unchanged (Kaneko and Doi, 1966).

Evidence for the existence of regulatory genes was obtained by Spizizen (1965). One apparently asporogenic mutant, which possessed defects (see above), produced some spores at 30° but not at 37°; another produced the antibacterial factor at 25–30° but not at 37°; by transformation it was shown that both mutants still contained intact gene elements capable of replacing defective regions of other mutants.

During sporulation both the vegetative cell genome and the spore genome appear to be transcribed. β-Galactosidase synthesis was induced in the mother cell at time t_3, but not in the prespore at this stage (Aubert and Millett, 1963b). As indicated above new syntheses of protein bodies and possibly the spore coats and exosporium are controlled by the mother-cell genome. This genetic activity of the mother cell may continue for some time after the metabolism of the spore has come to a halt.

Halvorson (1965) postulates two models for the control of sporulation (Fig. 27). Three properties govern the system: (i) sporulating cells contain a vegetative genome and a spore genome, (ii) transcription of the spore genome requires initial activation and proceeds in an ordered

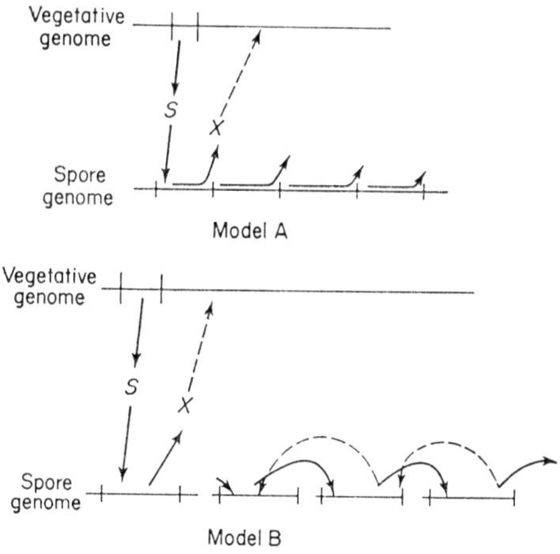

FIG. 27. Models for the regulation of sporulation. Model A postulates the unidirectional transcription of spore genome; model B, the sequential induction and repression of the spore genome. S indicates the sporulation initiation factor; × the inhibition of transcription of the vegetative genome; - - -→, inhibition; ——→, activation or transcription. From Halvorson, 1965.

manner, and (iii) transcription of the vegetative genome in the spore protoplast is inhibited by products of the spore genome. Halvorson suggests that the mechanisms for specifically inhibiting the vegetative genome may involve chemical modification (e.g. methylation of DNA, cross-linking the two DNA strands by antibiotics, etc.), inactivation by combination with specific spore protein, or inhibition of an RNA polymerase that is specific for the vegetative genome. He suggests the sporulation factor (Srinivasan and Halvorson, 1963) and the antibiotic produced early in sporulation (Balassa, 1964b) as possibilities for S and X respectively (see Fig. 27). The models provide for two possible mechanisms for the sequential ordering of gene products.

During spore morphogenesis, besides the ordered transcription occurring in the spore protoplast, there is probably an ordered differentiation occurring in the mother cell itself. Moreover, there occur physical or mechanical effects probably not directly under genetic control. These involve the variable amount of mother-cell cytoplasm enclosed at septation, the physical incorporation of mother-cell cytoplasm into the spore coats, and the tertiary physical and chemical changes in the closely associated layers and structures during the physicochemical changes associated with maturation. These non-genetically controlled variables could have profound effects on the properties of the spores.

c. *Induction of sporulation.* Activation of the spore genome may occur (i) under conditions of mass sporulation, (ii) during exponential growth, and (iii) in primary germ cells during their outgrowth. So far we have been concerned only with sporulation *en masse.* However, in minimal media, despite attempts to synchronize sporulation, some cells constantly form spores during exponential growth (Aubert *et al.*, 1961; Kerravala *et al.*, 1964). Schaeffer *et al.* (1965) studied the probability (P) at which a *B. subtilis* cell is induced to sporulate in the exponential phase of growth in minimal media. The value of P was constant for a particular set of conditions, varying with the medium and the nitrogen source. Except for a few instances, it appeared to be directly related to growth rate. The authors concluded that every cell at some step was faced with the choice of being able to continue growing or to sporulate and that this depended on the repression of the spore genome by the internal concentration of a nitrogen-containing catabolite, the concentration of which could vary from cell to cell. Presumably in minimal media at slow growth rates the concentration of this catabolite varies considerably due to amplification of micro-heterogeneities in the population. In these cells the intracellular concentrations or micro-environments probably resemble those of nearly every cell in a synchronized mass culture during sporogenesis.

Vinter and Slepecky (1965) found that the elongated germ cell of *B. cereus* spores (before first division) sporulates in a diluted germination medium; they termed this "microcycle sporogenesis". Cells less elongated or exponential-phase cells failed to sporulate. Microcycle sporogenesis has also been observed in *B. megaterium* (Vinter and Slepecky, 1965) and in germinating cells in a simple glucose–salts medium (Holmes and Levinson, 1966).

The factors inducing sporogenesis in each of these circumstances are probably similar. The following factors have been postulated as inducers of sporulation.

(i) Depletion of carbon- or nitrogen-containing substrates, growth factors or minerals resulting in the cessation of growth (Grelet, 1957). This is essentially the situation of healthy cells facing starvation after the depletion of nutrients (Knaysi, 1945).

(ii) Sporulation factors. Evidence for the existence of an endogenous factor ("sporogen") in cells committed to sporulation (stage I or II), which induced exponential-phase cells of *B. cereus* T to sporulate, was obtained by Srinivasan and Halvorson (1963). The factor is active for other species and also present in other species, but was not present in five non-sporing species (Srinivasan, 1965). The factor is heat-stable, ninhydrin-negative and resistant to the ribonucleases and trypsin. It manifested its activity in environmental conditions unsuitable for vegetative growth. Srinivasan suggests that it may derepress the spore genome. Wooley and Collier (1966) have isolated two peptide sporogens that are active for *Clostridium roseum*. These contained alanine, serine, glutamate, aspartate, lysine and an unidentified low mol. wt. compound. The possibility of hormone-like factors with somewhat similar rôles occurring in mixed cultures, spent media and in certain substances that stimulate sporulation has also been suggested (Murrell, 1961b).

(iii) Retardation factors. A heat-stable factor in supernatants from 11–12·5 hr. cultures of *Cl. butyricum* had no effect on the total yield of cells; but it decreased the growth rate and induced exponential-phase cells to sporulate (Bergère and Hermier, 1965a, b). The activity persisted in the presence of glucose and fresh nutrients. Retardation of the growth rate from 0·7 to 0·5 divisions/hr. in a chemostat induced mass sporulation in *B. megaterium* (Aubert and Millet, 1961). An Sp^- mutant continued growth when the growth rate was decreased to 0·2 divisions/hr.

(iv) Decreased protein synthesis. Doi and Igarashi (1964a) suggested that sporulation may be triggered by events that decrease overall protein synthesis. This could lead to the activation of proteases associated with sporogenesis (see p. 213), leading to a fresh supply of amino acids for m-RNA and new protein synthesis.

(v) Blocked DNA synthesis. The specific association of antibiotic production with the sporogenic property (Balassa *et al.*, 1963; Balassa, 1964b; Schaeffer *et al.*, 1963) led Schaeffer (1963) to suggest that bacilli may produce an antibiotic to block DNA synthesis. Edein, an antibiotic from *B. brevis*, does this, but has not been associated with sporogeny.

d. *Mechanism of induction.* Possibly all of the above factors activate the induction mechanism directly or indirectly via a common step. But what is the mechanism that induces the nuclear changes resulting in axial thread formation preceding membrane invagination and the ordered sequence of biochemical events? Jacob *et al.* (1960) suggested that the spore genome may exist as an episome in an autonomous state, and non-functional during vegetative growth; activation or integration of the episome would lead to sporogenesis. This would hardly necessitate the nuclear re-arrangements involved (see Part II, p. 136) especially if a recombination and diploid-to-haploid reduction occur as postulated by some authorities (see references given in Murrell, 1955). Some support for the episome theory was obtained by Rogolsky and Slepecky (1964). Acriflavine, which eliminates episomes in the autonomous state, converted a high percentage of early exponential-phase cells to asporogenous mutants. This may be due to the mutagenic action of acriflavine during this growth phase or to the sensitivity of the episome in a detached state during this period; the authors preferred the latter interpretation. Further support has recently been obtained by Bott and Davidoff-Abelson (1966) using acridine orange. The dispersion of the many unlinked genes on the gene map (Fig. 26) and the possible size of the spore genome (>100 genes, 6×10^7 daltons) may indicate that the episome theory is unlikely.

Activation of the spore genome or the latter part of the sequential induction system (Model B, Fig. 27) has been considered to occur by regulatory feed-back mechanisms, reversal of catabolite repression of the spore genome (Schaeffer *et al.*, 1965) or derepression, by substrate depletion, of part or parts of the chromosome. The last of the phenomena certainly seems to be involved in the induction of many enzyme systems associated with sporulation (Hanson *et al.*, 1963a, b; Szulmajster and Hanson, 1965; Blumenthal, 1965; Ramaley and Bernlohr, 1965). Even in microcycle sporogenesis, an acetate utilization system was activated or induced (W. K. Holmes, in preparation for press). Both a carbon and nitrogen source (glucose and glutamate) were involved in repression of synthesis of the TCA cycle enzymes in *B. subtilis* (Hanson *et al.*, 1964b).

From differences in the properties of m-RNA, Doi and Igarashi (1964c) inferred that, during step-down growth transition, there was a sudden derepression of many sites. Whether this resulted from an initial effect on one enzyme or regulator is not known. Studies on catabolite repression led Schaeffer *et al.* (1965) to infer that derepression of at least one enzyme was necessary; as there were probably several operons in the spore genome, the co-ordinated expression was most easily explained by sequential induction; an enzyme of the first operon might be the key enzyme in such a case.

VI. Maturation and the Physico-Chemical State of the Mature Resting Spore

A. Maturation

In Section II (p. 144) and elsewhere, changes in the appearance of spores during stage V and VI were described. These changes are associated with a decreased permeability to dilute stains, Ca^{2+} uptake, DPA formation, and the development of heat resistance. What are the basic chemical and physical changes responsible for the changes in these properties? During stages III and IV, the forespore starts to become partially refractile; the refractile zone enlarges and increases in refractivity to stage VI as the cortex thickens and whitens, as seen in electron micrographs. During stage IV the forespore is still permeable to dilute crystal violet solutions; dehydration is therefore unlikely to be responsible for the refractile change during this period. The apparent growth of the refractile zone, if real and not an optical effect, is important in interpreting the final physicochemical change. Is it due to increasing solute concentrations in the protoplast, compression of the protoplast, or contraction of the thickening cortex into a refractile shell about the protoplast?

The studies of Young and Fitz-James (1962) indicate clearly that Ca^{2+} and DPA are required for the final whitening or ripening process; Ca^{2+} uptake occurs at a linear rate during times t_4–t_6; the requirement for Ca^{2+} precedes, and is essential for DPA formation and retention in the spore (Fig. 21). Ca^{2+} deprivation for too long a period prevented uptake and presumably DPA synthesis. The above maturation changes were reversed. Ca^{2+}-deficient spores often showed cleavages between the cortex and coat, the cytoplasm retained its affinity for Pb^{2+} and the ribosomes remained resolved (Young and Fitz-James, 1962). The spores had a full complement of integuments.

Spores formed subsequent to chloramphenicol addition at stage IV lacked complete spore coats; they were, however, refractile, resistant to a mild heat treatment (80°, 10 min.), had a whitened cortex, and about 20% of the normal amount of DPA. They were unstable, and lysed several hours later (Ryter and Szulmajster, 1965). Thus the coats are essential for preserving the physicochemical state brought about by Ca^{2+} and dipicolinic acid. Further, the results of the cycloserine experiments (see p. 222) suggest that a small change in the chemical structure of the cortex can prevent the attainment of stability in the system, even when the spore possesses cortex and coats.

B. State of the Mature Spore

1. *State of the Protoplast*

Until germination is initiated, the protoplast has negligible respiratory activity (Church and Halvorson, 1957). Recent results using a radiochemical method indicate that, with their turnover rate, endogenous reserves would last for hundreds of years (Desser and Broda, 1965).

Although most of the enzymes in spores are solubilized and activated on disruption and have similar labilities to vegetative enzymes, a few are in a somewhat different state (Halvorson *et al.*, 1966). DNA polymerase required activation during germination by deoxyribonuclease and lipase (Yoshikawa, 1965). A glucose dehydrogenase, which was not intrinsically stable on extraction, could be stabilized a million-fold by lowering the pH value to 6·5 and increasing the salt concentration to $5M$ (corresponding to a water activity of about 0·75). These treatments appeared to convert the enzyme from the dimer form to the monomer. The stability of the monomer was equivalent to that of the enzyme in the intact spore (Sadoff *et al.*, 1965). Monovalent cations were most effective, and the conversion was reversible. This suggests that, if the protoplast in the presence of suitable solutes had a water activity of 0·75, the heat stability of this enzyme and perhaps that of many other labile proteins would be increased.

Some other enzymes have a structural stability while they are bound to particulate material; these include alanine racemase (Stewart and Halvorson, 1954), catalase, and *p*-phenylenediamine oxidase (Murrell, 1955). Bound ribosomes are also present (Aronson, 1965b). The stabilized proteins may occur in layers external to the protoplast. The absence of thymine dimer formation during u.v. irradiation indicates that the physical state of the DNA is different from that found in vegetative cells or in solutions (Donnellan and Setlow, 1965).

2. *Solute Concentrations and Space Considerations*

The high DNA/RNA and nucleic acid/cytoplasm ratios (Section IIIA, p. 162) are indicative of the high density that may exist in the protoplast. Further, if all of the Ca^{2+} and DPA are localized in the protoplast or in the cortex, they will result in very high solute concentrations; if the protoplast constitutes 30% of the spore volume, and the Ca^{2+} and dipicolinic acid are located therein, they may reach concentrations of respectively 9 and 30%. Similar concentrations could occur in the cortex if these components were localized there.

3. *Chemical Stability of the Spore State*

Although the spore may remain viable for decades and is practically inert, it is not perfectly stable. In stored suspension, more free solutes are found on storage (Table 10, p. 159), residual phosphorus-containing material is released (Fitz-James, 1955), there is a change in the rate at which they initiate germination (Knaysi, 1948; Powell, 1950; Murty and Halvorson, 1957; Church and Halvorson, 1957), and viability slowly declines (Evans and Curran, 1960; Marshall *et al.*, 1963). Marshall *et al.* (1963) also showed that the viability changes may occur during storage at various low water activities, greatest stability occurring at a_w values of 0·2–0·3.

4. *Water Content and Permeability*

Four pieces of evidence indicate that spores are probably permeable to water. (i) Lewis *et al.* (1960) calculate that a cell with the dimensions of a spore and composed of the most impermeable material known could not remain impermeable to water; (ii) when spore pellets are suspended in heavy or tritiated water, the labelled water equilibrates with practically all (96+%) of the water in the pellet (Murrell and Scott, 1958; Murrell, 1961a; Black and Gerhardt, 1962). This evidence fails to indicate whether a small anhydrous region surrounded by an impermeable barrier exists in the spore. Correction for hydrogen exchange, which does occur (W. G. Murrell, unpublished observations), decreases the estimate of water space and suggests that a very small volume of water does not equilibrate with external water. Experiments using [^{18}O] water are needed to answer this satisfactorily; (iii) spores contain about 70% water (Murrell, 1961a; Black and Gerhardt, 1962). The dried spore has a density of 1·3–1·5. The space not occupied by water consists of dry space (if any?) plus the molecular space occupied by spore substance, which must include a considerable amount of molecular space from which water is excluded by molecular restrictions such as in haemoglobin (Kendrew, 1963). This is the apparent dry matter volume. Allowing for and assuming a uniform high density of spore dry matter, calculations suggest that the possibility of any anhydrous region in the spore of the size of the spore protoplast is unlikely; (iv) spores can be equilibrated by controlling the vapour phase of the atmosphere to any desired level of dryness and this has profound effects on their heat resistance (Fig. 28; Murrell and Scott, 1966) and viability (Marshall *et al.*, 1963). This implies that water can be moved from the wet region of the spore and so alter greatly the stability of the protoplast. This movement of water can occur very rapidly; for example spores of *Bacillus megaterium*, equilibrated at 0·3 a_w to about 5–10% water content, when placed under conditions giving 0·9–1·0 a_w by the bithermal method begin to die at the rate characteristic of 0·9–1·0 a_w ($D_{110°} = 18$ sec.) within 20–30 sec.; the $D_{110°}$ at 0·3 a_w was about 20 hr. This strongly suggests that water is permeating into the protoplast rather than exerting a physical effect on the integuments which changes the state of the protoplast. Further, if the effect is simply on a region exterior to an anhydrous region, it is difficult to see why the effect should be so different for spores of different species (Fig. 28).

On the other hand, spores swell rapidly on the initiation of germination, to the extent of 20% of the packed cell volume, and this fact together with the accompanying loss of DPA and Ca^{2+} is believed to reflect a rapid uptake of water (Hitchins and Gould, 1962). However, the swelling could result from disruption and plasticity changes in the coats, accompanied by relaxation of the contracted state of the cortex without a gross change in the protoplast volume.

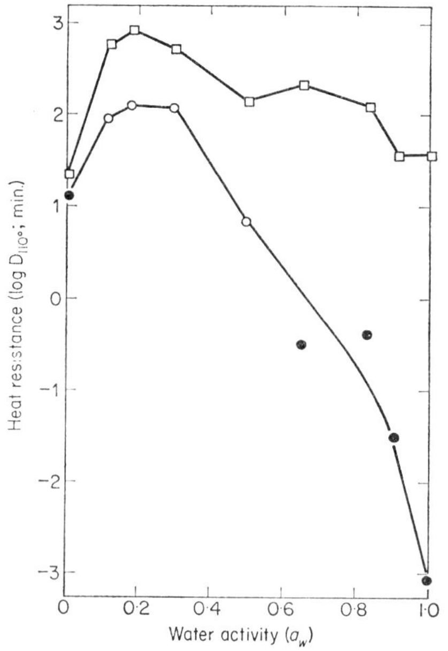

FIG. 28. Heat resistance at various water activities (a_w) at 110° of spores of *Clostridium botulinum* type E (circles) and *Bacillus stearothermophilus* (squares). Spores of the two species in mixed suspension were dried, equilibrated and heated together. Heat resistance is given as the logarithm of the decimal reduction time (D value). From Murrell and Scott, 1966.

5. *Water Sorption Properties*

Water sorption isotherms indicate that spores have a much lower water-binding capacity than vegetative cells (Waldham and Halvorson, 1954; Lewis, 1961; Marshall *et al.*, 1963; Grecz and Smith, 1966). This results from lower free-solute concentrations in the spores (Section IIIA, p. 159) and from molecular cross-linking of the usual hydrated groups (—CO—NH—, —COOH, NH_2).

6. *Solute Permeability*

About 40–50% of the spore volume in *B. cereus* T (including the exosporium) appears to be permeable to many low mol. wt. solutes

(Black *et al.*, 1960; Black and Gerhardt, 1961). With compounds of higher mol. wt., the amount of apparent penetration into the spore was related linearly to molecular size, and the upper limit occurred at a mol. wt. of about 160,000 (Gerhardt and Black, 1961). Lysozyme molecules do not appear to penetrate the spore coat until the coat-disulphide bonds are broken; the molecules then reach the cortex where they attack the glycopeptide (Gould *et al.*, 1963).

In *B. subtilis* only 6% and 29% pyruvate uptake were found at pH values of 4·6 and 3 respectively; exposure to pyruvate at a pH value of 1·1–1·3 for 5 min., however, allows a large uptake (> 60 μmoles/g./hr.), and, on transfer to phosphate buffer at pH 7·3, initiation of germination occurs (Falcone and Brasciani, 1963). The authors suggest that reversible changes in the electrochemical properties of the spore surface and the pyruvate molecule are responsible for the uptake. Uptake of [^{14}C] glutamate and glycine occurred in spores of *B. licheniformis*; even though these amino acids did not initiate germination they became associated almost entirely with the protoplast, and hardly at all in the integument fraction (Martin and Harper, 1965).

VII. Heat Resistance

Space does not permit a discussion of the resistance of spores to other treatments. Heat resistance is probably the most sensitive indicator of the resistant state of the spore, and the basis of heat resistance probably largely accounts for other types of resistance except perhaps radiation resistance, which appears to involve other factors (Vinter, 1957b, 1961).

The considerable progress in our knowledge of the spore still does not give us yet an adequate explanation of the basis of heat resistance in terms that explain the mechanism involved in (i) the vegetative to spore transition and (ii) the large range in heat resistance of different species. Vegetative cells of the sporing organisms vary in heat resistance from those that die at 45° (*Clostridium botulinum* E) to those able to grow at 75° or higher (*Bacillus stearothermophilus*). Each bacterium forms spores similar in structure and DPA contents, but the spores differ in heat resistance by a factor of about 10^5 in a dilute aqueous system (Fig. 28). Thus any hypothesis for the basis of heat resistance needs to explain the increase in heat resistance from the vegetative cell (V) to the spore (S) and S_1 to S_n in the following scheme, in which

$$\begin{array}{ccc} V_1 & \xrightarrow{\quad a_1 \quad} & S_1 \\ b\big\downarrow & & \big\downarrow c \\ V_n & \xrightarrow{\quad a_n \quad} & S_n \end{array} \qquad \text{(about 90,000).}$$

V_1 and S_1 represent the species of lowest resistance and V_n and S_n the species of highest resistance. Factors a_1 to a_n cannot be expressed quantitatively due to the large changes in temperature coefficients. Thermophilic species usually produce more resistant spores. No one has attempted to compute whether the factors a_1 to a_n are constant or whether S_1 to S_n reflect the V_1 to V_n differences.

The theory most commonly suggested to explain the heat resistance of spores in a dilute aqueous system involves impermeability to water coupled with an anhydrous internal region, presumably the protoplast. The results outlined in the previous section (p. 236) strongly suggest that neither of these postulates is true. The coats and deeper layers are permeable to water and many low mol. wt. solutes. Any anhydrous region is likely to be smaller than the protoplast. No mechanism for dehydrating the protoplast or for maintaining it in a dry state has been suggested. To comply with the water status and permeability properties, Black and Gerhardt (1962) have suggested that the protoplast occurs as an insoluble gel with macromolecules cross-linked through stable but chemically reversible bonds to form a high polymer matrix with entrapped free water. There is no direct evidence to support this theory.

Before this theory was formulated and for similar reasons, Lewis *et al.* (1960) postulated the contractile cortex theory. This envisaged the protoplast as stabilized by a decrease in its water content obtained by exertion of hydrostatic pressure following contraction of the cortex. This theory is compatible with water permeability. Water need not be entirely excluded and any water in the protoplast could exchange with labelled external water, as no impermeable barrier is postulated. The glycopeptide polymeric composition of the cortex with the many free carboxyl groups (Fig. 8, p. 156) and its similarity to typical acid-type ion exchangers suggest that the cortex would be ideal for a contraction mechanism. This led Warth *et al.* (1963b) to suggest that a decrease in the electrostatic repulsion of the carboxyl groups by a cation may provide the contraction mechanism. This theory was tested by examining the relationship between heat resistance and hexosamine and diaminopimelic acid contents as indicators of the backbone and the side-chain respectively of the glycopeptide, in *Bacillus* species with a 700-fold difference in heat resistance. Metal and DPA contents were also examined (Murrell and Warth, 1965). The highly significant relation between diaminopimelic acid content and the Mg/Ca ratio and heat resistance support the theory (Fig. 28). A multiple regression equation including values for hexosamine content, Mg/Ca ratio and Ca^{2+} content predicted a heat resistance with a multiple correlation coefficient of 0·96. If Ca^{2+} is chelated to DPA, as seems likely, then some other cation would be required to decrease the electrostatic repulsion of the carboxyl groups.

That the cortex is in a contracted or compressed state in the intact spore is indicated by the difference in thickness of this layer in Fig. 5d and Fig. 7a–c; in the former the thickness is $< 0{\cdot}1\ \mu$, and in the latter $> 0{\cdot}2\ \mu$ at the thinnest part. The presence of an infolded inner margin also suggests that when the spore is disrupted the cortex goes into the uncontracted state (see also Fig. 1, Murrell and Warth, 1965; Warth *et al.*, 1963b). Increased heat resistance under increased hydrostatic pressure (Johnson and Zobell, 1949) also supports the contractile cortex theory.

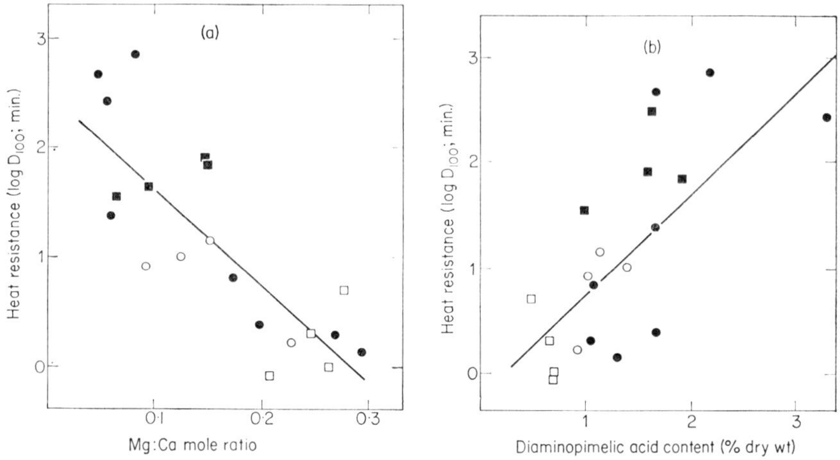

FIG. 29. Relationships between heat resistances and chemical composition of spores. Graph (a) shows the relationships between heat resistance and the magnesium:calcium mole ratios in different spores. Circles show data for species with normal morphology; squares, for species with extra integuments. Closed symbols are for spores grown at 50°; open symbols, spores grown at 30°. The regression equation for twenty preparations ($y = 2{\cdot}54 - 8{\cdot}95x$) was significant at the $P = 0{\cdot}001$ level. The heat resistance is given as the logarithm of the decimal reduction time. Graph (b) shows the α,ϵ-diaminopimelic acid contents of *Bacillus* spores in relation to their heat resistance. The regression equation ($y = -0{\cdot}296 + 0{\cdot}95x$) was significant at the $P = 0{\cdot}001$ level. Symbols are as in graph (a). Heat resistance is given as the logarithm of the decimal reduction time. From Murrell and Warth, 1965.

The theory could provide a unifying hypothesis to account not only for the increased resistance as a result of sporulation but also for the large differences between the heat resistances of species. Variations in the chemical composition of the glycopeptide (such as modification of the side-chain frequency (diaminopimelate concentration), structure and cross-linking) and in the cation concentration, could result in differences in the degree of contraction and the final water content of the protoplast, thereby leading to marked differences in heat resistance.

If the observations on heat resistance of spores equilibrated at controlled water activities (Fig. 28) mean that the water content of the protoplast is being varied, then these data suggest that the protoplasts of spores of *Cl. botulinum* type E in a dilute aqueous system should have a relatively high water content. A decrease in this water content results in a large increase in heat resistance. On the other hand, the spores of *B. stearothermophilus* show only a small response to the treatment suggesting that the water content of their protoplasts is already quite low when they are suspended in a dilute aqueous system. Spores of intermediate resistance had heat resistance versus a_w response curves intermediate between these extremes (Murrell and Scott, 1966).

The cortical system could also operate by the cortical glycopeptide behaving as an elastic ion-exchange gel which swells against relatively inextensible coats, and so applies pressure to the protoplast; the degree of swelling would depend on cation load and other factors (Alderton and Snell, 1963).

The importance of the cortex in heat resistance and stability is evident from studies with inhibitors (see p. 221). Coats, however, are essential for stability as calcium and DPA are not retained for long in coat-deficient spores (Fitz-James, 1963; Ryter and Szulmajster, 1965).

Other factors involved in heat resistance include the prevention of protein denaturation by ions and the high Ca-DPA concentrations. Nucleic acid polymers are known to protect proteins from denaturation. The high nucleic acid: cytoplasm ratio together with the high Ca-DPA concentrations, if the latter are located in the protoplast, may provide a highly dense and stable system within the integuments. In the words of Powell (1957) this may mean that "the protoplasm of the spore is in a highly condensed "waterproofed" system stabilized by the incorporation of calcium dipicolinate and possibly by the constitution of the spore coat". However, different degrees of solute packing would be necessary to explain species differences as the Ca^{2+} and DPA contents are of a similar order in all species.

VIII. Conclusions

This survey of a complicated biochemical product, the bacterial endospore, which results from the operation of many complex biochemical reactions reveals many aspects that need to be examined in more minute detail. Many interesting and sophisticated experiments are awaiting to be performed before our knowledge of the spore and how it is formed is complete. Studies up to now have only begun to unravel the mystery of this minute cell. Many of the reactions involved in its differentiation and in the genetic control mechanisms that operate are likely to be basic to biological systems in general.

Some of the more important aspects requiring further detailed precise study may be listed—the method of DNA replication and segregation at time t_0; the mechanism of derepression of the spore genome; metal–DPA relationships and the binding of DPA in the spore; location of Ca^{2+} and DPA; biosynthesis of DPA, cortex and coats; accurate physical and chemical studies on the state of the spore components in the resting spore and the mechanism of heat resistance. The mechanism of cortex formation requires elucidation. Does this involve both the inner and outer plasma membranes of the spore and is it laid down in the contracted state as it polymerizes in the presence of calcium and DPA; does this explain the increase in volume of the refractile region? Very precise correlations of Ca^{2+} uptake and DPA formation together with glycopeptide synthesis, coat formation and development of other properties in perfectly synchronized cultures are required. The interesting biochemical reactions involved in many of the stages of differentiation need unravelling and relating accurately to cytological development; the site of their occurrence needs pin-pointing.

If this study suggests the course of future research to a few readers then my time will not have been in vain.

IX. Acknowledgements

I would like to thank D. F. Ohye for the electron micrographs, and all authors and journals for kind permission to use their figures for illustration. Figures 4, 7, 10, 19 and 21 are reproduced with kind permission of the Rockefeller University Press from the *Journal of Biophysical and Biochemical Cytology* and the *Journal of Cell Biology*.

References

Alderton, G. and Snell, N. (1963). *Biochem. biophys. Res. Commun.* **10**, 139.

Alper, R., Lundgren, D. G., Marchessault, R. H. and Cote, W. A. (1963). *Biopolymers* **1**, 545.

Ames, B. N. and Hartman, P. (1963). *Cold Spr. Harb. Symp. quant. Biol.* **28**, 569.

Aronson, A. I. (1965a). *J. molec. Biol.* **11**, 576.

Aronson, A. I. (1965b). *J. molec. Biol.* **13**, 92.

Aronson, A. I. and Del Valle, M. R. (1964). *Biochim. biophys. Acta* **87**, 267.

Aubert, J. P. and Millet, J. (1961). *C. R. Acad. Sci., Paris* **253**, 1880.

Aubert, J. P. and Millet, J. (1963a). *C. R. Acad. Sci., Paris* **256**, 1866.

Aubert, J. P. and Millet, J. (1963b). *C. R. Acad. Sci., Paris* **256**, 5442.

Aubert, J. P., Millet, J. and Casroriadis-May, C. (1961). *C. R. Acad. Sci., Paris* **253**, 1731.

Bach, M. and Gilvarg, C. (1964). *Fed. Proc.* **23**, 313.

Bach, J. A. and Sadoff, H. L. (1962). *J. Bact.* **83**, 699.
Bailey, G. F., Karp, S. and Sacks, L. E. (1964). *J. Bact.* **89**, 984.
Baillie, A. and Norris, J. R. (1962). *J. appl. Bact.* **25**, vii.
Baillie, A. and Norris, J. R. (1963). *J. appl. Bact.* **26**, 102.
Baillie, A. and Norris, J. R. (1964). *J. Bact.* **87**, 1221.
Balassa, G. (1963a). *Biochim. biophys. Acta* **72**, 497.
Balassa, G. (1963b). *Biochim. biophys. Acta* **76**, 410.
Balassa, G. (1963c). *In* "Mécanismes de régulation des activités cellulaires chez les microorganismes", (J. Senez, ed.), p. 565, Gordon and Breach, New York.
Balassa, G. (1964a). *Biochim. biophys. Res. Commun.* **15**, 236.
Balassa, G. (1964b). *Biochim. biophys. Res. Commun.* **15**, 240.
Balassa, G. (1966a). *Ann. Inst. Past.* **110**, 17.
Balassa, G. (1966b). *Ann. Inst. Past.* **110**, 316.
Balassa, G. (1966c). *Ann. Inst. Past.* **110**, 175.
Balassa, G. and Contesse, G. (1965). *Ann. Inst. Past.* **109**, 683.
Balassa, G., Ionesco, H. and Schaeffer, P. (1963). *C. R. Acad. Sci., Paris* **257**, 986.
Benassi, G. and Bonanni, M. V. (1965). *Boll. Soc. Ital. Biol. Sper.* **41**, 228.
Benger, H. (1962). *Z. Hyg. Infektionskr.* **148**, 318.
Benger, H. (1966). *Z. Bakteriol.* (in Press).
Berger, J. A. and Marr, A. G. (1960). *J. gen. Microbiol.* **22**, 147.
Bergère, J. L. and Hermier, J. (1964). *Ann. Inst. Past.* **106**, 214.
Bergère, J. L. and Hermier, J. (1965a). *Ann. Inst. Past.* **109**, 80.
Bergère, J. L. and Hermier, J. (1965b). *Ann. Inst. Past.* **109**, 391.
Bernlohr, R. W. (1964). *J. biol. Chem.* **239**, 538.
Bernlohr, R. W. (1965a). *In* "Spores III", (L. L. Campbell and H. O. Halvorson, eds.), p. 75, Amer. Soc. Microbiol., Ann Arbor.
Bernlohr, R. W. (1965b). *Bact. Proc.* 36.
Bernlohr, R. W. and Novelli, G. D. (1959). *Nature, Lond.* **184**, 1256.
Bernlohr, R. W. and Novelli, G. D. (1960a). *Arch. Biochem. Biophys.* **87**, 232.
Bernlohr, R. W. and Novelli, G. D. (1960b). *Biochim. biophys. Acta* **41**, 541.
Bernlohr, R. W. and Novelli, G. D. (1963). *Arch. Biochem. Biophys.* **103**, 94.
Bernlohr, R. W. and Sievert, C. (1962). *Biochem. biophys. Res. Commun.* **9**, 32.
Bishop, H. L. and Doi, R. H. (1966). *J. Bact.* **91**, 695.
Black, S. H. and Arredondo, M. I. (1966). *Experientia* **22**, 77.
Black, S. H. and Gerhardt, P. (1961). *J. Bact.* **82**, 743.
Black, S. H. and Gerhardt, P. (1962). *J. Bact.* **83**, 960.
Black, S. H., Hashimoto, T. and Gerhardt, P. (1960). *Canad. J. Microbiol.* **6**, 213.
Blumenthal, H. J. (1965). *In* "Spores III", (L. L. Campbell and H. O. Halvorson, eds.), p. 222, Amer. Soc. Microbiol., Ann Arbor.
Bott, K. F. and Davidoff-Abelson, R. (1966). *J. Bact.* **92**, 229.
Bott, K. F. and Lundgren, D. G. (1964). *Radiation Res.* **21**, 195.
Bradley, D. E. and Franklin, J. G. (1958). *J. Bact.* **76**, 618.
Brenner, M., Gray, E. and Paulus, H. (1964). *Biochim. biophys. Acta* **90**, 401.
Buono, F., Testa, R. and Lundgren, D. G. (1966). *J. Bact.* **91**, 2291.
Canfield, R. A. and Szulmajster, J. (1964). *Nature, Lond.* **203**, 496.
Cavallo, G., Falcone, G. and Imperato, S. (1963). *Bact. Proc.* 25.
Charney, J., Fisher, W. P. and Hegarty, C. P. (1951). *J. Bact.* **62**, 145.
Chatterjie, A. N. and Park, J. T. (1964). *Proc. nat. Acad. Sci. Wash.* **51**, 9.
Church, B. D., Halvorson, H., Ramsey, D. S. and Hartman, R. S. (1956). *J. Bact.* **72**, 242.
Church, B. D. and Halvorson, H. (1957). *J. Bact.* **73**, 470.
Church, B. D. and Halvorson, H. (1959). *Nature, Lond.* **183**, 124.

Clarke, M. C. and Swaine, D. J. (1962). *Geochim. Cosmochim. Acta* **26**, 511.
Curran, H. R. (1957). *Publ. Amer. Inst. Biol. Sci.*, No. 5, p. 1.
Curran, H. R., Brunstetter, B. C. and Myers, A. T. (1943). *J. Bact.* **45**, 485.
Day, L. E. and Costilow, R. N. (1964a). *J. Bact.* **88**, 690.
Day, L. E. and Costilow, R. N. (1964b). *J. Bact.* **88**, 695.
Del Valle, M. R. and Aronson, A. I. (1962). *Biochem. biophys. Res. commun.* **9**, 421.
Dennis, E. S. and Wake, R. G. (1966). *J. molec. Biol.* **15**, 435.
Desser, H. and Broda, E. (1965). *Nature, Lond.* **206**, 1270.
Doi, R. H. (1965). *In* "Spores III", (L. L. Campbell and H. O. Halvorson, eds.), p. 111, Amer. Soc. Microbiol., Ann Arbor.
Doi, R. H. and Halvorson, H. (1961). *J. Bact.* **81**, 642.
Doi, R. H. and Igarashi, R. T. (1964a). *Nature, Lond.* **203**, 1092.
Doi, R. H. and Igarashi, R. T. (1964b). *J. Bact.* **87**, 323.
Doi, R. H. and Igarashi, R. T. (1964c). *Proc. nat. Acad. Sci. Wash.* **52**, 755.
Doi, R. H. and Igarashi, R. T. (1965). *J. Bact.* **90**, 384.
Donnellan, J. E. and Setlow, R. B. (1965). *Science* **149**, 308.
El-Bisi, H. M., Lechowich, R. V., Amaha, M. and Ordal, Z. J. (1962). *J. food Sci.* **27**, 219.
Ellar, D. J. and Lundgren, D. G. (1966). *J. cell. Biol.* (in Press).
Evans, F. R. and Curran, H. R. (1960). *J. Bact.* **79**, 361.
Falaschi, A., Spudich, J. and Kornberg, A. (1965). *In* "Spores III", (L. L. Campbell and H. O. Halvorson, eds), p. 88, Amer. Soc. Microbiol., Ann Arbor.
Falcone, G. and Brasciani, F. (1963). *Experientia* **19**, 152.
Falcone, G. and Cavallo, G. (1958). *Atti. r. Acad. Lincai* **24**, 759.
Farkas, J. and Kiss, I. (1965). *Acta Microbiol.* **12**, 15.
Finlayson, A. J. and Simpson, F. J. (1961). *Canad. J. Biochem. Physiol.* **39**, 1551.
Fitz-James, P. C. (1955). *Canad. J. Microbiol.* **1**, 502.
Fitz-James, P. C. (1957). *Publ. Amer. Inst. biol. Sci.*, No. 5, p. 85.
Fitz-James, P. C. (1960). *J. biophys. biochem. Cytol.* **8**, 507.
Fitz-James, P. C. (1962). *J. Bact.* **84**, 104.
Fitz-James, P. C. (1963). *In* Mécanismes de régulation des activités cellulaires chez les microorganismes", (J. Senez, ed.), p. 529, Gordon and Breach, New York.
Fitz-James, P. C., Robinow, C. F. and Bergold, G. H. (1954). *Biochim. biophys. Acta* **14**, 346.
Fitz-James, P. C. and Young, I. E. (1958). *J. biophys. biochem. Cytol.* **4**, 639.
Fitz-James, P. C. and Young, I. E. (1959). *J. Bact.* **78**, 743.
Fleming, H. P. (1964). Ph.D. Thesis: University of Illinois, Urbana, Illinois,
Fleming, H. P. and Ordal, Z. J. (1964). *J. Bact.* **88**, 1529.
Foerster, H. F. and Foster, J. W. (1966). *J. Bact.* **91**, 1333.
Foster, J. W. and Heiligman, F. (1949). *J. Bact.* **57**, 613.
Gerhardt, P. and Black, S. H. (1961). *J. Bact.* **82**, 750.
Gerhardt, P. and Ribi, E. (1964). *J. Bact.* **88**, 1774.
Gollakota, K. G. and Halvorson, H. O. (1960). *J. Bact.* **79**, 1.
Gollakota, K. G. and Halvorson, H. O. (1963). *J. Bact.* **85**, 1386.
Gorini, L. and Audrain, L. (1956). *Biochim. biophys. Acta* **19**, 289.
Gould, G. W., Georgala, D. L. and Hitchins, A. D. (1963). *Nature, Lond.* **200**, 385.
Gould, G. W. and Hitchins, A. D. (1963a). *J. gen. Microbiol.* **33**, 413.
Gould, G. W. and Hitchins, A. D. (1963b). *Nature, Lond.* **197**, 622.
Grecz, N. and Smith, R. F. (1966). *Bact. Proc.* 32.
Grecz, N., Upadhyay, J. and Tang, T. C. (1966). *Abstr. Biophys. Society*, p. 113.
Green, J. H. and Sadoff, H. L. (1965). *J. Bact.* **89**, 1499.
Grelet, N. (1952). *Ann. Inst. Past.* **82**, 310.

Grelet, N. (1955). *Ann. Inst. Past.* **88**, 60.
Grelet, N. (1957). *J. appl. Bact.* **20**, 315.
Gruft, H., Buckman, J. and Slepecky, R. A. (1965). *Bact. Proc.* 37.
Hachisuka, Y. and Kuno, T. (1963). *Int. Symp. Physiol. Ecol. Biochem. of Germination, Greifswald, Germany,* Sept. 1963.
Hachisuka, Y., Tochikubo, K. and Murachi, T. (1965). *Nature, Lond.* **207**, 220.
Hahn, F. E., Wisseman, C. L. and Hopps, H. E. (1954). *J. Bact.* **67**, 675.
Halvorson, H. O. (1957). *J. appl. Bact.* **20**, 305.
Halvorson, H. (1962). *In* "The Bacteria", (I. C. Gunsalus and R. Y. Stanier, eds.), Vol. 4, p. 223, Academic Press Inc., New York.
Halvorson, H. O. (1965). *In* "Function and Structure in Micro-organisms", *Symp. Soc. gen. Microbiol.* **15**, 343 (M. R. Pollock and M. H. Richmond, eds.), Cambridge University Press.
Halvorson, H. and Howitt, C. (1961). *In* "Spores II", (H. O. Halvorson, ed.), p. 149, Burgess Publ. Co., Minneapolis.
Halvorson, H. O., Vary, J. C. and Steinberg, W. (1966). *Annu. Rev. Microbiol.* **20**, 169.
Hancock, R. and Park, J. T. (1958). *Nature, Lond.* **181**, 1050.
Hannay, C. L. (1953). *Nature, Lond.* **172**, 1004.
Hannay, C. L. (1956). *In* "Bacterial Anatomy", *Symp. Soc. gen. Microbiol.* **6**, 318 (E. T. C. Spooner and B. A. D. Stocker, eds.).
Hannay, C. L. (1957). *J. biophys. biochem. Cytol.* **3**, 1001.
Hannay, C. L. (1961). *J. biophys. biochem. Cytol.* **9**, 285.
Hannay, C. L. and Fitz-James, P. C. (1955). *Canad. J. Microbiol.* **1**, 694.
Hanson, R. S., Blicharska, J. and Szulmajster, J. (1964a). *Biochem. biophys. Res. Commun.* **17**, 1.
Hanson, R. S., Blicharska, J., Arnaud, M. and Szulmajster, J. (1964b). *Biochem. biophys. Res. Commun.* **17**, 690.
Hanson, R. S., Srinivasan, V. R. and Halvorson, H. O. (1961). *Biochem. biophys. Res. Commun.* **5**, 457.
Hanson, R. S., Srinivasan, V. R. and Halvorson, H. O. (1963a). *J. Bact.* **85**, 451.
Hanson, R. S., Srinivasan, V. R. and Halvorson, H. O. (1963b). *J. Bact.* **86**, 45.
Hardwick, W. A. and Foster, J. W. (1952). *J. gen. Physiol.* **35**, 907.
Hardwick, W. A. and Foster, J. W. (1953). *J. Bact.* **65**, 335.
Harper, M. K., Curran, H. R. and Pallansch, M. J. (1964). *J. Bact.* **88**, 1338.
Hashimoto, T. and Gerhardt, P. (1960). *J. biophys. biochem. Cytol.* **7**, 195.
Hashimoto, T., Black, S. H. and Gerhardt, P. (1960). *Canad. J. Microbiol.* **6**, 203.
Hermier, J. (1964). *Ann. Inst. Past.* **106**, 214.
Hitchins, A. D. and Gould, G. W. (1962). *J. gen. Microbiol.* **28**, iii.
Hodson, P. H. and Beck, J. V. (1960). *J. Bact.* **79**, 661.
Hodson, P. H. and Foster, J. W. (1965). *J. Bact.* **90**, 1503.
Hodson, P. H. and Foster, J. W. (1966). *J. Bact.* **91**, 562.
Holbert, P. E. (1960). *J. biophys. biochem. Cytol.* **7**, 373.
Holmes, K. C. and Monro, R. E. (1965). *J. molec. Biol.* **14**, 572.
Holmes, P. K. and Levinson, H. S. (1966). *Bact. Proc.* 16.
Hunnell, J. W. and Ordall, Z. J. (1961). *In* "Spores II", (H. O. Halvorson, ed.), p. 101, Burgess Publishing Co., Minneapolis.
Jacob, F., Schaeffer, P. and Wollman, E. L. (1960). *In* "Microbial Genetics", *Symp. Soc. gen. Microbiol.* **10**, 67 (W. Hayes and R. C. Clowes, eds.), Cambridge University Press.
Jann, G. J. and Eichhorn, H. H. (1964). *Bact. Proc.* 13.
Janssen, F. W., Lund, A. J. and Anderson, L. E. (1958). *Science* **127**, 26.

Johnson, F. H. and Zobell, C. E. (1949). *J. Bact.* **57**, 359.
Kadota, H. and Iijima, K. (1965). *Agr. biol. Chem.* **29**, 80.
Kadota, H., Iijima, K. and Uchida, A. (1965). *Agr. biol. Chem.* **29**, 870.
Kaneko, I. and Doi, R. H. (1966). *Proc. nat. Acad. Sci. Wash.* (in Press).
Kanie, M., Fujimoto, S. and Foster, J. W. (1966). *J. Bact.* **91**, 570.
Kendrew, J. C. (1963). *Science* **139**, 1259.
Kerravala, Z. J., Srinivasan, V. R. and Halvorson, H. O. (1964). *J. Bact.* **88**, 374.
Keynan, A., Murrell, W. G. and Halvorson, H. O. (1962). *J. Bact.* **83**, 395.
Knaysi, G. (1945). *J. Bact.* **49**, 473.
Knaysi, G. (1948). *Bact. Rev.* **12**, 19.
Knaysi, G. (1965). *J. Bact.* **90**, 453.
Kolodziej, B. J. and Slepecky, R. A. (1964). *J. Bact.* **88**, 821.
Kominek, L. A. (1964). Ph.D. Thesis: University of Illinois, Urbana, Illinois.
Kominek, L. A. and Halvorson, H. O. (1965). *J. Bact.* **90**, 1251.
Kondo, M. and Foster, J. W. (1965). *Bact. Proc.* 36.
Kondo, M., Takeda, Y. and Yoneda, M. (1964). *Bikens Jour.* **7**, 153.
Krask, B. J. (1953). *J. Bact.* **66**, 374.
Kuratomi, K., Ohno, K. and Akabori, S. (1957). *J. Biochem. (Tokyo)* **44**, 183.
Laishley, E. J. and Bernlohr, R. W. (1966). *Bact. Proc.* 97.
Lawrence, N. L. and Halvorson, H. O. (1954). *J. Bact.* **68**, 334.
Lechowich, R. V. and Ordal, Z. J. (1962). *Canad. J. Microbiol.* **8**, 287.
Lee, W. H. and Ordal, Z. J. (1963). *J. Bact.* **85**, 207.
Leitzmann, C. and Bernlohr, R. W. (1965). *J. Bact.* **89**, 1506.
Levinson, H. S. and Hyatt, M. T. (1964). *J. Bact.* **87**, 876.
Levinson, H. S., Hyatt, M. T. and Moore, F. E. (1961). *Biochem. biophys. Res. Commun.* **5**, 417.
Levinson, H. S. and Sevag, M. G. (1954). *J. Bact.* **67**, 615.
Levinson, H. S., Sloan, J. D. and Hyatt, M. T. (1958). *J. Bact.* **75**, 291.
Lewis, J. C. (1961). *In* "Spores II", (H. O. Halvorson, ed.), p. 165, Burgess, Publishing Co., Minneapolis.
Lewis, J. C., Snell, N. S. and Burr, H. K. (1960). *Science* **132**, 544.
Long, S. K. and Williams, O. B. (1960). *J. Bact.* **79**, 629.
Lubin, M. and Ennis, H. L. (1965). *Biochim. biophys. Acta* **80**, 614.
Lund, A. (1958). *Hormel Institute Ann. Rept.* p. 16.
Lund, A. (1959). *Hormel Institute Ann. Rept.* p. 14.
Lund, A. (1961). *Hormel Institute Ann. Rept.* p. 20.
Lundgren, D. G. and Beskid, G. (1960). *Canad. J. Microbiol.* **6**, 135.
Lundgren, D. G. and Bott, K. F. (1963). *J. Bact.* **86**, 462.
Malveaux, F. J. and Cooney, J. J. (1964). *Bact. Proc.* 13.
Mandel, M. and Rowley, D. B. (1963). *J. Bact.* **85**, 1445.
Mandelstam, J. and Rogers, H. J. (1958). *Nature, Lond.* **181**, 956.
Mandelstam, J. and Rogers, H. J. (1959). *Biochem. J.* **72**, 654.
Marmur, J., Seaman, E. and Levine, J. (1963). *J. Bact.* **85**, 461.
Marshall, B. J., Murrell, W. G. and Scott, W. J. (1963). *J. gen. Microbiol.* **31**, 451.
Martin, H. H. and Foster, J. W. (1958). *J. Bact.* **76**, 167.
Martin, J. H. and Harper, W. J. (1965). *J. dairy Sci.* **48**, 282.
Masui, M., Kawasaki, N. and Tabata, F. (1960). *Osaka City Med. J.* **6**, 139.
Matches, J. R., Walker, H. W. and Ayres, J. C. (1964). *J. Bact.* **87**, 16.
Matz, L. and Gerhardt, P. (1964). *Bact. Proc.* 14.
Mayall, B. H. and Robinow, C. F. (1957). *J. appl. Bact.* **20**, 333.
Mayer, G. D. and Beers, R. J. (1964). *Bact. Proc.* 102.
Mazaneck, K., Kocur, M. and Martinec, T. (1965). *J. Bact.* **90**, 808.

Meisel-Mikolajczyk, F. (1965). *Bull. Acad. Polon. Sci.* **13**, 7.
Millet, J. (1963). *C. R. Acad. Sci., Paris* **257**, 784.
Millet, J. and Aubert, J. P. (1960). *Ann. Inst. Past.* **98**, 282.
Monro, R. E. (1961). *Biochem. J.* **81**, 225.
Murrell, W. G. (1955). "The Bacterial Endospore", University of Sydney Press.
Murrell, W. G. (1961a). *In* "Spores II", (H. O. Halvorson, ed.), p. 229, Burgess Publishing Co., Minneapolis.
Murrell, W. G. (1961b). *In* "Microbial Reaction to Environment", *Symp. Soc. gen. Microbiol.* **11**, 100 (G. G. Meynell and H. Gooder, eds.), Cambridge University Press.
Murrell, W. G. and Scott, W. J. (1958). *Abstr. VII Intern. Congr. Microbiol.* p. 26.
Murrell, W. G. and Scott, W. J. (1966). *J. gen. Microbiol.* **43**, 411.
Murrell, W. G. and Warth, A. D. (1965). *In* "Spores III", (L. L. Campbell and H. O. Halvorson, eds.), p. 1, Amer. Soc. Microbiol., Ann Arbor.
Murty, G. G. K. and Halvorson, H. O. (1957). *J. Bact.* **73**, 235.
Nakata, H. M. (1963). *J. Bact.* **86**, 577.
Nakata, H. M. (1966). *J. Bact.* **91**, 784.
Nakata, H. M. and Halvorson, H. O. (1960). *J. Bact.* **80**, 801.
Norris, J. R. and Baillie, A. (1962). *J. appl. Bact.* **25**, vii.
Norris, K. P. and Greenstreet, J. E. S. (1958). *J. gen. Microbiol.* **19**, 566.
Ohye, D. F. and Murrell, W. G. (1962). *J. cell. Biol.* **14**, 111.
Pelcher, E. A., Fleming, H. P. and Ordal, Z. J. (1963). *Canad. J. Microbiol.* **9**, 251.
Perkins, W. E. and Tsuji, K. (1962). *J. Bact.* **84**, 86.
Perry, J. F. and Foster, J. W. (1954). *J. gen. Physiol.* **37**, 401.
Perry, J. J. and Foster, J. W. (1955). *J. Bact.* **69**, 337.
Perry, J. J. and Foster, J. W. (1956). *J. Bact.* **72**, 295.
Pfennig, N. (1957). *Arch. Mikrobiol.* **26**, 345.
Powell, J. F. (1950). *J. gen. Microbiol.* **4**, 330.
Powell, J. F. (1953). *Biochem. J.* **54**, 210.
Powell, J. F. (1957). *J. appl. Bact.* **20**, 349.
Powell, J. F. and Hunter, J. R. (1953). *J. gen. Physiol.* **36**, 601.
Powell, J. F. and Strange, R. E. (1953). *Biochem. J.* **54**, 205.
Powell, J. F. and Strange, R. E. (1956). *Biochem. J.* **63**, 661.
Powell, J. F. and Strange, R. E. (1959). *Nature, Lond.* **184**, 878.
Ramaley, R. F. and Bernlohr, R. W. (1963). *Bact. Proc.* 128.
Ramaley, R. F. and Bernlohr, R. W. (1965). *J. molec. Biol.* **11**, 842.
Ramaley, R. F. and Bernlohr, R. W. (1966). *Arch. Biochem. Biophys.* **117**, 34.
Record, B. R. and Grinstead, K. H. (1954). *Biochem. J.* **58**, 85.
Riemann, H. (1963). Ph.D. Thesis: Copenhagen University.
Riemann, H. and Ordal, Z. J. (1961). *Science* **133**, 1703.
Robinow, C. F. (1953). *J. Bact.* **66**, 300.
Robinow, C. F. (1960). *In* "The Bacteria", (I. C. Gunsalus and R. Y. Stanier, eds.), Vol. 1, p. 207, Academic Press Inc., New York.
Rode, L. J. and Foster, J. W. (1960a). *Proc. nat. Acad. Sci. Wash.*, **46**, 118.
Rode, L. J. and Foster, J. W. (1960b). *J. Bact.* **79**, 650.
Rode, L. J. and Foster, J. W. (1960c). *Arch. Mikrobiol.* **36**, 67.
Rode, L. J. and Foster, J. W. (1960d). *Nature, Lond.* **188**, 1132.
Rogolsky, M. and Slepecky, R. A. (1964). *Biochem. biophys. Res. Commun.* **16**, 204.
Ryter, A. (1965). *Ann. Inst. Past.* **108**, 40.
Ryter, A., Bloom, B. and Aubert, J. P. (1966). *C. R. Acad. Sci., Paris* **262**, 1305.
Ryter, A., Ionesco, H. and Schaeffer, P. (1961). *C. R. Acad. Sci., Paris* **252**, 3675.
Ryter, A. and Jacob, F. (1964). *Ann. Inst. Past.* **107**, 384.

Ryter, A. and Szulmajster, J. (1965). *Ann. Inst. Past.* **108**, 640.
Sacks, L. E. and Alderton, G. (1961). *J. Bact.* **82**, 331.
Sadoff, H. L. (1961). *In* "Spores II", (H. O. Halvorson, ed.), p. 180, Burgess Publishing Co., Minneapolis.
Sadoff, H. L., Back, J. A. and Kools, J. W. (1965). *In* "Spores III", (L. L. Campbell and H. O. Halvorson, eds.), p. 97, Amer. Soc. Microbiol., Ann Arbor.
Salton, M. R. J. (1964). "The Bacterial Cell Wall", p. 256, Elsevier, Amsterdam.
Salton, M. R. J. and Marshall, B. (1959). *J. gen. Microbiol.* **21**, 415.
Schaeffer, P. (1961). Ph.D. Thesis: University of Paris.
Schaeffer, P. (1963). *Symp. Biol. Hungary.* **6**, 123.
Schaeffer, P. and Ionesco, H. (1960). *C. R. Acad. Sci., Paris* **251**, 3125.
Schaeffer, P., Ionesco, H., Ryter, A. and Balassa, G. (1963). *In* "Mécanismes de régulation des activités cellulaires chez les microorganismes", (J. Senez, ed.), p. 553, Gordon and Breach, New York.
Schaeffer, P., Millet, J. and Aubert, J. P. (1965). *Proc. nat. Acad. Sci. Wash.* **54**, 704.
Sébald, M. and Schaeffer, P. (1965). *C. R. Acad. Sci., Paris* **260**, 5398.
Slepecky, R. A. (1961). *In* "Spores II", (H. O. Halvorson, ed.), p. 171, Burgess Publishing Co., Minneapolis.
Slepecky, R. A. and Foster, J. W. (1959). *J. Bact.* **78**, 117.
Slepecky, R. A. and Law, J. H. (1961). *J. Bact.* **82**, 37.
Snoke, J. E. (1964). *Biochim. biophys. Res. Commun.* **14**, 571.
Spizizen, J. (1965). *In* "Spores III", (L. L. Campbell and H. O. Halvorson, eds.), p. 125, Amer. Soc. Microbiol., Ann Arbor.
Spotts, C. R. and Szulmajster, J. (1962). *Biochim. biophys. Acta* **61**, 635.
Srinivasan, V. R. (1965). *In* "Spores III", (L. L. Campbell and H. O. Halvorson, eds.), p. 64, Amer. Soc. Microbiol., Ann Arbor.
Srinivasan, V. R. and Halvorson, H. O. (1963). *Nature, Lond.* **197**, 100.
Stent, G. S. (1964). *Science* **144**, 816.
Stevenson, J., Miller, K., Strothman, R. and Slepecky, R. A. (1962). *Bact. Proc.* **47**.
Stewart, B. T. and Halvorson, H. O. (1953). *J. Bact.* **65**, 160.
Stewart, B. T. and Halvorson, H. O. (1954). *Arch. Biochem. Biophys.* **49**, 168.
Strahs, G. and Dickerson, R. E. (1965). *Spore Newsletter* **2**, 30.
Strange, R. E. (1956). *Biochem. J.* **64**, 23.
Strange, R. E. and Dark, F. A. (1956). *Biochem. J.* **62**, 459.
Strange, R. E. and Dark, F. A. (1957a). *J. gen. Microbiol.* **16**, 236.
Strange, R. E. and Dark, F. A. (1957b). *J. gen. Microbiol.* **17**, 525.
Strange, R. E. and Powell, J. F. (1954). *Biochem. J.* **58**, 80.
Strominger, J. L. (1962). *Fed. Proc.* **21**, 134.
Stuy, J. H. (1958). *J. Bact.* **76**, 179.
Sugiyama, H. (1951). *J. Bact.* **62**, 81.
Szulmajster, J. (1964). *Bull. Soc. chim. Biol.* **46**, 443.
Szulmajster, J. and Canfield, R. E. (1963). *In* "Mécanismes de régulation des activités cellulaires chez les microorganismes", (J. Senez, ed.), Gordon and Breach, New York.
Szulmajster, J. and Hanson, R. S. (1965). *In* "Spores III", (L. L. Campbell and H. O. Halvorson, eds.), p. 162, Amer. Soc. Microbiol., Ann Arbor.
Szulmajster, J. and Schaeffer, P. (1961a). *C. R. Acad. Sci. Paris* **252**, 220.
Szulmajster, J. and Schaeffer, P. (1961b). *Biochem. biophys. Res. Commun.* **6**, 217.
Takagi, A., Kawata, T., Yamamoto, S., Kubo, T. and Okito, S. (1960). *Jap. J. Microbiol.* **4**, 137.
Takagi, A., Nakamura, K. and Ueda, M. (1965). *Jap. J. Microbiol.* **9**, 131.

Takahashi, I. (1965a). *J. Bact.* **89**, 294.
Takahashi, I. (1965b). *J. Bact.* **89**, 1065.
Tanenbaum, S. W. and Kaneko, K. (1964). *Biochemistry* **3**, 1314.
Thomas, R. S. (1964). *J. cell Biol.* **23**, 113.
Thompson, R. S. and Leadbetter, E. R. (1963). *Arch. Mikrobiol.* **45**, 27.
Tichane, R. M. and Bennett, W. E. (1957). *J. Amer. chem. Soc.* **79**, 1293.
Tinelli, R. (1955a). *Ann. Inst. Past.* **88**, 212.
Tinelli, R. (1955b). *Ann. Inst. Past.* **88**, 364.
Tinelli, R. (1955c). *Ann. Inst. Past.* **88**, 642.
Tokuyasu, K. and Yamada, E. (1959a). *J. biophys. biochem. Cytol.* **5**, 123.
Tokuyasu, K. and Yamada, E. (1959b). *J. biophys. biochem. Cytol.* **5**, 129.
Tsuji, K. and Perkins, W. E. (1962). *J. Bact.* **84**, 81.
Vinter, V. (1957a). *Fol. Biol.* **3**, 193.
Vinter, V. (1957b). *In* "Spores II", (H. O. Halvorson, ed.), p. 127, Burgess Publishing Co., Minneapolis.
Vinter, V. (1959). *Fol. Microbiol.* **4**, 216.
Vinter, V. (1960). *Fol. Microbiol.* **5**, 217.
Vinter, V. (1961). *Nature, Lond.* **189**, 589.
Vinter, V. (1962a). *Nature, Lond.* **196**, 1336.
Vinter, V. (1962b). *Fol. Microbiol.* **7**, 275.
Vinter, V. (1962c). *Fol. Microbiol.* **7**, 115.
Vinter, V. (1962d). *Experientia* **18**, 409.
Vinter, V. (1963). *Fol. Microbiol.* **8**, 147.
Vinter, V. (1964). *Fol. Microbiol.* **9**, 58.
Vinter, V. (1965a). *Fol. Microbiol.* **10**, 280.
Vinter, V. (1965b). *Fol. Microbiol.* **10**, 288.
Vinter, V. and Slepecky, R. A. (1965). *J. Bact.* **90**, 803.
Virtanen, A. I. and Pulkki, L. (1933). *Arch. Mikrobiol.* **4**, 99.
Wake, R. G. (1963). *Biochem. biophys. Res. Commun.* **13**, 67.
Waldham, P. G. and Halvorson, H. O. (1954). *Appl. Microbiol.* **2**, 333.
Walker, H. W. and Matches, J. R. (1965). *J. food Sci.* **30**, 1029.
Walker, H. W., Matches, J. R. and Ayres, J. C. (1961). *J. Bact.* **82**, 960.
Warth, A. D. (1965). *Biochim. biophys. Acta* **101**, 315.
Warth, A. D., Ohye, D. F. and Murrell, W. G. (1962). *Abstr. VIII Intern. Cong. Microbiol.*, Montreal, p. 16.
Warth, A. D., Ohye, D. F. and Murrell, W. G. (1963a). *J. cell Biol.* **16**, 579.
Warth, A. D., Ohye, D. F. and Murrell, W. G. (1963b). *J. cell Biol.* **16**, 593.
Weinberg, E. D. (1964). *Appl. Microbiol.* **12**, 436.
Weinberg, E. D. and Tonnis, S. M. (1966). *Appl. Microbial.* **14**, 850.
Windle, J. J. and Sacks, L. E. (1963). *Biochim. biophys. Acta* **66**, 173.
Woese, C. R. (1961). *J. Bact.* **82**, 695.
Woese, C. R., Langridge, R. and Morowitz, H. J. (1960). *J. Bact.* **79**, 777.
Wooley, B. C. and Collier, R. E. (1966). *Bact. Proc.* 16.
Yoneda, M. and Kondo, M. (1959). *Bikens J.* **2**, 247.
Yoshida, N., Izumi, Y., Tani, I., Tanaka, S., Takaishi, K., Hashimoto, T. and Fukui, K. (1957). *J. Bact.* **74**, 94.
Yoshikawa, H. (1965). *Proc. nat. Acad. Sci. Wash.* **53**, 1476.
Young, I. E. (1958). *Canad. J. Microbiol.* **5**, 197.
Young, I. E. and Fitz-James, P. C. (1959a). *J. biophys. biochem. Cytol.* **6**, 467.
Young, I. E. and Fitz-James, P. C. (1959b). *J. biophys. biochem. Cytol.* **6**, 483.
Young, I. E. and Fitz-James, P. C. (1959c). *J. biophys. biochem. Cytol.* **6**, 499.
Young, I. E. and Fitz-James, P. C.(1962). *J. cell Biol.* **12**, 115.

AUTHOR INDEX

Numbers in italics refer to the pages on which references are listed at the end of each article.

A

Abdel-Kader, M. M., 27, 31, *36*, *37*
A'Brook, J., 31, *37*
Achenbach, N. E., 57, *93*
Adams, J. M., 89, *90*
Adelberg, E. A., 55, 57, *91*, *95*
Adler, J., 82, 86, *90*
Adye, J., 28, 30, 31, 32, *36*, *37*
Akabori, S., 149, *248*
Alderton, G., 171, 172, 243, *244*, *250*
Alexander, M., 63, *92*
Alföldi, L., 67, *90*
Allcroft, R., 25, 26, 31, *37*, *37*
Alper, R., 203, *244*
Amaha, M., 177, 182, *246*
Ames, B. N., 73, 80, 81, 82, 86, 87, *90*, *92*, 231, *244*
Anderson, L. E., 183, *247*
Anfinsen, C. B., 79, 80, 85, *95*
Aposhian, H. V., 60, *94*
Arber, W., 3, *21*
Armbrecht, B. H., 29, 31, *36*, *37*
Arnaud, M., 191, 192, 235, *247*
Aronson, A. I., 65, *90*, 210, 214, 227, 237, *244*, *246*
Arredondo, M. I., 157, *245*
Asao, T., 27, 31
Ashley, L. M., 26, *36*
Ashworth, L. J., Jr., 31, *36*, *37*
Attardi, G., 77, 79, 82, 86, *90*
Aubert, J. P., 194, 200, 211, 215, 223, 232, 233, 234, 235, *244*, *249*, *250*
Audrain, L., 150, *246*
Aurisicchio, A., 63, *95*
Austrian, B., 80, *90*, *95*
Austwick, P. K., 25, 29, *36*
Avi-Dor, Y., 4, *22*
Ayerst, G., 25, 29, *36*
Ayres, J. C., 166, 168, 169, *248*, *251*

B

Babinet, C., 63, *92*
Bach, J. A., 223, 237, *245*, *250*
Bach, M., 219, *244*
Bailey, G. F., 180, *245*
Baillie, A., 158, 223, 224, *245*, *249*
Balassa, G., 135, 137, 144, 162, 211, 212, 213, 214, 232, 234, *245*, *250*
Bampton, S. S., 31, *36*
Barer, R., 5, *21*
Barnes, J. M., 26, *36*
Barrett, C. B., 28, 31, *37*
Barton, A. A., 18, *21*
Beck, J. V., 161, *247*
Beckwith, J. R., 80, 81, 82, 86, 89, *90*, *91*, *94*, *95*
Beers, R. J., 197, *248*
Beerthuis, R. K., 27, 28, 31, *36*, *37*
Benassi, G., 158, *245*
Benger, H., 170, 174, 176, 218, 219, *245*
Bennett, W. E., 172, *251*
Bentzon, M. W., *94*
Beran, K., 18, *22*, *23*
Béranger, G., 5, *21*
Berg, P., 63, *91*
Berger, J. A., 181, *245*
Bergère, J. L., 185, 187, 191, 193, 207, 208, 234, *245*
Bergold, G. H., 162, 165, 167, 172, *246*
Bernlohr, R. W., 147, 148, 150, 165, 185, 187, 192, 193, 197, 198, 199, 200, 201, 202, 214, 223, 224, 229, 235, *245*, *248*, *249*
Berns, K. I., 51, *91*
Beskid, G., 186, 192, *248*
Bethune, J. L., 79, 80, *95*
Binkley, S. B., 64, *91*
Birch, A. J., 33, *36*
Bird, R., 59, *93*
Bishop, H. L., 158, 162, 164, *245*
Blach, S. H., *245*
Black, S. H., 157, 167, 176, 178, 200, 216, 238, 240, 241, *246*, *247*
Bladen, H. A., 74, 75, *91*
Blicharska, J., 191, 192, 235, *247*
Blinkova, A. A., 55, *91*
Bloom, B., 211, *249*
Blumenthal, H. J., 235, *245*
Boe, A. A., 10, *22*

Boezi, J., 63, *92*
Bohinski, R. C., 13, *21*
Bolding, J., 31, *37*
Bolle, A., 81, *94*
Bolton, E. T., 77, *91*
Bonanni, M. V., 158, *245*
Bonhoeffer, F., 52, 56, 60, *91*
Borek, E., 67, *91*
Borzani, W., 4, *21*
Bott, K. F., 149, 203, 235, *245*, *248*
Bouck, N., 55, *91*
Bourgeois, S., 68, 84, 85, *91*, *95*
Bradfield, J. R. G., 7, 17, *23*
Bradley, D. E., 143, *245*
Brasciani, F., 240, *246*
Brdrar, B., 20, *21*
Bremer, H., 63, 69, 74, 75, *91*
Brenner, M., 229, *245*
Brenner, S., 50, 55, 60, 61, 62, 67, 68, 77, 80, 81, 82, 84, 89, *90*, *91*, *92*, *93*, *94*, *95*
Bresler, S. E., 55, *91*
Bretz, H. W., 7, *21*
Brian, P. W., 31, *36*
Broda, E., 236, *246*
Brunstetter, B. C., 166, 168, 169, 170, *246*
Buchi, G., 27, *36*
Buckman, J., 170, *247*
Buono, F., 186, 194, 195, 196, 197, *245*
Burger, M., 88, *93*
Burr, H. K., 238, 241, *248*
Burrell, N. J., 31, *36*
Butler, W. H., 26, *36*
Buttin, G., 86, *91*
Byrne, R., 74, 75, *91*

C

Cairns, J., 51, 52, *91*
Cameron, D. W., 33 *36*
Campbell, A., 40, *91*
Campbell, A. D., 29, *37*
Campbell, J. J. R., 13, 14, *21*, *22*
Canfield, R. A., 209, 227, *245*
Canfield, R. E., 214, 225, 226, 227, *250*
Capecchi, M. R., 89, *90*
Carlton, W. W., 31, *37*
Carnaghan, R. B. A., 25, 26, 31, *36*, *37*
Caro, L., 55, 56, *92*
Casroriadis-May, C., 223, 233, *244*
Cavallo, G., 181, 223, *245*, *246*
Cecera, M. A., 75, *95*
Chamberlin, M., 63, *91*
Chambers, K., 33, *36*
Chang, S. B., 27, 31, *36*
Changeux, J.-P., 77, *94*
Chatterjee, B. R., 16, *21*
Charney, J., 170, *245*
Chatterjie, A. N., 222, *245*
Cheung, K. K., 27, *36*
Child, J. J., 34, *37*
Church, B. D., 164, 165, 176, 236, 237, *245*
Ciegler, A., 30, *36*
Clark, A. J., 88, *93*
Clark, B. F. C., 89, *91*
Clark, D. J., 17, *22*
Clarke, M. C., 169, *246*
Clowes, R. C., 67, *90*
Codner, R. C., 30, 31, *36*
Cohen, B., 9, 12, *21*
Cohn, M., 83, 84, 85, *91*, *95*
Collier, R. E., 234, *251*
Collins, V. G., 6, *21*
Contesse, G., 162, *245*
Cooney, J. J., 187, *248*
Costilow, R. N., 206, 207, 208, 216, *246*
Cote, W. A., 203, *244*
Cousin, D., 60, *93*
Cowan, C. I., 13, *22*
Craven, G. R., 79, 80, 85, *95*
Crawford, I. P., 80, 81, 82, *92*, *95*
Creek, R. D., 31, *37*
Crumpton, J. E., 4, 7, 17, *22*
Cukier, R., 74, *92*
Cummings, D. J., 59, *92*
Curran, H. R., 166, 168, 169, 170, 182, 237, *246*, *247*
Cuzin, F., 55, 60, 61, 62, *91*, *92*

D

Dagley, S., 67, *91*
Daniels, M. R., 26, *36*
Dark, F. A., 6, 9, 10, 11, 12, 13, 14, 18, *23*, 146, 147, 149, 150, 155, 165, 170, *250*
Das, H., 74, *91*
Davidoff-Abelson, R., 235, *245*
Davis, N. D., 31, *37*
Davison, P. F., 51, *93*
Dawes, E. A., 13, 14, 15, *21*
Dawkins, A. W., 31, *36*
Day, L. E., 206, 207, 208, 216, *246*
De Iongh, H., 28, 31, *36*
De Jong, K., 27, *37*
Delpy, L. P., 5, *21*
Del Valle, M. R., 210, 227, *244*, *246*
De Moss, J. A., 69, *94*
Dennert, G., 82, *92*
Dennis, E. S., 160, 211, *246*
Deppe, K., 82, *92*
De Silva, R., 6, *23*
Desser, H., 236, *246*
De Vogel, P., 29, 31, *36*
De Volt, H. M., 31, *37*
Dhaliwal, A. S., 10, *22*

Dickens, F., 26, *37*
Dickerson, R. E., 180, 181, *250*
Diener, U. L., 31, *37*
Doi, R. H., 45, *91*, 158, 162, 163, 164, 184, 214, 215, 231, 234, 235, *245*, *246*
Donnellan, J. E., 183, 237, *246*
Doty, P., 75, *95*
Drakulis, M., 20, *21*
Dubin, D. T., 76, *91*
Duncan, M. G., 14, *21*
Dutton, M. F., 34, *37*

E

Echols, H., 82, 87, *91*, *92*
Ecker, R. E., 43, 46, *91*
Edlin, G., 70, *91*
Eichhorn, H. H., 223, *247*
Eidlic, L., 68, *91*
El-Bisi, H. M., 177, 182, *246*
Elkort, A. T., 76, *91*
Ellar, D. J., 136, *246*
Ellis, J. J., 29, 30, 31, *37*
Engelhardt, D. L., 89, *95*
Englesberg, E., 80, 81, 86, 87, *91*, *93*
Ennis, H. L., 170, *248*
Evans, A., 63, *92*
Evans, F. R., 237, *246*
Ezekiel, D. H., 69, *91*

F

Falaschi, A., 158, *246*
Falcone, G., 181, 223, 240, *245*, *246*
Falk, I. S., 9, *23*
Fan, D. P., 71, 72, 73, 74, 76, *91*
Fangman, W. L., 68, *91*
Farber, E., 74, *95*
Farkas, J., 160, 181, *246*
Felsenfeld, G., 63, *92*
Fennell, D. I., 28, *37*
Fikhman, B. A., 5, *21*
Finlayson, A. J., 216, *246*
Fisher, W. P., 170, *245*
Fitz-James, P. C., 135, 136, 137, 138, 139, 140, 142, 144, 160, 161, 162, 164, 165, 166, 167, 171, 172, 177, 201, 202, 209, 210, 211, 212, 214, 215, 216, 217, 220, 221, 225, 227, 228, 230, 236, 237, 243, *246*, *247*, *251*
Fleming, H. P., 177, 178, 181, 200, *246*, *249*
Foerster, H. F., 179, *246*
Forchhammer, J., 71, *91*
Foster, J. W., 146, 147, 148, 158, 167, 170, 171, 172, 177, 179, 181, 184, 186, 190, 192, 200, 201, 211, 216, 217, 219, *246*, *247*, *248*, *249*, *250*

Fourie, L., 27, 31, *37*
Frank, M. E., 17, *21*
Fraenkel, D. G., 48, 64, *91*, *94*
Franklin, J. G., 143, *245*
Franklin, R. M., 63, *94*
Franzen, J. S., 64, *91*
Freter, R., 11, *21*
Friedman, L., 26, *37*
Fuchs, E., 75, *91*
Fuhs, G. W., 62, *91*
Fujimoto, S., 219, *248*
Fukui, K., 147, *251*
Fuller, W., 51, *91*
Fulton, C., 57, *91*
Furth, J. J., 63, *92*

G

Gabe, D. R., 6, *21*
Gabliks, J. Z., 26, *37*
Gaines, K., 63, 75, *91*
Gallant, J., 88, *91*, *92*
Gardner, W. K., Jr., 26, *36*
Garen, A., 87, 88, *91*, *92*
Garen, S., 87, *91*, *92*
Garry, B., 80, 87, *90*
Gartner, T. K., 82, *94*
Geiduschek, P., 63, *92*, *95*
Gellert, M., 63, *92*
Georgala, D. L., 240, *246*
Gerhardt, P., 145, 146, 166, 167, 168, 176, 178, 183, 200, 216, 238, 240, 241, *245*, *246*, *247*, *248*
Giacomoni, D., 63, *92*
Gibbons, N. E., 15, *23*
Gierer, A., 52, *91*
Gilliland, R. B., 4, 7, *21*
Gillespie, M. E., 46, *94*
Gilvarg, C., 219, *244*
Glansdorff, N., 88, *92*, *93*
Goldberg, I. H., 63, *92*
Goldstein, A., 47, 63, 74, 75, *91*, *92*
Goldstein, D. B., 47, 75, *92*
Gollakota, K. G., 176, 190, *246*
Goodman, H. M., 63, *92*
Gorini, L., 88, *92*, 150, *246*
Gorrill, R. H., 6, *21*
Gould, G. W., 181, 239, 240, *246*, *247*
Gray, E., 229, *245*
Graziosi, F., 63, *95*
Grecz, N., 162, 239, *246*
Green, J. H., 224, *246*
Green, M., 63, *92*
Greenstreet, J. E. S., 179, *249*
Grelet, N., 165, 193, 194, 233, *246*, *247*
Grinstead, K. H., 153, *249*

Gronlund, A. F., 14, *21*
Gros, F., 46, 65, 68, 72, 74, 77, 79, 82, 86, *90*, *92*, *95*
Gross, J. D., 55, 56, *92*
Grove, J. F., 31, *36*
Gruft, H., 170, *247*
Grundey, J. K., 31, *36*
Gundersen, V., 88, *92*

H

Hachisuka, Y., 183, 184, *247*
Hahn, F. E., 229, *247*
Hall, B. D., 63, 77, *92*
Hall, C., 79, *95*
Hall, H. H., 30, *36*
Halver, J. E., 26, *36*
Halvorson, H., 158, 164, 168, 170, 178, 183, 184, 220, 236, 237, *246*, *247*
Halvorson, H. O., 17, *23*, 158, 159, 176, 177, 178, 186, 187, 188, 190, 192, 193, 204, 205, 216, 223, 224, 225, 230, 231, 232, 234, 235, 237, 239, *245*, *246*, *247*, *248*, *249*, *251*
Hamilton, L. D., 51, *91*
Hanawalt, P. C., 52, *92*
Hancock, R., 221, *247*
Hannay, C. L., 144, 228, *247*
Hanson, R. S., 186, 187, 188, 190, 191, 192, 223, 224, 225, 235, *247*, *250*
Hardwick, W. A., 184, 186, 192, 200, 201, 211, *247*
Harkness, C., 31, *36*, *37*
Harold, F. M., 15, *21*
Harper, M. K., 182, *247*
Harper, W. J., 240, *248*
Harris, N. D., 3, *21*
Harrison, A. P., 9, 14, 16, *21*
Hartman, P., 231, *244*
Hartman, P. E., 73, 81, 82, 86, 87, *90*, *92*
Hartman, R. S., 164, *245*
Hartwell, L. H., 71, 74, *92*
Hashimoto, T., 147, 167, 168, 176, 178, 183, 200, 216, 240, *245*, *247*, *251*
Hayashi, M., 63, *92*
Hayes, D., 74, *92*
Hayes, W., 12, *21*
Hayflick, L., 17, *21*
Heathcote, J. G., 34, *37*
Hegarty, C. P., 170, *245*
Heliligman, F., 170, *246*
Helling, R. B., 82, *92*
Helmstetter, C. E., 59, *92*
Hemming, H. G., 31, *36*
Henning, U., 82, *92*
Hermier, J., 174, 185, 187, 191, 193, 207, 208, 234, *245*, *247*
Herzenberg, L. A., 87, *90*
Hesseltine, C. W., 29, 30, 31, *37*
Higa, A., 72, 73, 74, 76, *91*
Hill, C. W., 82, *92*
Hinshelwood, C. N., 9, *21*
Hitchins, A. D., 181, 239, 240, *246*, *247*
Hoare, D. S., 13, *22*
Hodges, F. A., 29, 31, *36*, *37*
Hodson, P. H., 161, 179, 219, *247*
Hoffman, F. J., 53, 54, *93*
Hoffman, H., 17, *21*
Hofschneider, P. H., 75, *91*
Holbert, P. E., 134, 143, *247*
Holker, J. S. E., 34, *37*
Holloway, P. W., 33, *36*
Holmes, K. C., *247*
Holmes, P. K., 148, 233, *247*
Hopps, H. E., 229, *247*
Howitt, C., 168, 178, 183, 220, *247*
Hunnell, J. W., 148, 149, 150, *247*
Hunter, J. R., 3, 4, 5, 7, 8, 9, 10, 11, 12, 14, 15, 16, 17, 18, 19, *22*, *23*, 200, *249*
Hurwitz, J., 63, *92*, *93*
Hyatt, M. T., 170, 171, 172, *248*

I

Igarashi, R. T., 45, *91*, 162, 163, 214, 215, 231, 234, 235, *246*
Iijima, K., 146, 149, 150, 151, *248*
Imamoto, F., 77, 79, 86, *92*, *94*
Imperato, S., 223, *245*
Imshenetskii, A. A., 16, *21*
Ionesco, H., 135, 137, 144, 191, 234, *245*, *249*, *250*
Irr, J., 80, 86, 87, *91*
Isaacson, P., 7, 17, *21*
Ishihama, A., 47, *94*
Ito, J., 79, 80, 81, 82, 86, *92*, *94*
Izumi, Y., 147, *251*

J

Jacob, F., 40, 51, 55, 60, 61, 62, 68, 76, 77, 79, 80, 81, 82, 83, 85, 86, *90*, *91*, *92*, *93*, *94*, *95*, 136, 137, 139, 234, *247*, *249*
Jacobs, S. E., 3, *21*
Jann, G. J., 223, *247*
Jannasch, H. W., 8, *21*
Janssen, F. W., 183, *247*
Jebb, W. H. H., 7, *21*
Jenkins, F. P., 26, *37*
Jensen, K. A., 17, *21*
Johnson, F. H., 242, *248*

Johnston, J. R., 18, *22*
Jones, H. E. H., 26, *37*
Jordan, E., 86, *93*, *95*
Julien, J., 45, *94*

K

Kadota, H., 146, 149, 150, 151, *248*
Kahan, E., 63, *93*
Kahan, F. M., 63, *93*
Kaiser, A. D., 82, 86, *90*
Kalckar, H. M., 86, *93*
Kaneko, I., 231, *248*
Kaneko, K., 217, 219, *251*
Kanie, M., 219, *248*
Kanner, L., 74, *91*
Karp, S., 180, *245*
Kawasaki, N., 160, *248*
Kawata, T., 142, 144, 228, *250*
Kaweh, M., 5, *21*
Kelly, C. D., 17, *21*
Kendrew, J. C., 238, *248*
Kennell, D., 46, *93*
Kenner, G. W., 33, *36*
Kepes, A., 71, 74, *93*
Kerravala, Z. J., 233, *248*
Ketterer, H., 4, *21*
Keuning, R., 27, *37*
Keynan, A., 177, *248*
Kida, S., 79, 86, *94*
Kiho, Y., 79, *93*
Kipling, C., 6, *21*
Kirschbaum, J. B., 63, 75, *92*
Kiss, I., 160, 181, *246*
Kjeldgaard, N. O., 40, 41, 42, 43, 46, 47, 48, 49, 59, 71, *91*, *93*, *94*
Knaysi, G., 16, *21*, 136, 174, 185, 186, 193, 233, 237, *248*
Koch, A. L., 5, *21*, 47, *93*
Kocur, M., 139, *248*
Koelensmid, W. A. A. B., 29, 31, *36*, *37*
Kogut, M.. 7, 17, *21*
Kohiyama, M., 60, *93*
Kolodziej, B., J., 170, *248*
Kominek, L. A., 176, 188, 203, 204, 205, 223, 224, *248*
Kondo, M., 146, 147, 148, 164, 165, 203, 217, 220, *248*, *251*
Konrad, M., 63, 74, 75, *91*
Kools, J. W., 237, *250*
Kornberg, A., 60, *94*, 158, *246*
Kos, E., 21, *21*
Krask, B. J., 197, *248*
Kubitschek, H. E., 6, *21*
Kubo, T., 142, 144, 228, *250*
Kuczynski, M., 4, *22*
Kuno, T., 183, *247*
Kurahashi, K., 86, *93*
Kuratomi, K., 149, *248*
Kurland, C. G., 43, 46, 65, 66, 67, 68, 76, *93*

L

Laffer, N. C., 31, *37*
Lagoda, A. A., 30, *36*
Laishley, E. J., 193, 200, *248*
Lamana, C., 15, *22*
Lamfrom, H., 60, *93*
Lancaster, M. C., 26, *37*
Langley, B. C., 31, *36*
Langridge, R., 162, 163, *251*
Lanzov, V. A., 55, *91*
Lark, C., 50, 59, 61, *93*
Lark, K. G., 40, 42, 43, 50, 53, 54, 59, 61, *93*, *94*
Larsen, H., 10, *22*
Law, J. H., 165, 203, 204, *250*
Lawrence, F. R., 14, *21*
Lawrence, N. L., 224, *248*
Leadbetter, E. R., 174, *251*
Lechowich, R. V., 172, ,176 177, 182, *246*, *248*
Lee, E. G. H., 32, *37*
Lee, N., 80, 81, 86, 87, *91*, *93*
Lee, W. H., 159, 160, *248*
Legator, M. S., 26, *37*
Leitzmann, C., 202, 214, *248*
Leive, L., 72, *93*
Levin, J. G., 74, 75, *91*
Levine, J., 163, *248*
Levinson, H. S., 170, 171, 172, 177, 233, *247*, *248*
Levinthal, C., 51, 72, 73, 74, 76, *91*, *93*
Lewis, J. C., 238, 241, *248*
Lewis, J. G., 239, *248*
Liebelova, J., 18, *22*
Lightbown, J. W., 7, 17, *21*
Lipe, R. S., 17, *22*
Long, L. Jr., 29, *37*
Long, S. K., 164, 165, *248*
Loper, J. C., 87, *92*
Lovett, S., 9, *22*
Lowe, D., 31, *36*
Lowney, L. I., 47, 75, *92*
Lubin, M., 170, *248*
Lund, A., 181, *248*
Lund, A. J., 183, *247*
Lundgren, D. G., 136, 149, 186, 192, 194, 195, 196, 197, 203, *244*, *245*, *246*, *248*

M

Maaløe, O., 40, 41, 42, 47, 48, 52, 58, 59, 64, 65, 66, 67, 68, 70, 76, *91*, *92*, *93*, *94*, *95*

Maas, R., 88, *93*
Maas, W. K., 88, *93*
McAuslan, B. R., 90, *93*
McCarthy, B., 15, *22*, 46, 77, *91, 93*
McCoy, J. M., 84, *94*
McDonald, D., 31, *37*
McGrew, S. B., 13, *22*
MacLeod, R. A., 19, *22*
McNiel, E. M., 6, *21*
McQuillen, K., 13, *22*
Magasanik, B., 40, 46, 71, 74, *92, 93, 94*
Mager, J., 4, *22*
Malamy, M., 63, *92*
Mallette, M. F., 13, *21, 22*
Malveaux, F. J., 187, *248*
Mandel, M., 160, *248*
Mandelstam, J., 14, *22*, 221, *248*
Marcker, K. A., 89, *91*
Marchessault, R. H., 203, *244*
Marcov, D., 18, *22*
Marmur, J., 163, *248*
Marr, A. G., 17, *22*, 181, *245*
Marshall, B. J., 147, 148, 149, 165, 237, 238, 239, *248*
Martin, F. M., 70, *93*
Martin, H. H., 190, 216, 217, 219, *248*
Martin, J. H., 190, 240, *248*
Martin, R. G., 78, 87, *90, 93*
Martinec, T., 139, *248*
Maruyama, Y., 54, *93*
Massie, H. R., 51, *93*
Masui, M., 160, *248*
Matches, J. R., 166, 168, 169, 177, *248, 251*
Mateles, R. I., 28, 30, 31, 32, *36, 37*
Mathews, M. M., 4, *22*
Matney, T. S., 57, *93*
Matsushiro, A., 79, 86, *92, 94*
Matthaei, J. H., 46, 73, *94*
Matz, L. 145, 146, 166, *248*
Mayall, B. H., 141, 167, 172, 183, *248*
Mayer, G. D., 197, *248*
Maynard Smith, J., 20, *22*
Mazaneck, K., 139, *248*
Meadow, P., 13, *22*
Meisel-Mikolajczyk, F., 165, *249*
Meselson, M., 52, *94*
Meynell, E., 6, *22*
Meynell, G. G., 6, *22*
Miller, K., 204, *250*
Millet, J., 194, 200, 215, 223, 224, 232, 233, 234, 235, *244, 249, 250*
Mishima, S., 79, *92*
Mitsui, H., 47, *94*
Mohr, V., 10, *22*
Moldave, K., 74, *94*
Monier, R., 45, *94*
Monod, J., 40, 76, 77, 79, 81, 82, 83, 85, 86, *92, 93, 94, 95*
Monro, R. E., 69, *95*, 201, *247, 249*
Moody, D. P., 34, *37*
Moore, F. E., *248*
Morgan, D. M., 7, *23*
Morikawa, N., 77, 79, *92*
Morowitz, H. J., 162, 163, *251*
Morris, D. W., 69, *94*
Mortimer, R. K., 18, *22*
Müller-Hill, B., 84, *94*
Munch-Petersen, A., 49, *94*
Murachi, T., 184, *247*
Murrell, W. G., 134, 138, 139, 141, 144, 145, 146, 147, 148, 149, 150, 153, 154, 155, 164, 168, 169, 170, 171, 172, 174, 175, 176, 177, 183, 185, 221, 222, 225, 228, 234, 235, 237, 238, 239, 241, 242, 243, *248, 249, 251*
Murty, G. G. K., 237, *249*
Myers, A. T., 166, 168, 169, 170, *246*

N

Nagata, T., 53, 55, *94*
Nakada, D., 14, *22*, 71, 74, *94*
Nakamoto, T., 63, *92*
Nakamura, K., 228, *250*
Nakata, H. M., 186, 188, 190, 192, 203, 204, *249*
Naono, S., 72, 74, 77, 79, 82, 86, *90, 92, 95*
Narkates, A. J., 12, *22*
Nathans, D., 69, *94*
Neidnardt, F. C., 40, 48, 50, 64, 68, *91, 94*
Nelson, A. A., 29, 31, *36, 37*
Nesbitt, B. F., 31, *37*
Ness, A. G., 9, 11, 14, *23*
Neuhard, J., 49, *94*
Neville, D., 63, *92*
Newberne, P. M., 31, *37*
Newton, W. A., 81, 89, *94*
Nilson, E. H., 17, *22*
Nirenberg, M. W., 46, 73, 74, 75, *91, 94*
Nishimura, T., 79, *92*
Noll, H., 74, *95*
Nomura, M., 66, *93, 94*
Norris, G. L. F., 31, *36*
Norris, J. R., 159, 223, 224, *245, 249*
Norris, K. P., 6, *22*, 179, *249*
Novelli, G. D., 185, 187, 192, 193, 223, 224, 229, *245*
Novick, A., 17, *22*, 84, *94*
Novick, R. P., 87, *94*
Nygaard, A. P., 63, *92*

O

Ochoa, S., 75, *94*, *95*
Ogg, J. E., 16, *22*, *23*
Ohno, K., 149, *248*
Ohye, D. F., 134, 138, 139, 140, 141, 144, 145, 146, 147, 148, 149, 150, 153, 155, 164, 169, 174, 183, 221, 222, 228, 241, 242, *249*, *251*
O'Kelly, J., 25, 31, *37*
Okito, S., 142, 144, 228, *250*
Olsen, R. H., 16, *22*
Ord, W. O., 27, 28, 31, *37*
Ordal, Z. J., 148, 149, 150, 159, 160, 172, 176, 177, 178, 182, 184, 200, *246*, *247*, *248*, *249*
Orgel, L. E., 84, 85, *91*
Orias, E., 82, *94*
Osawa, S., 47, 75, *94*
Osipova, I. V., 4, *22*
O'Sullivan, A., 42, 49, 53, *95*
Otaka, E., 75, *94*
Otsuji, N., 87, 88, *92*
Ozawa, A., 11, *21*

P

Pallansch, M. J., 182, *247*
Panda, B., 31, *37*
Pardee, A. B., 15, *22*, 65, 80, *90*, *94*, *95*
Park, J. T., 221, 222, *245*, *247*
Parrish, F. W., 29, *37*
Paulus, H., 229, *245*
Pelcher, E. A., 177, 178, 200, *249*
Perkins, W. E., 155, 208, *249*, *251*
Perrin, D., 83, 85, *95*
Perry, J. F., 179, 200, 216, *249*
Peterson, R. E., 30, *36*
Pfennig, N., 144, 159, *249*
Philp, J., 31, *37*
Philp, J. Mc. L., 26, *37*
Pine, M. J., 15, *22*
Pittillo, R. F., 12, *22*
Postgate, J. R., 3, 4, 5, 7, 8, 9, 10, 11, 12, 14, 15, 16, 17, 18, 19, *22*, *23*
Powell, E. O., 4, 6, 7, 17, *22*
Powell, J. F., 145, 153, 154, 155, 166, 170, 179, 181, 184, 200, 216, 217, 219, 223, 237, 243, *249*
Power, J., 80, 86, 87, *91*
Prestidge, L. S., 65, *94*
Preuss, A., 75, *91*
Previc, E. P., 46, *94*
Prickett, C. O., 31, *37*
Pritchard, R. H., 59, *94*
Pryadkina, M. D., 5, *21*
Pulkki, L., 144, *251*

Q

Quesnel, L. B., 7, 17, 19, *22*

R

Rabinowitz, M., 63, *92*
Rachmeler, M., 15, *22*
Rahn, O., 17, *21*
Ramaley, R. F., 198, 199, 224, 235, *249*
Ramsey, D. S., 164, *245*
Raper, K. B., 28, *37*
Rasmussen, K. W., 58, *93*
Razumovskaya, Z. G., 4, *22*
Record, B. R., 153, *249*
Reich, E., 63, *92*, *94*
Repko, T., 53, 54, *93*
Rhuland, L. E., 13, *22*
Ribbons, D. W., 13, 14, 15, *21*
Ribi, E., 144, 146, *246*
Rich, A., 63, 79, *92*, *93*, *95*
Richardson, C. C., 60, *94*
Richmond, M., 82, 87, *94*
Rickards, R. W., 33, *36*
Riemann, H., 155, 181, 184, *249*
Rizet, G., 18, *22*
Robinow, C. F., 134, 135, 136, 141, 144, 162, 165, 167, 172, 174, 183, *246*, *248*, *249*
Rode L. J., 158, 181, 182, *249*
Rogers, H. J., 221, *248*
Rogolsky, M., 235, *249*
Roman, A., 63, 75, *92*
Rosen, B., 70, *94*
Ross, K. F. A., 5, *21*
Rosset, R., 45, *94*
Roth, J. R., 87, *92*, *94*
Rouvière, J., 74, 77, 79, 82, 86, *90*, *92*
Rowley, D. B., 160, *248*
Ryan, A., 67, *91*
Ryan, F. J., 8, *22*
Ryter, A., 60, 61, *93*, *94*, 135, 136, 137, 139, 144, 160, 214, 227, 230, 234, 235, 243, *249*, *250*

S

Sacks, L. E., 171, 179, 180, *245*, *250*, *251*
Sadler, J. R., 84, *94*
Sadoff, H. L., 223, 224, 237, *245*, *246*, *250*
Salas, M., 75, *94*, *95*
Salmon, W. D., 31, *37*
Salton, M. R. J., 147, 148, 149, 165, *250*
Salunkhe, D. K., 10, *22*

Sarabhai, A. S., 81, *94*
Sargeant, K., 25, 30, 31, *36*, *37*
Sarnat, M. T., 63, *92*, *95*
Sato, K., 77, 79, 86, *92*, *94*
Schaechter, M., 40, 42, 43, 46, 47, 48, *91*, *93*, *94*, *95*
Schaeffer, P., 135, 137, 138, 144, 191, 223, 230, 233, 234, 235, *245*, *247*, *249*, *250*
Schaeffer, W., 26, *37*
Schaller, H., 60, *91*
Schatzberg, G., 4, *22*
Schecroun, J., 18, *22*
Schildkraut, C. L., 60, *94*
Schroeder, H. M., *37*
Schroeder, H. W., 31, *36*
Schulze, K. L., 17, *22*
Schumaier, G., 31, *37*
Schwartz, N. M., 82, *94*
Scott de B., 27, 31, *37*
Scott, W. J., 237, 238, 239, 243, *248*, *249*
Seaman, E., 163, *248*
Sébald, M., 223, *250*
Serman, D., 87, *92*
Setlow, R. B., 183, 237, *246*
Sevag, M. G., 170, *248*
Shatkin, A. J., 63, *94*
Shearer, C., 9, *22*
Sheridan, A., 25, 31, *37*
Shin, D. H., 74, *94*
Shon, M., 10, 12, *23*
Shotwell, O. L., 29, 30, 31, *37*
Sibatani, A., 75, *94*
Sierra, G., 15, *23*
Sievert, C., 147, 148, 150, 165, 229, *245*
Signer, E. R., 80, 86, *90*, *95*
Silver, S. D., 55, *95*
Sim, G. A., 27, *36*
Simmons, E. G., 29, *37*
Simpson, F. J., 216, *246*
Sistrom, W. R., 4, *22*
Skalka, A., 63, *92*
Slepecky, R. A., 165, 167, 170, 171, 172, 177, 178, 179, 203, 204, 233, 235, *247*, *248*, *249*, *250*, *251*
Sloan, J. D., 170, *248*
Smith, C. E., 63, *92*
Smith, D. D., 12, *23*
Smith, H. R., 29, 31, *36*, 37
Smith, I., 14, *22*
Smith, R. F., 239, *246*
Smith, M. A., 75, *94*, *95*
Smith, R. C., 64, *95*
Snell, N., 171, 172, *244*
Snell, N. S., 238, 241, 243, *248*
Snoke, J. E., 147, 148, *250*
Soffer, R. L., 72, *95*
Sparreboom, S., 27, *37*
Spiegelman, S., 63, 65, 77, *90*, *92*, *95*
Spizizen, J., 201, 231, *250*
Spotts, C. R., 211, 213, 214, 226, *250*
Spottswood, T., 88, *91*
Spudich, J., 158, *246*
Srinivasan, V. R., 186, 187, 188, 190, 192, 223, 224, 232, 234, 235, *247*, *248*, *250*
Staehlin, T., 74, *95*
Stahl, F. W., 52, *94*
Stanley, W. M., 75, *94*, *95*
Stapleton, R., 88, *92*
Steers, E., 79, 80, 85, *95*
Stent, G. S., 50, 63, 67, 68, 70, 73, 74, 75, 82, *90*, *91*, *93*, *95*
Steinberg, W., 159, 237, *247*
Stevenson, J., 204, *250*
Stewart, B. T., 158, 178, 192, 223, 237, *250*
Stodolsky, M., 63, *95*
Stonebridge, W. C., 31, *37*
Strahs, G., 180, 181, *250*
Strange, R. E., 4, 5, 6, 9, 10, 11, 12, 13, 14, 18, *23*, 145, 146, 147, 149, 150, 153, 154, 155, 165, 166, 170, 179, 216, 217, 219. 223, *249*
Streibelova, E., 18, *22*, *23*
Stretton, A. O. W., 81, 84, *91*, *94*, *95*
Strominger, J. L., 222, *250*
Strothman, R., 204, *250*
Strugger, S., 4, *23*
Stubblefield, R. D., 29, 30, 31, *37*
Stuy, J. H., 161, *250*
Sud, I. J., 42, 47, *95*
Sueoka, N., 42, 49, 53, *95*
Sugiyama, H., 164, *250*
Sundarajan, T. A., 75, *95*
Swaine, D. J., 169, *246*
Sykes, J., 3, *23*, 67, *91*
Sylvan, S., 15, *21*
Szolyvay, K., 82, *92*
Szulmajster, J., 187, 191, 192, 209, 211, 213, 214, 221, 223, 225, 226, 227, 230, 235, 236, 243, *245*, *247*, *250*

T

Tabata, F., 160, *248*
Takagi, A., 142, 144, 228, 229, *250*
Takahashi, I., 147, 230, *251*
Takaishi, K., *251*
Takeda, Y., 217, 220, *248*
Tanaka, S., 147, *251*
Tanenbaum, S. W., 217, 219, *251*
Tang, T. C., 162, *246*
Tani, I., 147, *251*
Tatum, E. L., 63, *94*
Taylor, A. L., 57, 78, 80, *95*

Tempest, D. W., 3, *23*
Temple Robinson, M. J., 33, *36*
Testa, R., 186, 194, 195, 196, 197, *245*
Thach, R. E., 75, *95*
Theil, E. C., 8, *23*
Thoman, M. S., 78, *95*
Thomas, C. A., 51, *91*
Thomas, R., 34, 35, *37*
Thomas, R. S., 143, 169, 173, 174, 183, *251*
Thompson, R. S., 174, *251*
Tichane, R. M., 172, *251*
Tinelli, R., 144, 145, 161, 165, 174, 193, 200, 202, 203, *251*,
Tissières, A., 68, *95*
Tkaczyk, S., 5, *21*
Tocchini-Valentini, G. P., 63, *92*, *95*
Tochilubo, K., 184, *247*
Tokuyasu, K., 228, *251*
Tomlinson, A. H., 7, *21*
Tonnis, S. M., 170, 229, *251*
Torriani, A., 87, *91*
Townsley, P. M., 32, *37*
Traut, R. R., 69, *95*
Tsuji, K., 155, 208, *249*, *251*
Turncock, G., 67, *95*

U

Uchida, A., 146, 149, 150, 151, *248*
Ueda, M., 228, *250*
Ullman, A., 85, 86, *93*
Underwood, J. G., 34, *37*
Upadhyay, J., 162, *246*

V

Vairo, M. L. R., 4, *21*
Valentine, R. C., 7, 17, *23*
Van der Merwe, K. J., 27, 31, *37*
Van der Zijden, A. S. M., 27, 31, *37*
Van Dorp, D. A., 27, *37*
Van Pelt, J. G., 28, *36*
Van Rhee, R., 29, *36*
Vary, J. C., 159, 237, *247*
Villa Trevino, S., 74, *95*
Vinter, V., 149, 175, 176, 182, 220, 221, 222, 225, 226, 233, 240, *251*
Virtanen, A. I., 144, *251*
Vles, R. O., 28, 31, *36*
Vogel, J. H., 88, *95*

W

Wade, H. E., 7, *23*
Wahba, A. J., 75, *94*, *95*
Wake, R. G., 160, 211, *246*, *251*
Walden, C. C., 32, *37*
Waldham, P. G., 239, *251*
Walker, H. W., 166, 168, 169, 177, *248*, *251*
Warner, J. R., 79, *95*
Warth, A. D., 134, 139, 144, ,141 145, 146, 147, 148, 149, 150, 153, 154, 155, 156, 157, 164, 166, 168, 169, 171, 172, 174, 175, 176, 177, 183, 185, 221, 222, 225, 241, 242, *249*, *251*
Watson, J. D., 66, *93*, *94*
Webster, B. R., 33
Webster, R. E., 89, *95*
Weibull, C., 6, *23*
Weinberg, E. D., 170 229, *251*
Weinberg, R., 82, *92*
Weiss, P., 63, *92*
Weiss, S. B., 63, *95*
Wettstein, F. O., 74, *95*
White, A. E., 67, *91*
White, P. B., 4, *23*
Wiame, J. M., 88, *93*
Wick, E. L., 27, 31, *36*
Wiesmayer, H., 86, *95*
Wiley, B. J., 29, *37*
Wild, D. G., 67, *91*, *95*
Wilkins, M. H. F., 51, *91*
Wilkinson, J. F., 14, *23*
Williams, O. B., 164, 165, *248*
Williams, R. P., 16, *21*
Willson, C., 46, 83, 85, *95*
Wilson, G. S., 6, 16, *23*
Wilson, H. R., 51, *91*
Windle, J. J., 179, *251*
Winslow, C. E. A., 9, *23*
Wisseman, C. L., 229, *247*
Withrow, A., 26, *37*
Woese, C., 72, *95*
Woese, C. R., 162, 163, *251*
Wogan, G. N., 26, 27, 31, *36*, *37*
Wollman, E., 51, *93*
Wollman, E. L., 234, *247*
Wooley, B. C., 234, *251*
Work, E., 13, *22*
Wyss, O., 12, *23*

Y

Yamada, E., 228, *251*
Yamamoto, S., 142, 144, 228, *250*
Yanagita, T., 54, *93*
Yaniv, M., 68, *95*
Yankofsky, S. A., 63, *95*
Yarmolinsky, M. B., 86, *95*
Yates, R. A., 80, *95*
Yegian, C., 69, 70, *91*, *93*
Yeo, R., 30, 31, *36*
Yoneda, M., 164, 165, 203, 217, 220, *248*, *251*

Yoshida, N., 147, *251*
Yoshikawa, H., 42, 49, 53, 60, *95*, 237, *251*
Young, I. E., 136, 137, 138, 140, 144, 160, 161, 162, 167, 171, 180, 181, 202, 209, 210, 211, 216, 217, 220, 225, 228, 236, *246*, *251*

Z

Zabin, I., 79, 80, *95*
Zamenhof, S., 8, *23*
Zelle, M. R., 16, *22*, *23*
Zhiltsova, G. K., 16, *21*
Ziegler, N. R., 17, *23*
Zillig, W., 75, *91*
Zimm, B. H., 51, *93*
Zinder, N. D., 89, *95*
Zipser, D., 81, 89, *94*
Zobell, C. E., 242, *248*
Zust, J. R., 29, *37*

SUBJECT INDEX

A

Acetate formation during sporulation of *B. cereus* T, 188
Acetate units, in the biosynthesis of aflatoxins, 33
Acetoacetyl-CoA reductase, synthesis of during sporulation of *B. cereus* T, 223
Acetobacter sp., viability of, 4
Acid washing, effect of on ion content of bacterial spores, 172
Acidic amino acids in ribosomal proteins from extreme halophiles, 108
Acidic nature of cell envelope protein in halobacteria, 118
Aconitate hydratase, synthesis of during sporulation of *B. cereus* T, 223
Actinomycin, action of, 63
  effect on m-RNA synthesis, 72
Activation of amino acids, 68
Adenosine deaminase, synthesis of during sporulation of *B. cereus*, 223
Adenylic acid as a nutrient for extreme halophiles, 99
Aerobacter aerogenes, cryptic growth of, 8
  effect of deuterium oxide on survival of, 9
  permeability of in relation to viability, 4
  RNA degradation in, 14
  shrinkage of on adding formaldehyde, 7
  survival of during starvation, 9
  viability of continuously grown, 17
Aflatoxin B_1, potency of, 26
  structural formula of, 27
Aflatoxin B_2, structural formula of, 27
Aflatoxin G_1, structural formula of, 27
Aflatoxin G_2, structural formula of, 27
Aflatoxin biosynthesis, possible pathway for, 35
Aflatoxins, biological effects of, 25
  biosynthesis of, 32
  chemical characteristics of, 27
  determination of, 28
  extraction of, 27
  fluorescence of, 27
  nature of, 25
  separation of, 27
  toxic properties of, 26
Aflatoxin-producing organisms, 28
Agents that release the Ca-DPA complex from bacterial spores, 181
Alanine, effect of on sporulation of *Cl. botulinum*, 208
Alanine as a precursor of bacterial spore DPA, 217
Alanine in amino acid pool of bacilli, 202
Alanine racemase, synthesis of during sporulation of *B. cereus*, 223
D-Alanine, incorporation of into bacterial spore glycopeptide, 222
Alkaline phosphatase, effect of salt on stability of, 105
  formation of, 70, 71, 72
  operon, 87, 88
  synthesis of during sporulation of *B. megaterium*, 224
Aluminium content of bacterial spores, 168
Amino-acid activating enzymes, 68
Amino acid composition of cell envelope proteins in halobacteria, 118
Amino acid contents of *B. licheniformis* during sporulation, 198
Amino acid requirements for spore formation in *B. cereus*, 195
Amino acids, effect of on DNA replication, 57, 58, 59
  presence of in DPA extracts of bacterial spores, 180
  requirement for DNA replication, 52, 53
  rôle of in the regulation of RNA synthesis, 64, 65, 66, 67, 68
Amino acids as carbon sources for extreme halophiles, 99
Amino acids as nitrogen sources for aflatoxin production, 32
Amino acids of spores, 144
Amino acids of spore coat, 147, 148, 149
Amino acids of spore cortex, 152–155
Amino acids of spore protoplast, 159
Ammonium ions, requirement for in extreme halophiles, 100
Amylase, synthesis of during sporulation of *B. licheniformis*, 224
Antigens, synthesis of during sporulation of *B. megaterium*, 223
Arabinose synthesis control, 80, 82, 86
Arginase, synthesis of during sporulation, 199

Arginase, synthesis of during sporulation of *B. licheniformis*, 224
Arginine, requirement for during sporulation of *Cl. botulinum*, 208
Arginine, catabolism of in *H. salinarium*, 111
Arginine deiminase of *H. salinarium*, 111
Arginine metabolism in *B. licheninormis*, reversal of during sporulation, 199
Arginine operon, 88
Ashing of bacterial spores, 174
Aspartate as a precursor of bacterial spore DPA, 217
Aspartate aminotransferase, effect of salt on stability of, 105
Aspartic β-semialdehyde, as a precursor of bacterial spore DPA, 219
Aspergillus effusus, aflatoxin production by, 29
Aspergillus flavus, aflatoxin production by, 25
  group, as aflatoxin producers, 28
Aspergillus flavus-oryzae, aflatoxin production by, 29
Aspergillus gymnosardae, aflatoxin production by, 29
Aspergillus micro-viridocitrinus, aflatoxin production by, 29
Aspergillus parasiticus, aflatoxin production by, 29
Aspergillus versicolor, sterigmatocystin production by, 34
Axial cord of forespore, 160
Axial thread of spore nucleus, 136, 137, 138
8-Azaguanine, impermeability of bacterial spores to, 210
Azide, effect of on substrate-accelerated death, 11

B

Bacillus anthracis, morphological changes in ageing cultures of, 16
Bacillus apiarus, potassium content of spores of, 169
  heat resistance and DPA content of spores of, 175
Bacillus brevis, carbohydrate metabolism of, 187
  metabolic changes during sporulation of, 192
Bacillus cereus, heat resistance and DPA contents of spores of, 175
  lipid content of spores of, 165
  PHB content of during spore formation, 203
  phosphorus content of spores of, 166
  protein synthesis in during spore formation, 226
Bacillus cereus T, hybridization of DNA and RNA from sporulating, 215
  relation of heat resistance to DPA synthesis in, 216
Bacillus cereus var. *albolactis*, carbohydrate metabolism of, 186
Bacillus cereus var. *lacticola*, carbohydrate metabolism of, 186
Bacillus cereus var. *mycoides*, carbohydrate metabolism of, 186
  heat resistance and DPA content of spores of, 175
  requirement for glutamate by during sporulation of, 194
Bacillus cereus var. *thuringiensis*, parasporal body of, 228
Bacillus coagulans, carbohydrate metabolism of, 187
  heat resistance and DPA content of spores of, 175
  metabolic changes during sporulation of, 192
  plates on spore coats of, 228
Bacillus licheniformis, carbohydrate metabolism of, 187
  heat resistance and DPA content of spores of, 175
  lipid content of spores of, 165
Bacillus megaterium, carbohydrate metabolism of, 186
  effect of growth rate transition on, 47, 48, 49
  heat resistance and DPA content of spores of, 175
  lipid content of spores of, 165
  PHB content of during spore formation, 203
  phosphorus content of spores of, 166
  protein synthesis in during spore formation in, 226
  relation between membrane lipid phosphorus and wall hexosamine in, 42
  rôle of retardation factors in sporogensis in, 234
Bacillus mycoides, effect of ageing on nucleoids in, 16
Bacillus polymyxa, lipid content of spores of, 165
Bacillus stearothermophilus, carbohydrate metabolism of, 187
  heat resistance and DPA content of spores of, 175
  lipid content of spores of, 165
  relation between heat resistance of and water activity in, 239
Bacillus subtilis, carbohydrate metabolism of, 187
  induction of DNA synthesis in synchronous cultures of, 60

C

D

Halobacterium salinarium, carbon replica of envelope fragment of, 115
cell yield of, 100
electron micrograph of thin section of, 115
glutamate uptake by, 108
leakage of DNA from, 109
nucleic acid content of, 109
protein content of cell envelope of, 118
salt content of, 102
Halococci, 98
generation time of, 100
Halophiles,
extreme,
distribution of, 99
media for cultivation of, 99
nutritional requirements of, 99
pigments of, 99
growth relations of extreme, 98
Halophilism, biochemical aspects of extreme, 97
Heat resistance, effect of water activity on in bacterial spores, 239
relation of to DPA content in bacterial spores, 174
Heat resistance of bacterial spores, 144 240
Hepatocarcinogenic nature of aflatoxins, 26
Heterogeneity of spore m-RNA in bacteria, 214
Hexagonal pattern on halobacterial cell envelope, 115
Hexosamines of spore cortex, 152–155
Histidine operon, 80, 81, 82, 87
Histological changes in animal tissues induced by aflatoxins, 26
Histone of spore protoplast, 162
Hybrid DNA-RNA, 77, 79
Hybridization of DNA and RNA from sporulating *B. subtilis*, 215

I

Immersion refractometry, use of in assessing microbial viability, 5
Indirect assessment of microbial viability, 4
Inducible enzymes, 77, 79, 82, 83, 84, 85, 86, 87
Induction of spore formation in bacteria, 230
Induction of sporulation in bacteria, mechanism of, 234
Initiators and DNA replication, 61
Inorganic composition of bacterial spores, 167
Inorganic constituents of bacterial spores, 168
Inorganic ions, rôles of in bacterial spores, 170
Inositol-requiring micro-organisms, suicide of, 12
Intracellular reserves, metabolism of during sporulation of bacilli, 200
Intermission, duration of during DNA replication, 42
Iron content of bacterial spores, 168
Irradiation, effect of in releasing the Ca-DPA complex from bacterial spores, 181
Isoleucine, presence of in DPA complex from bacterial spores, 180

K

Ketoaminopimelate, as a precursor of DPA, 219
Kojic acid as a possible precursor of aflatoxins, 34

L

Lactate dehydrogenase from *H. salinarium*, effect of salt on, 106
Lactose operon, 79, 80, 81, 82, 83–86
LD_{50} values for aflatoxins, 26
Leakage of pool materials from micro-organisms, as a measure of viability, 5
Lipid content of exosporium, 145, 146
Lipid content of spore coat, 147
Lipid contents of spores, 165
Lipids in cell envelopes of halobacteria, 117
Lipids of spores, 164
Lipoamide dehydrogenase activity of sporulating *B. subtilis*, 191
Lipoprotein particles in cell envelopes in halobacteria, 121
Location of DPA in bacterial spores, 183
Location of metals in bacterial spores, 173
Loss of DPA during bacterial spore germination, 238
Lysine, incorporation of into bacterial spore protein, 225
Lysis of extreme halophiles, 112
Lysis of *H. cutirubrum*, 113
Lysis products, growth on, 8
Lysozyme, action of on spore coat, 147, 148, 149
action of on spore cortex, 152, 153, 154

M

Magnesium, effect of on ribosome synthesis, 46

Magnesium contents of bacterial spores, 169
Magnesium-deficient bacteria, mortality of, 10
Magnesium ions, requirement for in extreme halophiles, 100
Maintenance energy, 13
and cryptic growth, 13
Maintenance of cell envelope in halobacteria, 112
Malate dehydrogenase, effect of salt on activity of, 104
synthesis of during sporulation of *B. cereus* T, 224
Malignant tumours induced by aflatoxins, 26
Malonyl-CoA system for fatty acid synthesis, absence of in *H. cutirubrum*, 117
Manganese, ability to replace calcium in bacterial spores, 178
Manganese contents of bacterial spores, 169
Manganese chelate, presence of in bacterial spores, 179
Manganous ions, stimulatory effect of on extreme halophiles, 100
Mannitol limitation, effect of on mortality of *A. aerogenes*, 10
Mapping of sporulation genes in bacteria, 230
Mating, DNA and genome replication during, 55, 56, 57
Maturation of spore, 144
Maturation of the physico-chemical state of the mature resting spore, 235
Mature spore, state of, 236
Media, effect of on aflatoxin production, 30, 31
Mesosome, 138, 139
Mesosomes and DNA replication, 61, 62
Messenger-RNA, decay of, 70–73.
effect of ribosomes on synthesis of, 74
effect on r-RNA synthesis, 74
regulation of formation of, 70–73
Messenger-RNA content of bacteria as affected by growth rate, 43, 46
Messenger-RNA synthesis during bacterial sporulation, 213
Messenger-RNA synthesis, regulation of, 88, 89
Metabolic changes during spore formation by bacilli, 185
Metabolic end-products, accumulation of prior to spore formation by bacilli, 185
Metabolic injury in micro-organisms, 3
Metabolic pathways in extreme halophiles, 110
Metabolism of extreme halophiles, 101
Metal content of bacterial spores, variation in, 171
Metal content of spores in relation to DPA content, 167
Methionine, incorporation of into aflatoxins, 32
incorporation of into bacterial spore protein, 225
Methods used in studying microbial survival under stress, 2
Methyl dipicolinate, isolation of from *B. cereus* spores, 179
Methylene blue staining, use of in assessing microbial viability, 4
Melting temperature of DNAs from extreme halophiles, 109
Mevalonate as a possible precursor of aflatoxins, 34
Mevalonate system for synthesizing isoprenoid chains in *H. cutirubrum*, 117
Microbial populations, starvation of, 9
Micrococcus, extremely halophilic strains of, 98
Micrococcus denitrificans, mobilization of poly-β-hydroxybutyrate in, 15
Micro-culture, measurement of viability by, 7
Microcycle sporogenesis in *B. megaterium*, 233
Micro-organisms producing aflatoxins, 28
Microsomal particles of spore protoplast, 162–164
Minimum stress, effect of on microbial viability, 1
Mitotic rate, inhibition of by aflatoxins, 26
Models of regulation, 88–90
Moist heat, effect of in releasing the Ca-DPA complex from bacterial spores, 181
Molecular architecture of cell envelopes in halobacteria, 122
Molecular weight of DNA, 50, 51
Monoethyl-DPA, extraction of from *Bacillus* spores, 179
Moribund micro-organisms, 3
Morphogensis of spores, genetic control of in bacteria, 230
Morphological changes in old bacterial cultures, 16
Mortality during growth, 16
Mother cell of spore, 135
Mould-damaged peanut meals, aflatoxins in, 25
Movements of spore nucleus, 136
Multicistronic messages, 77, 78, 80
Muramic acid, 145

Muramic acid, absence of from cell envelopes of halobacteria, 120
Muramic acid of spore coat, 149
Mutants for polarity, 81–83
Mutants with relaxed RNA control, 50
Mycotoxins, contamination of food by, 26

N

$NADH_2$ dehydrogenase, effect of salt on activity of, 103
$NADH_2$ oxidase activity of *B. subtilis* during sporulation, 191
$NADH_2$ oxidase (particulate), synthesis of during sporulation of *B. subtilis*, 223
Nickel, ability of to replace calcium in bacterial spores, 178
Nitrogen metabolism of clostridia during sporulation, 208
Nitrogen source, effect of on aflatoxin production, 32
Nitrogenous metabolism during sporulation of bacilli, 194
Non-specific reserve materials, metabolism of during sporulation in bacilli, 201
Novobiocin, biosynthetic origin of atoms in, 33
Nucleases, protection from, 75, 76
Nucleic acid changes during bacterial sporulation, 209
Nucleic acids in cell envelopes of halobacteria, 120
Nucleic acids in extreme halophiles, 108
Nucleoid of spore, 136
Nucleoids, numbers in *Salmonella* sp., 42, 50
Nucleoside triphosphates, levels of in cell, 64, 70, 71
  level of in growth rate transitions, 49
Nucleotides as nutrients for extreme halophiles, 99
Nucleotides in pools of bacilli during sporulation, 202
Nucleus of spore, 136
Nutritional factors affecting the DPA contents of bacterial spores, 176

O

Oats as a substrate for aflatoxin production, 31
Old cultures, survival of, 16
Operator genes, 76, 78, 79, 82–88
Operator model, the, 76–83
Operon, for alkaline phosphatase, 87, 88
Operon, for arginine, 88
  for galactose, 82, 86
  for histidine, 80, 81, 82, 87
  for lactose, 79, 80, 81, 82, 83–86
  for tryptophan, 79, 80, 81, 82
Operons, 76, 77, 78, 79, 80, 81, 83–88
Optical effect, use of in assessing microbial viability, 4
Organic acids, formation of by bacilli in relation to spore formation, 185
Ornithine carbamoyltransferase activity of *B. licheniformis* during sporulation, 193
L-Ornithine 2-oxo acid aminotransferase, synthesis of during sporulation of *B. licheniformis*, 224
Oxygen, role of in *Bacillus* metabolism, 193
Oxygen demand during sporulation of *B. cereus* T, 188

P

Parasporal bodies in bacterial spores, 228
Particles in cell envelopes of halobacteria, 122
Pathology of aflatoxin poisoning, 26
Peanuts as a substrate for aflatoxin production, 31
Penicillin sensitivity of halobacteria, 120
Penicillinase induction, 82, 87
Penicillium citreo-viride, DPA synthesis in, 219
Penicillium puberulum, aflatoxin production by, 29
Permeability of bacterial spores, 238
Permeability of micro-organisms, relation of to death, 4
pH value, rôle of in survival of bacteria, 9
  importance of in spore formation, 185
Phenylalanine, effect of on DPA content of *Bacillus cereus* spores, 176
  incorporation of into aflatoxins, 32
  incorporation of into bacterial spore protein, 225
Phosphate, impermeability of bacterial spores to, 210
Phosphate-limited *A. aerogenes*, acceleration of death of by K^+, 10
Phosphatides in cell envelope of *H. cutirubrum*, 117
Phospholipids in *B. polymyxa* spores, 166
Phospholipids of spores, 164
Phosphorus contents of bacterial spores, 166
α-Picolinic acid, effect of on DPA content of *Bacillus cereus* spores, 176

Physical factors affecting the DPA contents of bacterial spores, 176
Physico-chemical state of the mature resting spore, maturation of, 235
Plasma membrane, 136, 137, 138, 139, 140, 142
Plate counts in assessment of microbial viability, 5
Podospora anserina, clonal senescence in, 18
Polarity mutants, 81–83
Poly-β-hydroxybutyrate, content of in bacilli during sporulation, 202
  mobilization of in *Micrococcus denitrificans*, 15
Poly-β-hydroxybutyrate formation during sporulation of *B. cereus* T, 188
Poly-β-hydroxybutyrate in spores, 164
Polycistronic messages, 78, 79, 81, 82
Polysaccharide reserves, rôle of in cell survival, 14
Polysomes, 46, 74
  size of, 79
Polysomes in bacterial spores, 214
Pool amino acids in bacilli during sporulation, 197
Pool constituents, utilization of during sporulation in bacilli, 201
Pool materials, leakage of from microorganisms as a measure of viability, 5
Population effect on bacterial survival, 16
Potassium content of bacterial spores, 169
Potassium ions, content of in extreme halophiles, 101
  requirement for extreme halophiles, 100
Preribosomal particles, 67, 75
Primulin staining of yeasts, 18
Proflavine, effect of on m-RNA synthesis, 72
Proline as a precursor of bacterial spore DPA, 217
Propionate production by *Cl. botulinum* during sporulation, 207
Protease, synthesis of during sporulation of *B. licheniformis*, 223
Protease activity of *B. licheniformis* during sporulation, 193
Protein, regulation of synthesis of, 46, 47
  relation to growth rate, 40, 41
Protein content of cell envelopes in halobacteria, 118
Protein content of exosporium, 145, 146
Protein from ribosomes in extreme halophiles, composition of, 107
Protein synthesis, effect of on RNA synthesis, 64, 65
Protein synthesis during sporulation, 225
Protein synthesis in growth rate transitions, 48, 49
Protein to RNA relationship, 46
Proteolytic ability of extreme halophiles, 111
Proteolytic enzymes in bacterial spores, activation of by manganous ions, 170
Protoplast in the spore, state of, 236
Protoplast of spore, 134, 139, 157–164
Pseudomonas fluorescens, effect of deuterium oxide on survival of, 9
Pseudomonas ovalis, substrate-accelerated death of, 8
Pseudomonas sp., viability of magnesium-limited, 3
Purification of enzymes from extreme halophiles, 105
Puromycin, action of, 74
  effect of on RNA synthesis, 64, 67, 69
Pyrimidine synthesis control, 80
Pyrroline-5-carboxylate reductase, synthesis of during sporulation, 199
  synthesis of during sporulation of *B. licheniformis*, 224
Pyruvate production during sporulation by *B. cereus* T, 188

R

Radioactive-labelled aflatoxins, preparation of, 32
Rainbow trout, effect of aflatoxins on, 26
Rate of synthesis of RNA molecule, 75
Ratio of calcium to DPA in bacterial spores, 175
Regulation models, 88–90
Regulation of enzyme synthesis, 76–90
Regulation of m-RNA formation, 70–73
Regulation of RNA synthesis, 62–76
  effect of ribosomes on, 73–76
Regulation of synthesis of DNA, 42
Regulation of synthesis of protein, 46, 47
Regulator genes, 76, 83, 88
Regulatory genes, rôle of in bacterial sporulation, 231
Relaxed RNA mutants, 50
Relaxed RNA synthesis, 67, 74
Replication cycle, the, 52–55
Replicators and DNA replication, 61
Replicons and DNA replication, 61, 62
Repressible enzymes, 77, 78, 87, 88
Repressor for alkaline phosphatase operon, 87, 88
Repressor for lactose operon, 83, 84
Reserve materials, rôle of in cell survival, 14
Respiratory activity of spore protoplast, 236

Resting bacteria, turnover of cell constituents in, 14
Resting spore, maturation of the physicochemical state in the mature, 235
Retardation factors, rôle of in bacterial sporogenesis, 234
Ribonucleic acid, extraction of from bacteria, 43, 45
regulation of synthesis of, 43–46
relation to growth rate, 40, 41
rôle of in bacterial sporulation, 211
synthesis of in *B. cereus*, 211
synthesis of in *B. subtilis*, 211
Ribonucleic acid from *H. salinarium*, base composition of, 109
Ribonucleic acid of spore protoplast, 161–164
Ribonucleic acid phosphorus, changes in during sporulation of *B. cereus*, 210
Ribonucleic acid synthesis, regulation of, 62–76
rôle of amino acid activation in, 68, 69
rôle of amino acids in, 64–68
rôle of ribosomes in, 66, 67, 73–76
rôle of t-RNA in, 66–70
Ribosidase, synthesis of during sporulation of *B. cereus*, 223
Ribosomal protein, effect of on regulation of RNA synthesis, 66, 67
Ribosomal-RNA, action as messenger RNA, 75
effect of on regulation of RNA synthesis, 66, 67
preferential degradation of during bacterial sporulation, 212
Ribosomal-RNA content of bacteria, as affected by growth rate, 43, 45, 46
Ribosome synthesis in growth rate transitions, 48, 49
Ribosomes, content of in bacteria as affected by growth rate, 43, 44
degradation of in Mg^{2+}-starved *E. coli*, 15
effect of on regulation of RNA synthesis, 66, 67
efficiency of, 46, 47
in spore maturation, 144
numbers of in bacterial spores, 215
regulating effect on RNA synthesis, 73–76
sedimentation analysis of from sporulating *B. cereus* var. *alesti*, 212
Ribosomes in extreme halophiles, effect of salt on, 107
Ribosomes of spore protoplast, 162–164
Rice as a substrate for aflatoxin production, 31
Ripening of spores, 236
Rise of extreme halophilism, comments on, 126
RNA, rate of synthesis in growth rate transitions, 47, 48, 49, 50
RNA to protein relationship, 46
Rôle of DPA in bacterial spores, 183
Rye as a substrate for aflatoxin production, 31

S

Salmonella sp., number of nucleoids in, 42
Salmonella typhimurium, effect of growth rate transition on, 47, 48, 49
molecular weight of DNA from, 50
relation between growth rate and polymer synthesis in, 40, 41
relation between surface area: mass ratio and growth rate in, 42
ribosomal content of, 43, 44
Salt, role of in maintenance of cell envelope in halobacteria, 112
Salt concentration inside extreme halophiles, 101
Salt linkages in cell envelope of halobacteria, 113
Salts, functions of in maintaining cell envelope in halobacteria, 123
Salts as stabilizers of enzymes in extreme halophiles, 105
Sarcina, extremely halophilic strains of, 98
Sarcina morrhuae, salt content of, 102
Satellite DNA in extreme halophiles, 110
Scarring of yeasts, 18
Sedimentation coefficients of ribosomes in extreme halophiles, 107
Senescence of micro-organisms, 17
Separation of aflatoxins, 27
Serine as a precursor of bacterial spore DPA, 217
Silicon contents of bacterial spores, 169
Slide culture, measurement of viability by, 7
Sodium chloride, specificity of requirement for in extreme halophiles, 100
Sodium chloride concentration, effect of on growth of extreme halophiles, 98
Sodium contents of bacterial spores, 169
Solute concentrations in bacterial spores, 237
Solute permeability in bacterial spores, 239
Solute uptake in extreme halophiles, effect of salt on, 108
Solutes, utilization of in pool during sporulation in bacilli, 201
Sonication, effect of in releasing the Ca-DPA complex from bacterial spores, 181

Sorghum as a substrate for aflatoxin production, 31
Soybeans as a substrate for aflatoxin production, 31
Space considerations in bacterial spores, 237
Specific m-RNA formation, 77, 78, 79
Specificity of calcium requirement for sporulation, 177
Spectrophotometric method for determining aflatoxins, 28
Spirillum serpens, substrate-accelerated death of, 8
Spodography of bacterial spores, 173
Spontaneous death of vegetative microorganisms, 1
Sporal m-RNA, 215
  in bacteria, genetic loci for, 230
Spore, bacterial, development of impermeability in, 209
  calcium content of, 137
  chemical composition of, 144–184
  cytological development of, 135–144
  lipids of, 164
Spore coat, 134, 135, 140, 142, 143, 146, 147, 151, 152
  action of lysozyme on, 147, 148, 149
  amino acids of, 147, 148, 149
  chemical composition of, 146–151
  glycopeptide of, 146, 148
  lipid of, 147
  muramic acid of, 149
  penetration of by solutes, 240
  resistance of, 150
  X-ray pattern of, 151
Spore components in bacteria, synthesis of, 216
Spore core, 135
Spore cortex, 134, 139, 140, 141, 144, 151–155
  chemical composition of, 151–155
  formation of, 140–141
Spore DNA, 137
Spore exosporium, 134, 140, 144, 145, 146
Spore formation, biochemistry of, 184
Spore formation by bacilli, metabolic changes during, 185
Spore formation in bacilli, time scale for, 185
Spore formation in bacteria, induction of, 230
Spore heat-resistance, 144
Spore markers on genetic map of *B. subtilis*, 230
Spore maturation, 144
Spore mother cell, 135, 139, 140, 142, 143
Spore nucleoid, 135, 137, 138, 139, 144
Spore nucleus, 136
Spore protoplast, 134, 139, 157–164
  composition of soluble fraction in, 157–158
  DNA of, 160–162
  enzymes of, 158–159
  free solutes of, 159–160
  ribosomes of, 162–164
  RNA of, 161–164
Spore ripening, 144
Spore septum formation, 138, 139
Spore-specific proteins, 225
Spore wall, 135
Sporogen, nature of, 234
Sporogenesis, factors inducing in bacteria, 233
Sporosarcina ureae, dipicolinic acid in spores of, 174
Sporulation,
  bacterial,
    control of, 232
    DNA synthesis during, 209
  induction of in bacteria, 233
Sporulation factor, 232
Sporulation-less mutants of *B. subtilis*, 231
Sporulation stages, 135
Stability of ribosomes in extreme halophiles, 107
Stages of sporulation, 135
Staining, use of in assessing microbial viability, 4
Starvation of microbial populations, 9
Starvation-resistant mutants, 14
Step time in DNA replication, 42, 50
Sterigmatocystin, possible rôle of in aflatoxin biosynthesis, 34
Stimulation of electron transport in spores, effect of DPA on, 184
Streptomycin treatment of *E. coli*, 17
Strontium, ability to replace calcium in bacterial spores, 178
Structure of cell envelope in halobacteria, 114
Substrate-accelerated death, 8, 10
  strain specificity of, 11
Succinate dehydrogenase, effect of salt on activity of, 103
  synthesis of during sporulation of *B. cereus* T, 223
N-Succinylglutamate, synthesis of during sporulation of *B. megaterium*, 223
Sucrose as a carbon source for aflatoxin production, 32
Sugars in hydrolysates of cell envelopes in halobacteria, 119
Suicidal mutants and substrate-accelerated death, 12
Supressors, 84, 85, 87

Sulphur reserves, utilization of during starvation, 15
Survival of micro-organisms under minimum stress, 1
Surface patterns on halobacteria, 114
Swelling of spores during germination, 238
Synchronized cultures, use of in studies on the biochemistry of spore formation, 184
Synchronous cultures, induction of DNA synthesis in, 59, 60

T

Taurine in spore coat, 148
Teichoic acid of exosporium, 145, 146
Temperature, effect of in releasing the Ca-DPA complex from bacterial spores, 181
  effect of on ion contents of *Bacillus* spores, 172
Temperature-sensitive mutants, 60
Tetracyclines, effect of on DPA content of *Bacillus cereus* spores, 176
Tetrazolium dyes, use of in assessing microbial viability, 5
Thymine, effect of deficiency of on DNA replication, 62
Thymine limitation, effect of on induction of DNA synthesis, 57, 58, 59
Thymine-less death, 12
Thymine-requiring micro-organisms, suicide of, 12
Time for amino acid insertion in polypeptide, 75
Time for nucleotide insertion in RNA, 75
Time of DPA synthesis in sporulating bacteria, 216
Time scale for spore formation in bacilli, 185
Titration of protein in cell envelope of *H. halobium*, 119
Total counts in assessment of microbial viability, 5
Toxicology of aflatoxin poisoning, 26
α-Toxin, synthesis of during sporulation of *Cl. histolyticum*, 223
Trace element contents of bacterial spores, 169
Training of non-halophiles to the halophilic habit, 126
Transfer-RNA, effect of on regulation of RNA synthesis, 66–70
Transfer-RNA content of bacteria as affected by growth rate, 43, 45
Transfer-RNA in bacterial spores, 214
Transition of growth rates, effect of, 47–50
Tricarboxylic acid, involvement of in sporulation in bacilli, 190
Tryptophan, incorporation of into aflatoxins, 32
  operon, 79, 80, 81, 82, 86
Tyrosine, degradation of to acetate, 33
  incorporation of into aflatoxins, 32
  presence of in DPA complex from bacterial spores, 180

U

Ultraviolet spectra of bacterial spores, 179
Unit membranes in halobacteria, 116
Unstressed populations of micro-organisms, death of, 2
Uridylic acid as a nutrient for extreme halophiles, 99

V

Valerate production during sporulation of *Cl. botulinum*, 207
Valine, presence of in DPA complex from bacterial spores, 180
Viability, measurement of by microculture, 7
Viability measurements on microorganisms, 1
Viability of micro-organisms, definition of, 2
Volatile acid production during sporulation of *Cl. botulinum*, 206

W

Water activity, effect of on heat resistance of bacterial spores, 239
  effect of on viability of bacterial spores, 238
Water content of bacterial spores, 238
Water sorption properties of bacterial spores, 239
Water-binding capacity of bacterial spores, 239
Wheat as a substrate for aflatoxin production, 31
Whitening of spores, 236

X

X-ray analysis of aflatoxins, 27
X-ray pattern of spore coat, 151

Y

Yeasts, senescence in individuals in a population of, 18
Yields of aflatoxin, 31

Z

Zinc, ability to replace calcium in bacterial spores, 178
Zinc ions, effect of on aflatoxin production, 31